Experimental Embryology of Vascular Plants

Edited by B.M. Johri

With 81 Figures

Springer-Verlag
Berlin Heidelberg New York 1982

Professor Brij Mohan Johri
Department of Botany
University of Delhi
Delhi 110007
India

For explanation of the cover motif see legend to Fig. 5. p. 141.

ISBN-13: 978-3-642-67800-4   e-ISBN-13: 978-3-642-67798-4
DOI: 10.1007/978-3-642-67798-4

Library of Congress Cataloging in Publication Data. Main entry under title: Experimental embryology
of vascular plants. Includes bibliographies and indices. 1. Botany–Embryology, 2. Embryology,
Experimental. 3. Plant tissue culture. 4. Plant cell culture. I. Johri, B. M. QK665.E96. 582.033.
80.23524.

2131/3130-543210

Dedicated to
Professor Panchanan Maheshwari FRS
(1904–1966)

Foreword

A long time ago botany used to be regarded as the *scientia amabilis*, the friendly science, eminently suitable for leisured amateurs. Since then, and particularly in this century, it has grown tremendously in its importance and in its intimate contacts with various other disciplines of science, some of which, like plant genetics and plant physiology, at one time indeed used to be included under the broad term botany. In spite of the fact that such subjects have expanded into major scientific fields of their own, botany, the mother science, continues to maintain its central place: this is because it deals with plants which constitute one of the most vital life-supporting systems of this planet. Furthermore, interacting and benefiting from advances made in other sciences, it has steadily progressed in a number of areas. Experimental embryology of vascular plants is one such field where spectacular advances have been made in recent years. The time is therefore particularly opportune for the publication of an authoritative book on the subject.

It is very appropriate that the book has been planned and edited by Professor B. M. Johri, one of India's foremost botanists, whose contributions in embryology, plant morphology and morphogenesis are internationally known. He was closely associated over a number of years with Professor P. Maheshwari, the great botanist and embryologist, to whom the book is dedicated. In a research and teaching career spanning half a century, Professor Johri has made outstanding contributions to science, and earned the reputation of being a gifted teacher. Among other distinctions, he has been President of the Indian Botanical Society and the International Society of Plant Morphologists, Vice-President of the Third International Congress of Plant Tissue and Cell Culture, and of the Twelfth International Botanical Congress. He was editor of the well-known journal, *Phytomorphology*, for many years. He has been the recipient of numerous awards including the prestigious Birbal Sahni Gold Medal. He has travelled widely. I have mentioned all this only to indicate Professor Johri's special qualifications for editing this publication. It gives me particular pleasure to write this Foreword because I have known Professor Johri for many years and have admired him for his total devotion to science, his erudition, and the meticulous care with which he plans and carries out his investigations. He is also known for insisting on high standards in the presentation of the results of research.

The present book on experimental embryology of pteridophytes, gymnosperms and angiosperms, covers a wide range of topics including flower culture, anther culture, ovary, ovule and nucellus culture, pollenpistill interaction and control of fertilization, endosperm culture, embryo culture, and protoplast culture. The relevant aspects of vital importance have been emphasized. The authors of the individual chapters are well-known authorities in their respective fields. Their contributions, together with Professor Johri's short but authoritative and penetrating Introduction, make this book invaluable not only for the undergraduate and post-graduate students for whom it is primarily intended, but also for all those interested in understanding the subject and making use of the progress so far achieved. I am confident that the book will be welcomed. I wish Professor Johri and his co-authors great sucess.

New Delhi   B. P. PAL
April 10, 1982   Ph.D. (Cantab.), F.R.S.
                                   Hon. F.L.S., F.N.A., F.A.Sc.

Dr. Pal is a former Director, Indian Agricultural Research Institute; Director-General, Indian Council of Agricultural Research; Past President, Indian National Science Academy; Past General-President, Indian Science Congress Association; and Ex-Chairman, National Committee on Environmental Planning and Coordination.

Preface

Experimental Embryology of Vascular Plants is essentially meant for undergraduate and post-graduate students. Therefore, no effort has been made to review the literature exhaustively. The urgency for the preparation of such a text arose because of the significant advances in experimental embryology during the last 25 years, and the paucity of books presenting the scattered information as a coherent theme.

The refinements in techniques and availability of a large number of new phyto-chemicals have accelerated the pace of research, and many new areas of in vitro studies on plant growth, development, and differentiation have emerged.

In the early years usually root segments of tomato, discs from the cambial region of carrot root, and pith tissue from the stem of tobacco were grown in rather simple nutrient media supplemented with growth promoters, individually or in combinations, in several concentrations. Gradually, other plant tissues and organs were also cultured in vitro. Such studies were not restricted to angiosperms, but gymnospermous, bryophytic and pteridophytic tissues were also grown in cultures. This led to experimental studies on regeneration, modifications in life cycles, apospory, and apogamy. The single cell culture on liquid and agarified media, mass culture of cell suspensions, and protoplast culture have made new impacts on in vitro investigations.

The totipotency of the plant cell, as envisaged by Haberlandt, at the beginning of the 20th century, has been fully vindicated. If, as yet, we are not able to obtain the desired results in all instances, it is due to lack of proper techniques, and insufficient knowledge about nutrient media and other physical and chemical conditions, which are essential for proper growth of cells, tissues and organs.

In 1951 appeared a paper by J. P. Nitsch, from the Biological Laboratories of the Harvard University, on the culture of unpollinated and pollinated ovaries of gherkin (a cucurbit) and some other plants. This investigation, conducted under the guidance of Professor R. H. Wetmore, paved the way for cell, tissue, and organ cultures of reproductive organs in angiosperms.

Originally, it was planned to include chapters only on "Flower Culture", "Anther Culture", "Ovary, Ovule and Nucellus Culture", "Pollen-pistil Interaction and Control of Fertilization", "Endosperm Culture", "Embryo Culture", and "Protoplast Culture". For the sake of a more comprehensive treatment of the subject, accounts of experimental embryology of pteridophytes and gymnosperms have also been added. In the chapter on "Experimental Embryology of Pteridophytes", some aspects of ultrastructure and histochemistry are included; however, to keep the text to a desirable limit such an approach has not been adopted for gymnosperms and angiosperms.

The manuscripts on *Experimental Embryology of Vascular Plants* were submitted to the publisher in January 1979. I suggested to Mr. N. K. Mehra, Representative in India for Springer-Verlag, that he should explore the possibility of bringing out a cheap edition in India. In spite of his exhaustive efforts Mr. Mehra could not find a satisfactory printer. Finally, some time in June/July 1981, the manuscripts were returned to Springer-Verlag, Heidelberg.

In view of the unusual delay in publication it is appropriate that, wherever necessary, the contributor should update the information. Thus, recent studies have been added to most chapters.

I wish to emphasize that this book is not exclusively devoted to in vitro culture of cells and tissues of reproductive organs. In vivo development has been considered at appropriate places for comparison. A great deal of new information has recently been gained about the wall structure of pollen grain and stigma involved in compatibility and incompatibility reactions. Much attention has been paid to this aspect. Several chapters have been contributed by my colleagues in this Department, and have been prepared in collaboration with me.

A textbook of this nature may well have shortcomings, and I would greatly appreciate if the readers would bring them to my attention. I have no doubt, however, that the undergraduate and post-graduate students will find the information in various chapters useful.

University of Delhi   B. M. JOHRI
January, 1982

Acknowledgements

Experimental Embryology of Vascular Plants had been planned when the investigations on experimental embryology at the Department of Botany, University of Delhi, had progressed significantly during 1955–1966 under Professor P. Maheshwari's leadership and foresight.

I have had the pleasure and privilege of working with Professor Maheshwari, and had many occasions to discuss various aspects of tissue culture with colleagues during my visits abroad in the last 25 years. Among them I wish to mention: E. Ball, M. Bopp, A. Braun, R. Butenko, R. Gautheret, D. Hess, A. C. Hildebrandt, H. W. Knobloch, G. Melchers, G. Morel, M. G. Mullins, U. Näff, J. P. Nitsch, C. Nitsch, I. Potrykus, A. N. Rao, F. Skoog, F. C. Steward, H. E. Street, J. G. Torrey, P. R. White, Y. Yamada and M. Zenkteler.

Discussions with my co-workers H. Y. Mohan Ram and N. S. Rangaswamy, and with my son M. M. Johri, have also been very fruitful.

I am very grateful to all the contributors who readily accepted my invitation and prepared various chapters, and also updated them.

Mr. Ajay Mathur, Miss Archana Sahai and Dr. K. G. Ramawat read the proofs, and Dr. Sunanda Ghosh and Miss Sahai checked the indices. I greatly appreciate their assistance.

I owe a special word of thanks to Dr. Konrad F. Springer and Dr. Dieter Czeschlik of Springer-Verlag, Heidelberg, for their interest in this book, to Miss E. Schuhmacher (Springer) for very ably managing its production, and to Mr. N. K. Mehra, Representative of Springer-Verlag in India, who served as an excellent communication link between Dr. Springer and myself.

This work could not have been completed without the facilities available to me at the University of Delhi, for which I wish to express my gratitude to Professor S. C. Maheshwari, Professor H. Y. Mohan Ram, and Professor R. N. Kapil who have been Chairman of the Department of Botany after my retirement in 1974.

I should like to express my warm appreciation for my wife Raj, for her understanding, patience and support during the preparation of this book.

B. M. JOHRI

Contents

7. Pollen-Pistil Interaction and Control of Fertilization

8. Endosperm Culture

Contributors

DeMaggio AE, Department of Biological Sciences, Darthmouth College, Hanover, New Hempshire 03755, USA

George, Leela, Bio-organic Division, Plant Morphogenesis and Tissue Culture Section, Bhabha Atomic Research Centre, Trombay, Bombay 400085, India

Johri BM, Department of Botany, University of Delhi, Delhi 110007, India

Kitchlue, Sushma, Department of Botany, Miranda House, University of Delhi, Delhi 110007, India

Konar RN, Department of Botany, University of Delhi, Delhi 110007, India (deceased December 30, 1979)

Narayanaswamy S, Bio-Organic Division, Plant Morphogenesis and Tissue Culture Section, Bhabha Atomic Research Centre, Trombay, Bombay 400085, India (Present address: 24 Theagaraja Gramani Street, Theagaraja Nagar, Madras 600017, India)

Norstog K, Department of Biological Sciences, Northern Illinois University, Dekalb, Illinois 60115, USA (Present address: Fairchild Tropical Garden, 10901 Old Cutler Road, Miami, Florida 33156, USA)

Raghavan V, Department of Botany, Ohio State University, Columbus, Ohio 43210, USA

Rangan TS, Department of Botany, University of Florida, Gainsville, Florida 32611, USA

Rao PS, Bio-organic Division, Plant Morphogenesis and Tissue Culture Section, Bhabha Atomic Research Centre, Trombay, Bombay 400085, India

Shivanna KR, Department of Botany, University of Delhi, Delhi 110007, India

Srivastava PS, Department of Botany, SGTB Khalsa College, University of Delhi, Delhi 110007, India

1. Introduction

B. M. JOHRI

Retrospectively, experimental embryology can be said to have its roots in the third century B.C. when the Mesopotamians had a special ceremony for the pollination of the date palm. A man climbed up a male tree, brought down the inflorescence, and handed it over to the high priest who touched the female inflorescences with it. This ensured a good supply of dates.

Camerarius (1694) observed that in the female plants of *Mercurialis annua*, if completely isolated from male plants, the fruits did not contain a fertile seed. He continued his experiments with *Ricinus* and *Zea mays*, and showed that an interaction between stamens and carpels was necessary for the production of seeds.

Kölreuter (1761) artificially pollinated *Nicotiana, Dianthus, Matthiola*, and *Hyoscyamus* and produced hybrids. While the exact role of pollination and fertilization had not yet been discovered, Kölreuter did not envisage that his simple technique of artificial pollination would, in due course of time, become the most important tool for plant breeding and crop improvement.

P. Maheshwari (1950) summarizes the earlier studies on experimental embryology in vivo and in vitro. The in vitro investigations owe their origin to the pioneer studies of Haberlandt (1902; see also Krikorian and Berquam 1969) with "mature" cells. Although he did not succeed in his efforts, he prophesied that the cultivation of isolated cells in nutrient solutions should make possible an experimental approach to many important problems (see White 1954). This prophecy has been more than fulfilled during the last five decades.

We now have several formulations of nutrient media which support the in vitro growth of reproductive tissues, and very significant results have been and are being obtained. Cell, tissue, and organ culture is now a well-established discipline with a wide range of contacts with other disciplines including physiology, biochemistry, and genetics.

In recent years several treatises dealing with different aspects of cell, tissue, and organ culture have been published. Raghavan (1976) deals with experimental embryogenesis in vascular plants. *Plant Cell, Tissue, and Organ Culture, Applied and Fundamental Aspects*, edited by Reinert and Bajaj (1977), includes chapters on anther culture, culture of isolated microspores, ovule culture, in vitro pollination and fertilization, endosperm culture, embryo culture, and embryogenesis is discussed in some other chapters too. *Plant Tissue and Cell Culture*, edited by Street (1977), has chapters on anther and pollen culture, aspects of organization – organogenesis, embryogenesis, and cytodifferentiation; relevant information concerning experimental embryology is included in several other chapters. Gautheret (1977) edited the Georges Morel Volume, which has much information pertaining to experimental embryology, especially on anther culture and production of haploid plants. The Science Press, Peking, published the *Proceedings of the Symposium on*

Plant Tissue Culture, jointly organized by Academia Sinica, China, and the Australian Government, held at Peking in 1978. This volume includes recent works on plant tissue culture, and most of them emphasize the possibilities of exploitation of plant tissue culture methods as practiced in China. *Frontiers of Plant Tissue Culture, 1978*, edited by Thorpe (1978), includes the papers presented at the Fourth International Congress on Plant Tissue and Cell Culture held at Calgary, Canada. It emphasizes the progress made, especially during the last decade, through the use of plant tissue culture, in the knowledge of plant growth and development. The impact of plant tissue culture on industry and agriculture, genetic manipulation, regulation of growth and morphogenesis, and primary and secondary metabolism are the diverse but related themes discussed in *Frontiers*. *Fundamentals of Plant Tissue Culture* by B. Grout and K. Short (1979) is an introductory text useful to senior school and undergraduate students. It should serve as a sourcebook of methodologies for media preparations and techniques of aseptic culture. A recent book *Plant Cell and Tissue Culture – Principles and Applications* (Sharp et al. 1979) deals with the current global status of food and energy, physiology of growth and morphogenesis, and genetics and agricultural applications. In these edited books various chapters have been prepared by specialists who have provided first-hand information on organ, tissue, and cell culture. However, there is not a single book, so far, giving a connected account of studies on experimental embryology, which should also be useful to undergraduate and post-graduate students.

Scope of the Present Work

The chapters that follow are devoted to experimental embryology of pteridophytes, gymnosperms, and angiosperms. Each chapter is self-contained and, therefore, some repetition is inevitable. The text and legends have not been loaded with too many details of media for which reference may be made to the citations in the text and references. The contributors have successfully attempted to bring out the landmarks in the development of our knowledge of experimental embryology, especially during the last 25 years. The significant achievements and many of the unresolved problems have been pointed out. Johri (1962, 1965, 1971 a, b, 1974) has discussed most aspects of experimental embryology in several reviews. However, progress in this area has been so rapid that a comprehensive review is no more possible. Hence the need for this multi-authored edited book on the experimental embryology of vascular plants.

Some Landmarks

The most spectacular achievement seems to be the in vitro development of haploid plants from pollen (Chap. 5). The fact that the various national and international symposia held on anther/pollen culture during the last 10 years outnumber those on any other singular aspect of in vitro studies, indicates the rich potentials of this area of investigation.

Soon after the discovery of origin of haploid plants from pollen through anther culture (Guha and Maheshwari 1964), the next step was to culture isolated pollen grains. In this direction C. Nitsch (1976) has given an excellent account of the technique and procedure for pre-treating the anthers and isolating pollen grains.

C. Nitsch (1977) discusses the advantages of isolated microspore culture over anther culture. The degenerating anther tissue may leach out growth-inhibiting substances which are completely eliminated in pollen culture. Also, the possibility of diploid callus from anther tissue is avoided. The production of a large number of haploids, through pollen culture, and homozygous diploids is of great importance to the geneticists.

Success with the culture of pollinated ovaries with proembryos and free nuclear endosperm made it possible to culture ovaries and ovules with the zygote and primary endosperm nucleus. The next step was the in vitro culture and pollination of ovaries and ovules. Test tube pollination and fertilization of gynoecia, placenta with ovules, and isolated ovules (Chap. 7) is a great step forward in removing barriers to crossability, and obtaining intergeneric and interspecific hybrids (see also Chap. 6). This technique is not yet being usefully exploited but its potentiality cannot be underestimated.

Endosperm tissue – long considered to be "dead" and incapable of undergoing morphogenesis – has been successfully cultured and organogenesis achieved (the first report was by Johri and Bhojwani 1965). Subsequently, a number of diverse taxa have yielded callus of unlimited growth, and differentiation leading to triploid roots, shoots, and plantlets in about a dozen taxa (Chap. 8).

It would be of interest to mention that the techniques for the in vitro production of haploids, control of fertilization, and endosperm culture were initially perfected by investigators at the Department of Botany, University of Delhi, under the inspiring guidance of Prof. P. Maheshwari, FRS, to whom this book is dedicated.

Another aspect which deserves to be mentioned is protoplast culture. Strictly speaking, this may not directly pertain to experimental embryology. However, embryogenesis does occur in protoplasts of a number of taxa (Chap. 10). Therefore, it appeared desirable to include a chapter on protoplasts. Some other aspects of protoplast culture have also been considered for the sake of a comprehensive coverage.

Nutrient Media

Selection of a suitable nutrient medium to support optimal growth, development, and differentiation of various cells, tissues, and organs, in minimal time, is a time-consuming task. If the media already in use do not prove satisfactory, various combinations have to be tested until the attainment of a desired medium. Raghavan (1976) gives the details of 18 media which have been used by different workers, from time to time, for culturing vegetative and reproductive tissues. For growing reproductives tissues the media commonly used are those prescribed by P.R. White, J.P. Nitsch and C. Nitsch, T. Murashige and F. Skoog, as well as modifications of these and other media. Various concentrations of casein hydrolysate, yeast extract, coconut milk, auxins, kinins, gibberellic acid, combinations of amino acids, and other chemicals and plant extracts are also used as supplements. In recent years attempts have been made to avoid the use of coconut milk or other plant extracts, due to the fact that it becomes extremely difficult to identify the active fraction which evokes a particular response. Therefore, efforts are being made to devise simple, well-defined media so that meaningful correlations can be established.

Polyembryony

Johri (1971 b) discussed various aspects of embryogenesis in tissue cultures. The sequence of development is usually comparable to the development in vivo. The origin of the supernumerary embryoids can sometimes be traced to a single initial of the tissue, or the proliferated callus. It may be noted that squash mounts of callus may show configurations simulating embryogenic stages. These should be regarded as embryoids *only* if further development takes place leading to differentiation of various organs of the embryo. If organogenesis fails to occur, there is no justification to interpret such cell assemblages as embryoids. According to Haccius (1971) the embryo is a bipolar structure, and differentiation of a shoot and radicular pole occurs at opposite ends. Also, it should not be connected with the vascular tissue of the explant during its initiation and development.

Polyembryony can occur by budding, or cleavage, of the zygotic embryonal mass. Such budding and cleavage is much more common in proembryos than embryos of advanced stage. The addition of casein hydrolysate to the medium promotes embryogenesis and polyembryony in cultures.

Johri (1965) stated that polyembryony may be classified into three categories:

1. Zygotic Polyembryony. Budding and/or cleavage of proembryo, or differentiation of supernumerary embryos from embryonal callus – *Anethum, Cuscuta, Dendrophthoe, Exocarpus,* and *Gossypium.*

2. Nucellar Polyembryony. Callused nucellar tissue – *Citrus microcarpa.*

3. Embryoids
a) Callused floral primordia – *Ranunculus sceleratus* (see also Chap. 4).
b) Stem of in vitro plantlets – *Ranunculus sceleratus* (see also Chap. 4).
c) Callused anthers – *Datura innoxia* (see also Chap. 5).
d) Callused leaf segment – *Kalanchoë pinnata.*
e) Callused root explant – *Daucus carota.*

Recently, embryoids have also been obtained from callused endosperm of *Petroselinum hortense*[1] (see Chap. 8), and isolated protoplasts (from ovular callus) of *Citrus sinensis* (see Chap. 10).

Polyembryony (occurring in vivo) is usually classified into "true" and "false" types, depending on whether the embryos arise in the same embryo sac or in dif-

---

1 This appears doubtful and, probably, embryoids may have originated from callused embryonal tissue

ferent embryo sacs in the same ovule (P. Maheshwari 1950). Judging from in vitro studies, it would be appropriate to classify polyembryony into two categories: spontaneous and induced. Spontaneous type could be further distinguished into "true" and "false" types. Induced polyembryony should include "zygotic" and "non-zygotic" polyembryony mentioned above.

The capacity for embryogenesis is widely distributed and, possibly, it is an inherent property of somatic plant cells (see Reinert et al. 1977). Narayanaswamy (1977) has summarized various aspects of regeneration of plants in tissue cultures, and lists various categories of plants in which embryos develop in cultures. Additions to these lists continue as the work progresses in various laboratories.

It may be noted that phenomenal success has no doubt been achieved in the cultures of cells, tissues, and organs of diverse taxa, but much more remains to be done in this direction to understand the intricacies of morphogenesis. With better understanding of the unsolved problems, and further refinements in the techniques, new vistas for future progress will open up.

Vasil (1980) has edited two volumes of *Perspectives in Plant Cell and Tissue Culture*. Besides several other topics, there are chapters on androgenic haploids, intraovarian and in vitro pollination, endosperm culture, embryo culture, and isolation, culture, and fusion of protoplasts. Another edited book (Thorpe 1981) is *Plant Tissue Cultures: Methods and Applications in Agriculture*. The areas of interest discussed include nutrition, media, characteristics of plant cell and tissue cultures, basic technical aspects of androgenesis, growth and behaviour of cell cultures, embryogenesis and organogenesis, in vitro fertilization, embryo culture, and isolation, fusion and culture of protoplasts. *Cloning Agricultural Plants via In Vitro Techniques* (Conger 1981) essentially deals with micropropagation of ornamental species, fruit crops, vegetable crops, agronomic crops, and trees. There is useful information on media and media components, age of explants, anther culture, haploid propagation, organogenesis, embryogenesis, embryo culture, and protoplast culture.

References

Anonymous (1978) Proceedings of symposium on plant tissue culture. Academica Sinica, Science Press, Peking
Camerarius RJ (1694) De sexu plantarum epistola. Tübingen, Germany
Conger BV (ed) (1981) Cloning agricultural plants via in vitro techniques. CRC Press Inc, Boca Raton, Florida
Gautheret RJ (ed) (1977) La culture des tissus et des cellules des végétaux. Résultants généraux et réalisations pratiques. Travaux dédiés à la memoire de Georges Morel. Masson, Paris
Grout B, Short K (1979) Fundamentals of plant tissue culture. Neo Plants Ltd, Stoke
Guha S, Maheshwari SC (1964) In vitro production of embryos from anthers of *Datura*. Nature (London) 204:497

Haberlandt G (1902) Kulturversuche mit isolierten Pflanzenzellen. Sitzungsber Akad Wiss
    Wien Math, Naturwiss Kl 111:69–92
Haccius B (1971) Zur derzeitigen Situation der Angiospermenembryologie. Bot Jahrb
    91:309–329
Johri BM (1962) Controlled growth of ovary and ovule. In: Maheshwari P, Johri BM, Vasil
    IK (eds) Proceedings of the Summer School of Botany, Darjeeling. Ministry Sci Res
    Cult Affairs, New Delhi, pp 94–105
Johri BM (1965) Chemical induction of polyembryony. In: Ramakrishnan CV (ed) Tissue
    culture. W. Junk, The Hague, pp 330–334
Johri BM (1971 a) Differentiation in plant tissue cultures. Presidential address, Sec Bot,
    58 th Indian Sci Congr Bangalore II:159–186
Johri BM (1971 b) Embryogenesis in tissue cultures. In: Les cultures de tissus de plantes.
    Colloq Int CNRS Paris 193:269–280
Johri BM (1974) Experimental morphogenesis. Invitation lectures delivered at the Univer-
    sity of Mysore. Prasaranga, Univ Mysore, pp 1–64
Johri BM, Bhojwani SS (1965) Growth responses of mature endosperm in cultures. Nature
    (London) 208:1345–1347
Kölreuter JC (1761–1766) Vorläufige Nachricht von einigen das Geschlecht der Pflanzen be-
    treffenden Versuchen und Beobachtungen
Krikorian AD, Berquam DL (1969) Plant cell and tissue cultures: The role of Haberlandt.
    Bot Rev 35:59–88
Maheshwari P (1950) An introduction to the embryology of angiosperms. McGraw Hill,
    New York
Narayanaswamy S (1977) Regeneration of plants from tissue culture. In: Reinert J, Bajaj
    YPS (eds) Applied and fundamental aspects of plant cell, tissue, and organ culture.
    Springer, Berlin Heidelberg New York, pp 179–206
Nitsch C (1976) Protocol for the practice of pollen culture. In: Dudits D, Farkas GL, Maliga
    P (eds) Cell genetics in higher plants. Proc Int Training Course, Akadémiai Kiadó, Bu-
    dapest, pp 221–226
Nitsch C (1977) Culture of isolated microspores. In: Reinert J, Bajaj YPS (eds) Applied and
    fundamental aspects of plant cell, tissue, and organ culture. Springer, Berlin Heidelberg
    New York, pp 268–278
Raghavan V (1976) Experimental embryogenesis in vascular plants. Academic Press, Lon-
    don New York
Reinert J, Bajaj YPS (eds) (1977) Applied and fundamental aspects of plant cell, tissue, and
    organ culture. Springer, Berlin Heidelberg New York
Reinert J, Bajaj YPS, Zbell B (1977) Aspects of organization – organogenesis, embryogen-
    esis, cytodifferentiation. In: Street HE (ed) Plant tissue and cell culture. 2 nd edn. Bot
    Monographs, vol 11. Blackwell, Oxford, pp 389–427
Sharp WB, Larson PO, Paddock EF, Raghavan V (eds) (1979) Plant cell and tissue culture
    – Principles and applications. Ohio State Univ Press, Columbus
Street HE (ed) (1977) Plant tissue and cell culture. 2nd edn. Bot Monographs, vol 11. Black-
    well, Oxford
Thorpe TA (ed) (1978) Frontiers of plant tissue culture. Proc 4 th Int Congr Plant tissue and
    cell culture. Univ Calgary, Calgary
Thorpe TA (ed) (1981) Plant tissue culture: Methods and applications in agriculture. Aca-
    demic Press, New York London
Vasil IK (ed) (1980) Perspectives in plant cell and tissue culture. Int Rev Cytol, Suppl 11A,
    11B. Academic Press, New York London
White PR (1954) The cultivation of animal and plant cells. Thames and Hudson, London

2. Experimental Embryology of Pteridophytes

A.E. DeMaggio

In the years since the publication of the book *Embryogenesis in Plants* (Wardlaw 1955) there has been no complete review of embryology in pteridophytes. Yet, during this time, there has been a considerable increase in the amount of experimental work done in fern embryology. Embryology is defined here in its broadest sense to include all stages in reproduction, as Maheshwari (1964) has suggested. Our understanding of the reproductive biology of ferns has thus grown substantially in the past 25 years, and certain principles of embryological development have been uncovered which now can be applied generally to this group of plants. This chapter summarizes morphological, cytological, and biochemical studies conducted at various stages in fern embryology. It begins with a summary of the development of sex organs and the production of sperm and egg, and then considers those events that result in the growth of the embryo and development of the sporophyte.

Initiation of Antheridia

The antheridia are the first of the sex organs to develop on fern gametophytes. They are generally formed on the basal portion of the cordate prothallus, and often arise from superficial cells along the margin. Some ferns have the ability to produce an antheridium-inducing hormone which is capable of diffusing through the substrate and inducing nearby prothallia to form antheridia. Hormones that are capable of stimulating the formation of the male sex organ in ferns are called *antheridogens*. The antheridogen from *Anemia phyllitides* has been chemically characterized and found to be structurally related to the gibberellins (Endo et al. 1972). This is an extremely interesting discovery, especially in view of the known property of the gibberellins also to stimulate antheridial formation. The findings that gibberellins and antheridogens are structurally related and behave in much the same way to induce antheridia on gametophytes led some workers to propose a close relationship between these compounds during prothallial maturation.

The beginning of antheridial development is noted by the protrusion of a superficial cell which divides periclinally, separating a hemispherical antheridal initial from the basal cell. The antheridal initial subsequently divides anticlinally several times before another periclinal division produces a primary spermatogenous cell. At an early stage this cell is surrounded by sterile tissue which, through continued divisions, forms the antheridium. In many ferns the antheridium wall consists of three cells (four if the cap cell is divided), basal cell, ring cell, and cap cell. This type of antheridium is considered to be more advanced than the type having several to many cells in the wall (Atkinson 1973). The primary spermatogenous cell divides in regular fashion and soon a mass of cells with large, distinct nuclei is formed within the antheridium. Divisions of the spermatogenous cells appear to be simultaneous at least for a part of the time. As the sperm cells continue to develop the nuclei begin to flatten and the protoplasmic material in the cells begins to condense.

Spermatogenesis

In many ways spermatogenous cells are cytologically similar to meristematic cells. They possess a large centrally located nucleus with a prominent nucleolus and a very dense cytoplasm containing vesicles and organelles. In *Pteridium* (Bell and Duckett 1976) 32 spermatocytes form following five successive and synchronous divisions of the primary spermatogenous cell. During differentiation and maturation of the spermatocytes the blepharoplast is produced. This structure will ultimately give rise to the motile apparatus. Details of blepharoplast initiation are uncertain. However, in some archegoniates it appears to originate in the spermatocyte mother cell. In *Marsilea* it is a densely stained spherical or slightly oblong organelle 0.5–1.0 µm in diameter (Hepler 1976). During differentiation of the spermatocyte the blepharoplast consists of tubular structures or cylinders (Fig. 1 A). Microtubules radiate from it in all directions and appear to terminate close to its perimeter. These characteristics led Hepler (1976) to view the organelle as a microtubule-organizing center. The blepharoplast arises de novo and is transformed into numerous procentrioles. These later acquire the triplet tubular morphology characteristic of basal body centrioles. During the time that the cylinders

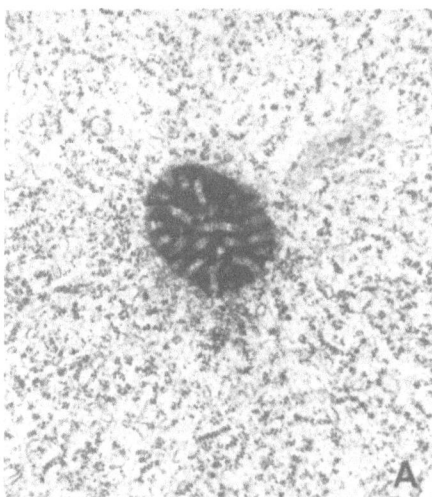 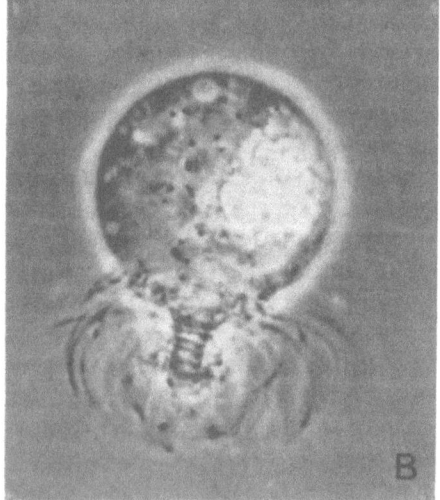

Fig. 1. A Mature blepharoplast of *Marsilea*. The lightly-stained interpenetrating channels are conspicuous. Segments of microtubules radiate in all directions from the blepharoplast region. The cytoplasm is rich in helical polyribosomes. (After Hepler 1976). **B** Mature sperm of *Marsilea vestita*. The sperm consists of a large vesicle containing starch, lipid bodies, and mitochondria. The nucleus is contained in the flagella-bearing coil. The sperm swims coil first. (After Laetsch 1967)

are forming, a rectangular body consisting of microtubules and plates appears in the blepharoplast. This is the multilayered structure (MLS). It subsequently enlarges and comes in contact with the nucleus. As the nucleus begins to lose its spherical shape, the nucleus and the MLS coil into helices. The centrioles which made up the shell of the blepharoplast differentiate into basal bodies, increase in length, and acquire characteristics of plant flagella (see also Laetsch 1967).

The mature spermatozoid seems to be similar in the leptosporangiate ferns studied. All are asymmetrical cells roughly 5 μm long by 3 μm wide, bounded by a plasma membrane and having the form of a sinistrally coiled helix. The spermatozoid consists principally of the MLS, mitochondria, nucleus, and a ribbon of aligned microtubules running the length of the gamete. The flagella occupy the anterior portion of the cell. It is generally believed that the shape of the helical gamete is maintained by the backbone or microtubular ribbon, which acts as an elastic cytoskeleton (Fig. 1 B).

Initiation of Archegonia

In Nature archegonia usually arise on the shaded or lower surface of the prothallus behind the apical notch or meristematic region of the heart-shaped gametophyte. They develop from superficial cells which can be identified as archegonial initials by the dense cytoplasm. The initial cell first produces a row of three cells from which the archegonium and its contents will be derived (Fig. 2). During develop-

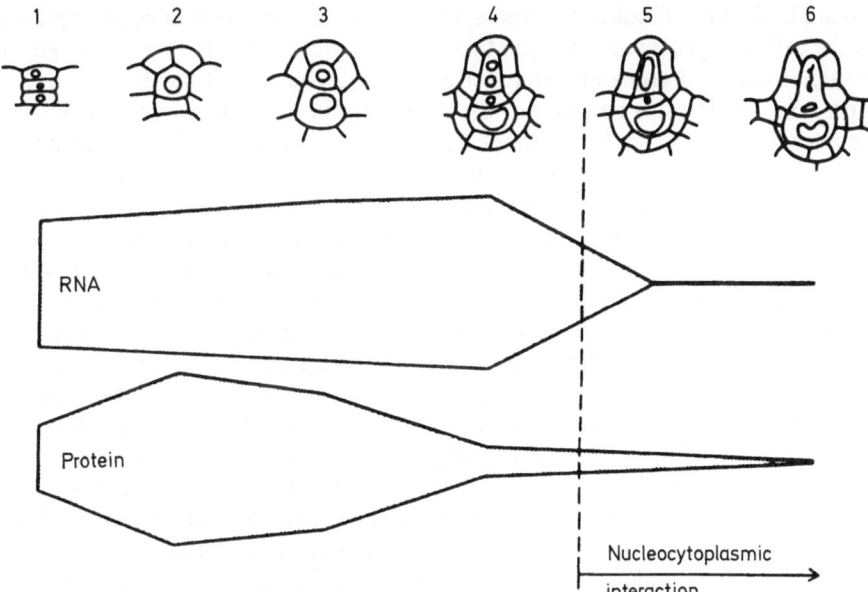

Fig. 2. A comparison of the amounts of RNA and protein synthesis during oogenesis. *1,2* Growth of primary cell. *3* Formation of central cell. *4–6* Formation and maturation of egg. Based on pulse-labeling and quantitative autoradiography. (After Bell and Duckett 1976)

ment the outer cell divides and, ultimately, produces the protruding neck of the archegonium. The innermost cell divides next, producing derivatives which form the basal cells of the jacket or venter. And, finally, the central or primary cell in the row of three produces the neck canal cells, the ventral canal cell, and the egg. The cytoplasm of the primary cell remains at the base of the archegonial canal, and during division this cytoplasmic mass becomes essentially the cytoplasm of the egg. Growth of the primary cell is thus considered the first significant event in oogenesis.

Oogenesis

Details of oogenesis in ferns are known for a number of species, but the ultrastructural and cytological changes occurring during this stage of development have only been studied extensively in *Pteridium* and *Dryopteris* (see Bell and Duckett 1976, for review).

Cytological Characteristics

The primary cell when first formed is box-like in appearance but, as differentiation progresses, it often enlarges to five-times its original size. The young cell possesses a large nucleus with irregularly shaped nucleoli and an organelle-rich cytoplasm containing numerous ribosomes and profiles of polysomes. Bell (1975) suggests that these features are related to the accompanying synthesis of RNA and protein, and are indicative of molecular events involved in the formation of egg cytoplasm. As the cell enlarges, many changes take place in the cytoplasm. Plastids begin to dedifferentiate and grana and starch are lost. Mitochondria show little change in appearance during this period, but the cytoplasm becomes very dense. The large vacuoles disappear, and their disappearance is accompanied by the appearance of a large number of small vesicles. The origin of these vesicles is unknown, and there is no evidence that they arise from Golgi or endoplasmic reticulum. As the cell continues to enlarge, most of the cytoplasm remains at the base of the archegonial canal. Subsequent division of the primary cell produces two cells of unequal size, a small neck canal cell and a large central cell. The neck canal cell is situated towards the archegonial neck and has only a small amount of cytoplasm, while the central cell inherits much of the cytoplasmic mass of the primary cell. Ultrastructural studies have shown that the cytological changes initiated in the primary cell continue with little interruption in the central cell. Plastids continue to dedifferentiate, and the numerous small vesicles begin to aggregate. It has been suggested that these vesicles are lysosomal, involved in the digestion of cytoplasm. However, characterization of the enzymes present in the vesicles has not yet been performed, and the physiological role of the vesicles during oogenesis remains obscure.

The central cell undergoes an unequal division producing a highly cytoplasmic egg occupying the base of the archegonium, and next to it, in the archegonial neck, a ventral canal cell with only a small amount of cytoplasm. The detailed cytological examination of oogenesis in *Pteridium* indicates that there is a remarkable continuity of cytoplasm from the primary cell to the central cell and eventually to the egg.

Biochemical Characteristics

Using quantitative ultraviolet microscopy, and light and electron microscopic autoradiography, Bell and his colleagues (Bell and Mühlethaler 1964; Sigee and Bell 1968, 1971) show that DNA is present in rather striking quantities in the cytoplasm of the central cell and egg of *Pteridium*. It is first detectable in the cytoplasm of the primary cell of the axial row, and continues to increase during cell differentiation. As this cell increases in size the number of mitochondria and proplastids multiplies. The increased cytoplasmic DNA can, thus, be attributed directly to these organelles. However, as the central cell and egg are formed, even greater amounts of DNA appear in the cytoplasm. But this is not accompanied by a conspicuous increase in the number of organelles. It was suggested that at this stage increased absorption measurements resulted from partial depolymerization of DNA, but this has not been confirmed. The mitochondria and proplastids in the egg seem to possess unusually large quantities of DNA, and there is evidence for the presence of DNA unassociated with organelles in the ground cytoplasm. Some investigators believe this DNA could have been derived from organelles which earlier disintegrated. Unusual amounts of cytoplasmic DNA have not been observed in somatic cells of the gametophyte, and high levels of DNA appear to be characteristic only of those cells involved in oogenesis.

During initiation of oogenesis and the formation of primary cell, labeling experiments with radioactive uridine indicate the synthesis of considerable quantities of RNA in nucleolus and cytoplasm (Jayasekera and Bell 1971). This pattern of RNA synthesis continues through the formation of the central cell and egg. The synthesis of RNA is already at a high level at the beginning of oogenesis (Fig. 2), and is thought to be correlated with the formation of ribosomes that takes place at this stage. Support for this view was provided by Cave and Bell (1974), who found that the peak of RNA synthesis coincides with that stage in development where ribosome frequency is also at its highest. RNA continues to be synthesized in the newly formed egg cell, but here it is accompanied by a dramatic reduction in protein synthesis. Figure 2 illustrates that protein synthesis is not uniform throughout oogenesis. There is considerable synthesis of protein during growth of the primary cell, and only little synthesis at later stages.

Maturation of the Egg

Once formed, the egg cell undergoes a series of significant cytological and biochemical changes as it matures. The initial observations of structural changes taking place during egg maturation depicted the degeneration of plastids and mitochondria and the evagination of the nucleus into cytoplasm. It was suggested that mitochondria and plastids degenerate and are eliminated from the egg cytoplasm. New organelles were assumed to be created de novo from the nuclear evaginations. The elimination and replacement of these organelles implied the creation of a new population of organelles in each sexual cycle (Bell 1970). In subsequent electron microscopic examination of oogenesis and egg maturation in other archegoniate plants, stages corresponding to nuclear evagination have been observed, but the elimination and regeneration of organelles has not been confirmed.

In continuing studies Bell (1975) observed that during the very early stages of egg maturation vesicles appear to be eliminated from the cytoplasm. These seem to move towards the periphery of the egg where, it is thought, they contribute their contents outside the plasmalemma for the formation of the characteristic lipoidal membrane surrounding the egg. In two species of *Dryopteris*, as well as in *Pteridium aquilinum*, the nucleus is seen to enlarge, become irregular in outline, and send out evaginations into the cytoplasm. At this time many nucleolar bodies are formed and move into the evagination. Mitochondria and plastids during this stage are poorly differentiated.

As the egg matures, the membrane surrounding it is conspicuous. The cytoplasm is penetrated by nuclear evaginations, sheets of endoplasmic reticulum, and many free ribosomes. Plastids are quite different from those observed in earlier stages and have ill-defined membranes with few internal lamellae. Mitochondria are variable in shape, some contain opaque aggregates and others have few swollen membranes.

As maturation proceeds, the ventral canal cell situated adjacent to the egg degenerates and the neck canal cell also breaks down. These events occur at the time when a considerable increase of cytoplasmic DNA can be measured in the egg. Bell and Duckett (1976) suggest that nucleotides from the degenerating cells may be incorporated into the maturing egg and account for the increased DNA. However, the experimental results reported by Sigee (1972) indicate that the high concentration of cytoplasmic DNA is derived largely by replication within the cytoplasm.

The mature egg appears to be isolated from the surrounding cells of the gametophyte by the presence of the egg membrane. In addition, the obliteration of plasmodesmata which were present during earlier stages of oogenesis seems to further isolate the egg from neighboring cells.

The extension of the nucleus into the cytoplasm of the egg is a unique feature of oogenesis in those homosporous ferns that have been studied. Unfortunately, we know very little about the role of this intimate nucleo-cytoplasmic interaction during oogenesis. Bell (1975) has argued that the nucleo-cytoplasmic interaction is an essential part of the process by which the nucleus prepares the egg cytoplasm for sporophytic growth. He proposes that some of the nuclear evaginations split off and differentiate into new organelles – mitochondria and plastids – which populate the cytoplasm of the mature egg. After fertilization these organelles become the organelles of the sporophyte. In some archegoniates, for example the liverwort *Sphaerocarpus*, the nuclear evaginations in the egg are reported to breakdown and release their contents into the cytoplasm. In *Pteridium* and *Dryopteris* a similar phenomenon could occur, but evidence that the nuclear evaginations give rise to new organelles for a sporophytic cytoplasm is not convincing. In fact, Sigee (1972) reports he could detect no evidence for complete elimination of organelles during oogenesis in *Pteridium* as had been previously suggested (Bell 1963). He observed that apparently normal mitochondria and plastids were present during the time when some organelles were disintegrating. In addition, his electron micrographs and autoradiographs provide no evidence to support the view that new organelles are formed from nuclear evaginations. At the present time we do not know the significance of these cytological events during oogenesis.

Fertilization

Dehiscence of antheridia generally occurs in ferns only minutes after the gameto-phytes have been in contact with water. Water entering the antheridia raises the internal pressure and causes the antheridial walls to swell. A short time later the cap of the antheridium or the opercular cell opens. Discharge of the coiled sperm follows quickly and a number of them, often still surrounded by mucilageneous material, are ejected from the antheridium. They become active as the mu-cilageneous sheath surrounding them is dissolved, and begin a rapid rotary motion. Soon afterwards they quickly swim away in their characteristic spiral manner. When released from the antheridia, *Lygodium* sperms swim at rates of 150 μ/s in large helices and often change direction abruptly. *Marsilea* sperms (Fig. 1 B), on the other hand, swim more slowly at rates of 120 μ/s, travelling a constant unidirec-tional path (Bilderback et al. 1974). Both *Lygodium* and *Marsilea* sperm appear in high speed cinematographs to beat three-dimensionally in a continuously travel-ling helical wave. The wave is propagated from base to tip of the flagella and rates of 65 and 30 beat cycles/s have been measured in *Lygodium* and *Marsilea* respec-tively. It is interesting to note that while the flagella function in a coordinated man-ner, all of them do not beat at the same time. After the flagella have completed a specific number of beat cycles, the body of the sperm makes a complete rotation. It is likely that this type of flagella locomotion is characteristic of sperm from most ferns.

Attraction of the Sperm

Swimming sperms appear to be attracted to the archegonia by the mucilaginous, protoplasmic material released during archegonial opening. Ordinarily, within 1 h after coming in contact with water, mature archegonia begin to open. The terminal cells of the archegonial neck swell and gradually separate providing a central pore or opening to the neck canal. The contents of the neck canal, now in active motion, are expelled often rather suddenly through the opening. In *Phlebodium* this mate-rial and the ventral canal cell are forcibly ejected from the opening under consid-erable pressure (Ward 1954 a). Once most of the neck canal contents have been eliminated, the terminal cells of the neck begin to part in an orderly fashion. The ventral canal cell is often eliminated during this time if it was not ejected earlier. The expelled material, remnants of the neck and ventral canal cells, usually remain in the vicinity of the opened archegonia but do not obstruct the passageway to the egg.

Some attention has been given to identifying those substances that diffuse from the archegonia during dehiscence and are responsible for the chemotactic response of ferns. In a variety of investigations it has been repeatedly demonstrated that C_4 acids, especially malic acid and its salts, have the ability to attract sperm of many ferns. This substance, or some related compound, may be excreted by cells of the axial row, establishing a gradient toward which the sperms swim. Bell and Duckett (1976) suggest that because cells of the axial row become increasingly isolated from the rest of the gametophyte, they become partly anaerobic and unable to completely oxidize C_4 acids. Traces of these acids would remain in the cells and diffuse into the external medium as the archegonia open.

We obviously need to learn much more about those substances capable of attracting the sperm before our description of fertilization events is complete. It is clear, however, that sperms swimming in the vicinity of opened archegonia are attracted to the organ. There they orient themselves toward the open passageway and, after a short delay, enter the canal.

Penetration of the Egg

The sperms entering the neck of the archegonium are elongated and do not retain their characteristic helical form. This has been observed in both light and electron microscopic studies of fertilization in several species of ferns. Accompanying the change in shape, the sperms show signs of disorganization. Nuclei and flagella appear disrupted and the microtubular band is often seen separating from the nucleus. The vesicular cytoplasm containing plastids, starch, and mitochondria is lost from the sperms as they become extended during passage through the archegonial neck. In the venter, sperms regain their helical form although the exact nature of the helix is often changed. Several investigators have concluded that the skeletal component primarily responsible for maintaining the spiral construction of the sperm is the microtubule band lying outside the nucleus. This band possesses considerable strength and probably serves as an elastic skeleton holding the cell together in the absence of a rigid wall. The cores of the helical sperm undulating in the venter of the archegonium are devoid of cytoplasm, and a loss of sperm cytoplasm appears to be a common feature of fertilization in a variety of archegoniates.

The actual mechanism by which the sperm penetrates the egg is still in question. Duckett and Bell (1971, 1972) doubt that flagella are effective in the narrow neck canal. They attribute penetration to the helical motion of the sperm, which is directly controlled by the flexing movement of the microtubular band. While cytological studies of sperm penetration and movement through the cytoplasm of the egg are scanty, some interesting details are available. In *Pteridium* a clear area within the egg, but separated from the egg cytoplasm, has been observed to contain one and often more coiled sperms. This has been referred to as a "fertilization cavity," but it is important to note it is apparently absent during fertilization of *Marsilea* eggs. Considerable disintegration of the sperm characterizes its entrance into the egg cytoplasm. The microtubular band is observed to separate from the nucleus and mitochondria, and both organelles loose their spiral shape. Other components of the sperm also can be seen within the egg cytoplasm. Some are clearly still associated with the microtubular band and others are in various states of disarray. Entrance of the sperm also causes physical disruption of the egg cytoplasm, and this is thought to be responsible for rapid structural changes in the egg. Since no special fertilization membrane is found to block entrance of additional sperm into the egg cytoplasm, prevention of polyspermy is attributed to changes taking place in the egg cytoplasm after sperm entry.

The skeleton of the sperm remains in the cytoplasm of the egg until some time after union of the sperm and egg nuclei. All organelles of the sperm are presumed to vanish gradually, digested by enzymes in the egg cytoplasm. By 12 h after fertilization prominent nuclear evaginations are absent from the young fern embryo, and the nucleus is now regular in outline and normal in appearance.

Pattern of Development in Fern Embryos

The general pattern of embryo development in ferns is similar for both eusporangiate and leptosporangiate members. Minor differences in the partitioning of early embryos are known to take place. However, the only significant variations appear to be in the timing at which different parts of the embryo develop into shoot, leaf, root, and foot.

The embryology of the osmundaceous fern *Todea barbara* has been thoroughly studied (DeMaggio 1961), and is used here to illustrate the segmentation pattern and sequence of organ development characteristic of many ferns. Once fertilization has been completed the young embryo becomes turgid, and fills the archegonial venter. At this time the cytoplasm is unevenly distributed, much of it lying in the basal portion of the embryo, and the embryo appears highly vacuolate. During the next few days the cytoplasmic contents become more evenly distributed, the embryo increases in size, and the jacket cells or calyptra enclosing the egg begins periclinal divisions (Fig. 3 A). The first division occurs 5–6 days after fertilization, and the wall forms in the plane of the archegonium at right angles to the axis of the prothallus (Fig. 3 B). A day later the second division takes place, again in the axis of the archegonium but this time parallel to the prothallial axis, and partitions the embryo into quadrants. Octants are generally formed 2 days later as synchronous divisions in the quadrants lead to the formation of walls perpendicular to the archegonial axis and to the two previously formed walls. During this time the embryo has continued to enlarge and maintains its spherical shape. Although the fern embryo, even at this early stage in development, is in contact with the surrounding calyptra, there is apparently no physical union of cells. Earlier workers in fern embryology held the position that quadrant and octant walls were of special morphological significance in the further development of the young embryo. In some cases the formation of a specific organ was ascribed to a particular descriptive region of the embryo. More recent investigators (Vladesco 1935; Ward 1954b; Wardlaw 1955; DeMaggio 1961) have shown that many ferns do not rigidly follow a uniform segmentation pattern, and the formation of various organs is often not limited to any one embryonic region but may involve neighboring regions. During the next 3 or 4 days two more sets of synchronized divisions take place in each cell of the embryo and partition it into 16 and 32 cells respectively. The regularity with which the post-octant stages of embryo development take place in *Todea* is not found in some other ferns, for example *Gymnogramme sulphurea* (Vladesco 1935) and *Phlebodium aureum* (Ward 1954 b).

Further divisions of the embryo are not synchronous and, in the next few days, differentiation and organ formation begin in various regions of the embryo. The first organ to be clearly distinguished is the foot (Fig. 3 C). Cells from the two quadrants of the embryo farthest away from the archegonial neck are involved in its formation. These cells undergo a series of anticlinal and periclinal divisions, enlarge

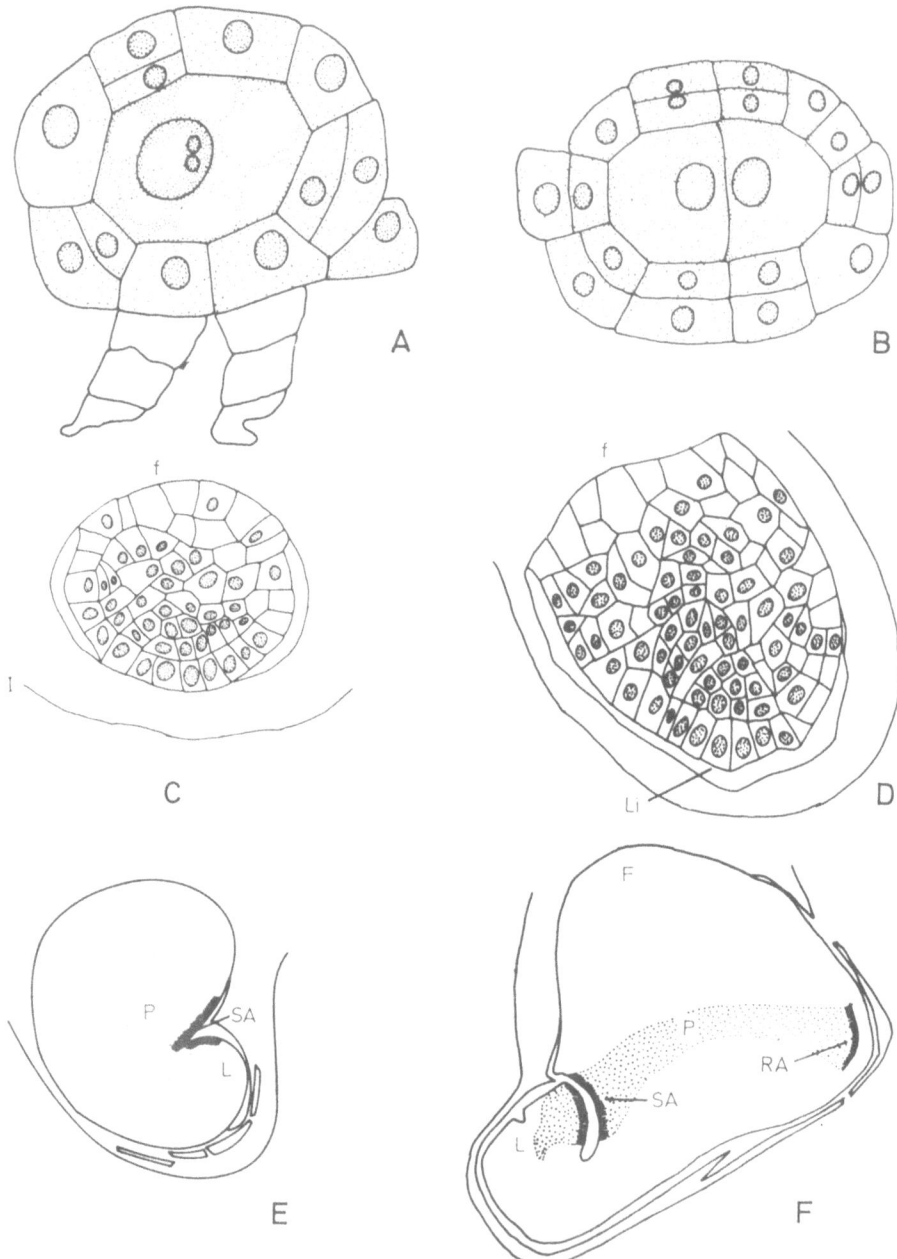

Fig. 3 A–F. Embryo development in *Todea barbara*. **A** Longitudinal section of venter with 5-day-embryo. Divisions are evident in jacket cells. **B** Transverse section of two-celled embryo, 6 days old. **C** 18-day-embryo showing well developed foot *f* and meristematic outer region. **D** 20-day-embryo with prominent leaf initials *Li*. **E** 28-day-embryo. The growing leaf had curled over the incipient shoot apical region *SA*, and procambial tissue *P* extends into the developing leaf *L* and stem. **F** 30-day-embryo with well developed shoot and root apices *SA*, *RA*. (After Wetmore and Steeves 1971)

considerably, and press tightly against the calyptra. The foot is considered to be the absorbing organ for the young embryo and, in a number of ferns, its cells intrude between cells of the parent gametophyte. Recent electron microscopic study has shown that foot cells exhibit many of the same characteristics as the specialized "transfer cells" now known to exist in a variety of plants. In *Todea*, no protrusion of the foot cells into the prothallial cells is evident, and the boundary between the starch-filled cells of the foot and parent prothallial cells can be distinguished until very late in embryo development. While the foot is in the process of forming, a number of cell divisions have taken place in the other two quadrants of the embryo and leaf initials appear in a position lateral and distal to the foot (Fig. 3 D, E). Continued divisions in this area of the embryo produce the protruding leaf primordium. For the next 10 days the leaf, surrounded by the expanding calyptra, continues to grow, curling over the top of the embryo in the earliest stages of crozier development. Procambial tissue that initially differentiated in the center of the embryo now develops continuously and acropetally into the primary leaf.

Stem and root initials are formed within 1 or 2 days after the differentiation of leaf initials (Fig. 3 E, F). The stem arises midway between leaf and foot, and is derived from the activity of apical cells that have formed through anticlinal divisions of the surface cells. As the shoot develops, procambial tissue differentiates acropetally into it, the tissue also joining with that of the growing leaf. At approximately the same time the root begins to form in a region diametrically opposite the stem. A distinct apical cell is evident 20 days after fertilization. Cellular differentiation takes place quite rapidly, and in a few days cells of the root cap and cortex are visible. As the root develops, strands of procambial tissue extend acropetally into it from the central region of the embryo. Towards the later stages of embryogenesis, each of the enlarging organs – leaf, stem, and root – is supplied by an interconnected and continuous arrangement of provascular tissue. At about 30 days after fertilization the root begins to break through the calyptra, and approximately 2 days later the primary leaf straightens and grows out over the prostrate prothallus. The young sporophyte is now prepared to begin an independent existence.

Regulation of Developmental Patterns in Fern Embryos

Over the years much effort has been made to understand how the precise pattern of embryo development in ferns is regulated. Studies have been largely stimulated by the desire to explain the alternation of generations, which is particularly prominent in ferns. Lang (1909) was perhaps one of the first investigators who, in his attempt to understand the nature of the differences between fern gametophyte and sporophyte, focused attention on the fern embryo. In his ontogenetic theory of alternation of generations, he pointed out that embryo and spore begin their development under different environmental conditions. He attributed causal significance to the "nutritive and correlative influences" which the prothallus has on the developing zygote. In recent times Wetmore and his colleagues (Wetmore 1959; Wetmore et al. 1963; Wetmore and Steeves 1971) further developed these ideas and formulated some specific questions which served to stimulate experimental research on fern embryos: To what degree is the containment of an archegonium concerned in the orderly cell division of fern embryos? Is the hereditary pattern of physical organization of the archegonium more directly correlated with the pattern of early embryology than is the action of genes? And, is it possible that the pattern of early embryology in plants is more influenced by the impact of the intimate milieu of the zygote, physical as well as nutritional, than immediately and directly by the genome?

Recent results from a variety of experimental studies provide only partial answers to these questions. Much of our present knowledge has been derived from results of surgical studies, some of which will be described here. The indirect nature of these experimental investigations makes it difficult to provide complete and unequivocal answers to the questions.

Surgical Experiments

Some of the first surgical experiments on fern embryos were designed and carried out by Ward and Wetmore (1954). Using prothallia of *Phlebodium aureum*, they cut away parts of the archegonium and jacket cells by making incisions into the prothallus with microscalpels. Their primary objective was to determine whether or not release of probable restraining influences of the archegonium and calyptra was reflected in changes in organization and development of the embryo. Figure 4 illustrates the position of certain incisions and the effects of these operations on subsequent embryo development. Embryos on treated prothallia ordinarily continue developing, and often emerge from the surrounding calyptra at an earlier stage of development than do normal embryos. In addition, surgically treated embryos have smaller and more numerous cells than ordinary embryos. They are generally cylindrical in form, and characteristically lack roots until very late in development. In many instances (Fig. 4F) the incised prothallial section, when either inverted or left in its usual position, produces spindle-like outgrowths which later develop leafy appendages. These experiments leave little doubt that surgical treatments interfere not only with the timing but also with the sequence of events taking place during embryogenesis. Moreover, the observation that some treatments caused precocious release of embryos from surrounding tissue provides at least indirect evidence for restraint as a morphogenetic factor.

It is interesting to note that the most marked effects on embryo growth observed by Ward and Wetmore (1954) resulted from simply removing the neck of the archegonium (Fig. 4G, H). In these instances tuberous, leaf-bearing structures form. These unusual growths are thought to arise as a result of interference with the timing of events during development, and not through significant change in the ultimate developmental pattern. Similar growths from fern zygotes are also seen after removing the archegonial neck from prothallia of *Thelypteris palustris* (Jayasekera and Bell 1959). Here, cylindrical, tuberous outgrowths emerge on which leaves and shoot apices later develop. The altered pattern of embryogenesis in this case is also attributed to the removal of mechanical pressure imposed by the archegonium on the swelling zygote.

When Ward and Wetmore (1954) made incisions that separated the embryo from the apical region of the prothallus, further growth of the embryo was significantly delayed. In other experiments (Jayasekera and Bell 1959) embryos were iso-

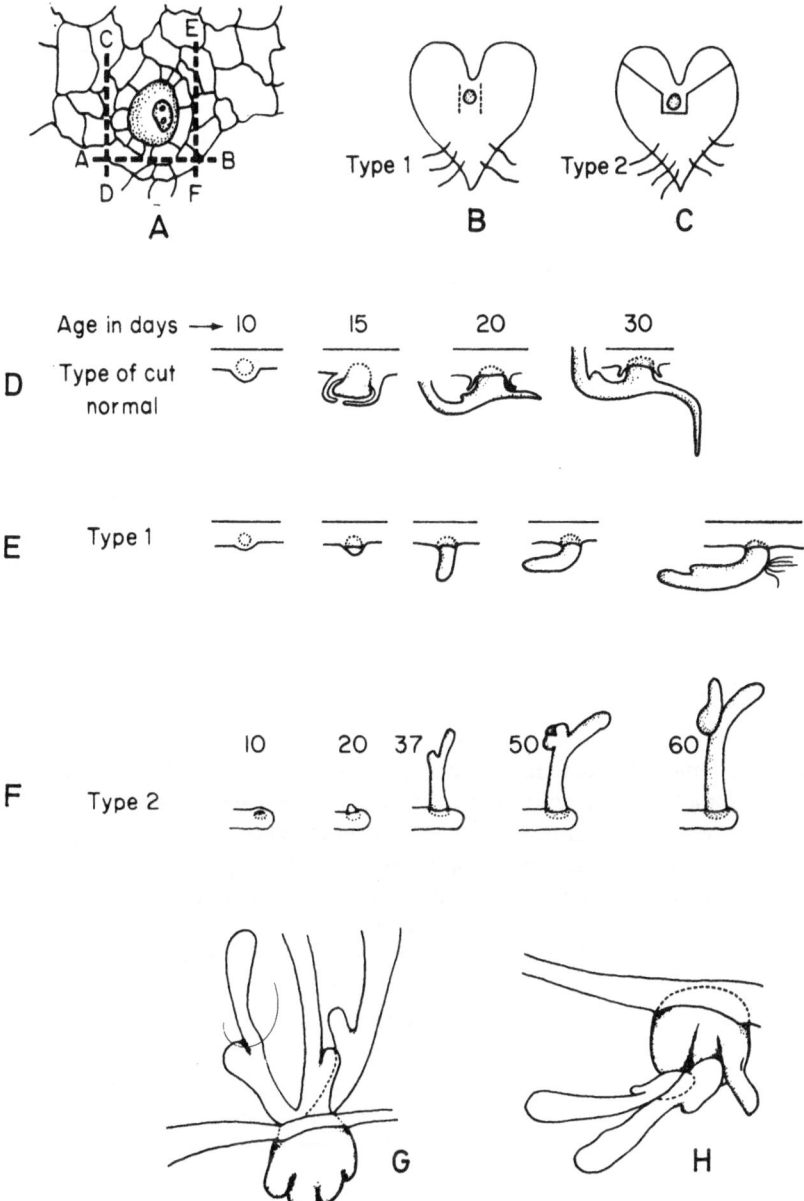

Fig. 4A–H. Experiments on the effect of surgical treatments on development of *Phlebodium aureum* embryos. **A** Longitudinal section of zygote and surrounding gametophytic tissues showing the pattern of vertical and horizontal incisions. An incision along the *line A–B* removes the archegonial neck. **B, C** Gametophytes showing the positions of vertical incisions that are designated *Type 1* and *2* below. In both types the archegonial neck is also removed. **D** Development of an unoperated embryo as the control for experimental treatments. **E, F** Development of embryos after *Type 1* and *2* operations. **G, H** Development of embryos after removal of archegonial neck only. (After Steeves and Sussex 1972)

lated on a pad of prothallial tissue and differentiation and organ development were similarly delayed. These experiments signal yet another important relationship between prothallus and embryo development: that under ordinary conditions, substances diffusing from the prothallus can influence the embryo. There is rather good evidence that the hormone auxin (indoleacetic acid) is produced in meristematic cells of the prothallus and has an effect on growth of the sporophyte (Albaum 1938). In experiments with fern embryos it has been observed that removing the zygote from the influence of the prothallial meristem often limits its development. The argument has been made that auxin moving from the meristem is essential for orderly embryogenesis and, in its absence, embryo development is retarded. This view is substantiated to some extent by experiments showing that when auxin is added to isolated pads of tissue bearing zygotes, the usual pattern of embryogenesis takes place.

In experiments where the zygote with adhering tissue is removed from the prothallus the nutritional balance of the young embryo must be severely damaged. What influence this has during the course of embryo development is not known with certainty. It is obvious that those nutrients ordinarily provided by the prothallus for sporophyte growth will need to be supplied. In addition to mineral salts, the most important substances limiting growth of the embryo appear to be vitamins and sugar (Durand-Rivières 1964).

Growth of Isolated Embryos

Developing embryos of *Marsilea vestita* are known to be strongly influenced by the immediate environment of the female gametophyte (Andres 1964). The early organization and symmetry of the embryo is influenced by the calyptra and surrounding tissue. When very young embryos are totally removed from the gametophyte, they are unable to survive even when supplied with a nutritive medium. These embryos are still dependent on the parent plant for metabolites essential for their further development. However, after embryos have attained bilateral symmetry, that is, following initiation of the leaf primordium, they can be isolated and grown independently of the female gametophyte. At this stage the leaf is sufficiently developed to begin manufacturing materials needed for sustaining the embryo. The fact that younger embryos are unable to be nurtured in vitro reflects both the complexities of the archegonial environment at this stage and our limited knowledge of the requirements for embryo development. One point is clear, that the stimulation for early embryo development is derived from the maternal plant.

The extent of variation between the requirements for growth of young as opposed to more mature embryos is evident in studies carried out on isolated embryos of *Todea barbara* (DeMaggio and Wetmore 1961; DeMaggio 1963). In *Todea* it is possible to remove embryos from their position in the archegonial venter beginning shortly after fertilization and continuing up to the time mature organs are formed. Older embryos excised 20 days after fertilization are already fairly well differentiated before being planted in a nutrient medium. They possess leaf, stem, and root meristems and have undergone considerable internal cellular differentiation. These embryos are easily grown to maturity on a simple nutrient medium containing mineral salts and sugar (Fig. 5 A, B). The situation was quite different

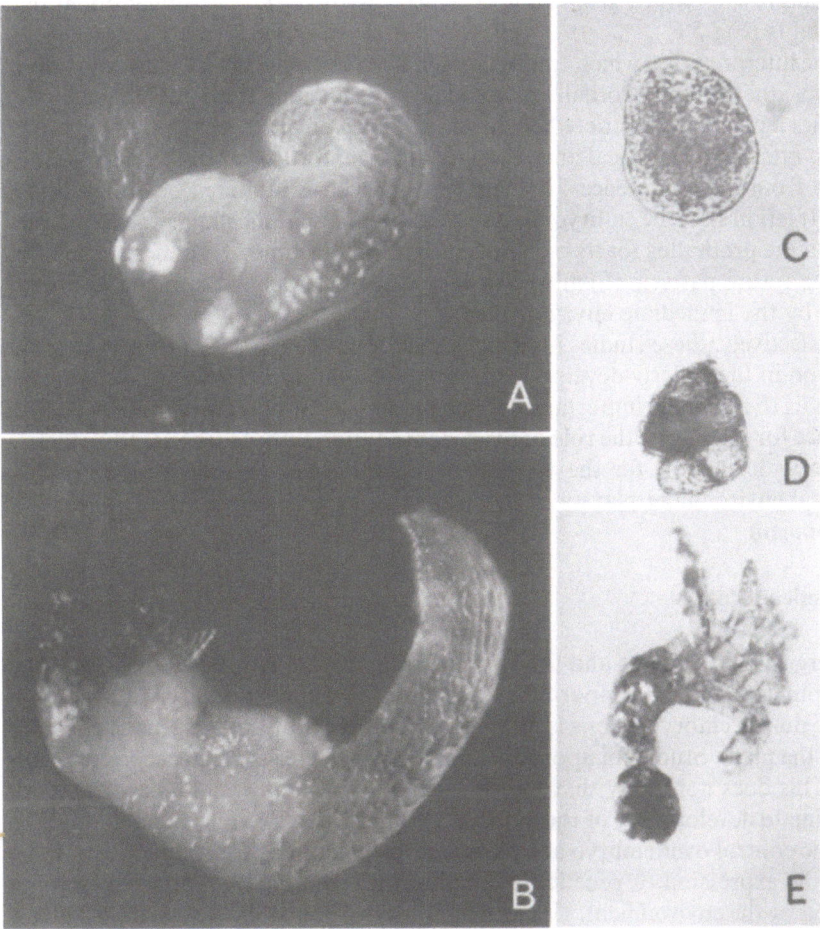

Fig. 5 A–E. Development of isolated embryos of *Todea barbara*. **A, B** Embryos isolated 20 days after fertilization. **A** After 1 month in culture. **B** 4 months in culture, showing well-developed primary leaf. **C, D, E** Embryos isolated before the first division. **C** Embryo isolated 4 days after fertilization. **D** Embryo grown for 2 weeks in culture. **E** 4 months in culture, showing branched thalloid emergence. (After DeMaggio and Wetmore 1961)

when only slightly younger embryos (17-day-old) were planted in a similar nutrient medium. In all instances they fail to continue their development unless growth-stimulating substances present in coconut water are supplied. The requirement for growth of these embryos is obviously more complex and more difficult to satisfy than for older embryos. Nevertheless, the materials required for continuing development of isolated embryos can be provided with some additions to the nutrient medium. This is not true when attempts are made to grow very young (4- to 5-day-old) embryos (Fig. 5 C). These undergo a number of cell divisions, enlarge slowly, and then do not immediately continue developing. For several weeks little activity can be detected (Fig. 5 D). When growth does resume, the embryos produce outgrowths which give rise to irregular, asymmetric, two-dimensional thalloid plants (Fig. 5E).

One interpretation which can be derived from these results is that in very young embryos the parent prothallus furnishes the nutrition and the archegonium provides a physically restricted environment, both of which are needed for the zygote to attain physiological and morphological maturity. When the embryo is removed from these influences, its pattern of growth is inhibited or drastically altered. If left in situ, the embryo matures and becomes multicellular and less dependent on the prothallus for its metabolic needs. By this time organ meristems form, and the future pattern of embryo development is largely determined and unaffected by the immediate environment.

Collectively, these studies indicate a role for physical pressure, hormones, and nutrition in the orderly development of the fern embryo. They do not, however, permit us to assess the importance of one factor over another, nor do they provide evidence for evaluating the role of the genome during embryogenesis. Nonetheless, they provide support for the view that the physical environment as well as the chemical environment plays a determinate role in regulating patterns of embryo development.

Biochemical Studies

The surgical experiments and the embryo isolation and culture studies described in the previous section support the thesis that the normal sequence of events occurring during embryogenesis is not predetermined simply by the genetic constitution of the plant. Studies of apogamy and apospory in ferns are in accord with this view. This does not imply that the genome is not fundamentally responsible for the ultimate development of the embryo; certainly it is. But our understanding of genomic control over embryo activities is severely limited. We need to learn more about the expression of genetic potential of the plant embryo, and how it is influenced by the environment. Unfortunately, this is probably the most difficult aspect of embryogenesis to study, partly because the plant embryo is enclosed and in immediate contact with parental tissue and partly because, as pointed out earlier, the immediate environment contributes significantly to the development of the embryo. Some biochemical work has been done utilizing plant embryos, although these studies have been limited to only a few species of higher plants (Marcus 1969; Ihle and Dure 1972; Walbot et al. 1972). There has been considerable progress in this area recently, and fundamental information has been acquired concerning the role of messenger RNAs and transfer RNAs and the synthesis of specific proteins.

The inaccessibility of most fern embryos precludes similar studies from being performed. Perhaps, in time, newer methods and techniques will be developed which will make biochemical studies of fern embryos possible. Meanwhile, cytochemical and autoradiographic techniques have been employed to obtain information about the content and synthetic activity of various macromolecules during embryo development. At the present time not enough studies have been conducted to be certain the results are general, or to permit valid interpretations to be made. However, a recent study of embryo development in *Marsilea vestita* illustrates the wealth of biochemical information that can be obtained using these simple procedures (Kuligowski-Andres 1975). During the initial stages in embryo development the amount of DNA and RNA present in the zygote is very small. As one might expect, increasing complexity of the zygote generally is correlated with an increase in the amount of cellular DNA and RNA. Not all of the cellular DNA increases. The cytoplasmic DNA (principally chloroplast DNA) remains constant while the nuclear DNA increases. Labeling studies using radioactive DNA precursors have confirmed this finding. They also have shown that in early embryogenesis DNA synthesis is greater in the prothallial cells than in the zygote. This situation changes during later stages where synthesis of DNA in the cells of the embryo increases as organs develop and differentiate. Another interesting feature of these studies is the observed increase in both nucleolar and cytoplasmic RNA content which occurs in cells of the embryo during development. This increase is directly associated with a dramatic increase in the number of nucleoli per nucleus in the cells. While we know nothing concerning the rates of synthesis of the various RNA species during this period, it is well known that nucleoli are the major sites of ribosomal RNA synthesis. It is therefore tempting to speculate that these stages in development are associated with increased synthesis of ribosomal RNA. During animal oogenesis selective replication of the ribosomal genes is the specialized mechanism underlying the massive synthesis of all the ribosomal RNA needed to support later embryo development (Davidson 1968). It would be interesting indeed if a similar mechanism was operating during certain stages of embryogenesis in *Marsilea*. From the preceding account of the experimental studies using ferns, it is obvious that a good deal more work is needed before we begin to uncover the molecular complexities underlying embryogenesis in this group of plants.

References

Albaum H (1938) Inhibitions due to growth hormones in fern prothallia and sporophytes. Am J Bot 25:124–133

Andres J (1964) Recherches préliminaires sur la culture in vitro des embryons de la fougère aquatique *Marsilea vestita* (Marcileacées). CR Acad Sci Paris 258:5956–5959

Atkinson L (1973) The gametophyte and family relationships. In: The phylogeny and classification of the ferns. Suppl 1. J Linn Soc Bot 67:73–90

Bell PR (1963) The cytochemical and ultrastructural peculiarities of the fern egg. J Linn Soc Bot 58:353–359

Bell PR (1970) The archegoniate revolution. Sci Prog, Oxford, 58:27–45

Bell PR (1975) Physical interactions of nucleus and cytoplasm in plant cells. Endeavour 34:19–22

Bell PR, Duckett JG (1976) Gametogenesis and fertilization in *Pteridium*. J Linn Soc Bot 73:47–78

Bell PR, Mühlethaler K (1964) Evidence for the presence of DNA in the organelles of the egg cells in *Pteridium aquilinum*. J Mol Biol 8:853–862

Bilderback DE, Jahn TL, Fonseca JR (1974) The locomotor behavior of *Lygodium* and *Marsilea* sperm. Am J Bot 61:888–890

Cave CF, Bell PR (1974) The synthesis of RNA and protein during oogenesis in *Pteridium aquilinum*. Cytobiologie 9:331–343

Davidson EH (1968) Gene activity in early development. Academic Press, London New York

DeMaggio AE (1961) Morphogenetic studies on the fern *Todea barbara* (L.) Moore. 2. Development of the embryo. Phytomorphology 11:64–76

DeMaggio AE (1963) Morphogenetic factors influencing the development of fern embryos. J Linn Soc Bot 58:361–376

DeMaggio AE, Wetmore RH (1961) Morphogenetic studies on the fern *Todea barbara*. 3. Experimental embryology. Am J Bot 48:551–565

Duckett JG, Bell PR (1971) Studies on fertilization in archegoniate plants. 1. Changes in the structure of the spermatozoids of *Pteridium aquilinum* (L.) Kuhn during entry into the archegonium. Cytobiologie 4:421–436

Duckett JG, Bell PR (1972) Studies on fertilization in archegoniate plants. 2. Egg penetration in *Pteridium aquilinum* (L.) Kuhn. Cytobiologie 6:35–50

Durand-Rivières R (1964) Contribution à la morphogènèse du sporophyte de *Pteris longifolia* L. Ann Sci Nat Bot 12:1–86

Endo MK, Nakanishi K, Näf U, McKeoss W, Walker R (1972) Isolation of the antheridogen of *Anemia phyllitides*. Physiol Plant 26:183–185

Hepler PK (1976) The blepharoplast of *Marsilea*: Its de novo formation and spindle association. J Cell Sci 21:361–390

Ihle JN, Dure L (1972) The developmental biochemistry of cottonseed embryogenesis and germination. 3. Regulation of the biosynthesis of enzymes utilized in germination. J Biol Chem 247:5048–5055

Jayasekera RDE, Bell PR (1959) The effect of various experimental treatments on the development of the embryo of the fern *Thelypteris palustris*. Planta 54:1–14

Jayasekera RDE, Bell PR (1971) The synthesis and distribution of RNA in developing archegonia of *Pteridium aquilinum*. Planta 101:76–87

Kuligowski-Andres J (1975) Contribution à l'étude d'une fougère, le *Marsilea vestita* (Marsiléacées), du stade embryon au stade sporophyte adulte. Ann Sci Nat Bot 16:151–216, 249–308

Laetsch WM (1967) Techniques in developmental biology. TY Crowell Co

Lang WH (1909) A theory of alternation of generations in archegoniate plants based upon the ontogeny. New Phytol 8:104–116

Maheshwari P (1964) Embryology in relation to taxonomy. In: Turrill WB (ed) Vistas in botany. Recent researches in plant taxonomy, vol IV. Pergamon Press, Oxford, pp 55–97

Marcus A (1969) Seed germination and the capacity for protein synthesis. Symp Soc Exp Biol 23:143–160

Sigee DC (1972) The origin of cytoplasmic DNA in the mature egg cell of *Pteridium aquilinum*. Protoplasma 75:323–324

Sigee DC, Bell PR (1968) Deoxyribonucleic acid in the cytoplasm of the female reproductive cells of *Pteridium aquilinum*. Exp Cell Res 49:105–115

Sigee DC, Bell PR (1971) The cytoplasmic incorporation of tritiated thymidine during oogenesis in *Pteridium aquilinum*. J Cell Sci 8:467–487

Steeves TA, Sussex IM (1972) Patterns in plant development. Prentice-Hall Inc., New Jersey

Vladesco MA (1935) Recherches morphologiques et expérimentales sur l'embryogénie et l'organogénie des fougères leptosporangiées. Rev Gen Bot 47:513–528

Walbot V, Brady T, Clutter M, Sussex IM (1972) Macromolecular synthesis during plant embryogeny: Rates of RNA synthesis in *Phaseolus coccineus* embryos and suspensors. Dev Biol 29:104–111

Ward M (1954a) Fertilization in *Phlebodium aureum* J Sm. Phytomorphology 4:1–17

Ward M (1954b) The development of the embryo of *Phlebodium aureum* J Sm. Phytomorphology 4:18–26

Ward M, Wetmore RH (1954) Experimental control of development in the embryo of the fern *Phlebodium aureum*. Am J Bot 41:428–434

Wardlaw CW (1955) Embryogenesis in plants. Wiley and Sons Inc., New York

Wetmore RH (1959) Morphogenesis in plants. Am Sci 47:326–340

Wetmore RH, Steeves TA (1971) Morphological introduction to growth and development. In: Steward FC (ed) Plant physiology: A treatise, vol. VI A. Academic Press, New York London, pp 3–166

Wetmore RH, DeMaggio AE, Morel G (1963) A. morphogenetic look at the alternation of generations. J Indian Bot Soc 42 A:306–320

3. Experimental Embryology of Gymnosperms

K. Norstog

Experimental plant morphology has become an important botanical discipline. Among its major areas is experimental embryology. Composing that branch, in turn, are subdivisions relating to experimental approaches (via microsurgery, embryo culture, etc.) as well as those treating special groups of plants. Not the least of the latter are the gymnosperms, in which interest may be attributed to two features: their undeniable fascination to botanists because they stand between cryptogams and flowering plants in their reproductive development; and their very great economic importance in forestry and horticulture.

While it is possible to find isolated reports of experimental embryology in gymnosperms dating nearly a century past, the beginning of the modern era of experimental embryology of gymnosperms can be credited in large measure to the pioneering work of Prof. Carl D. LaRue of the University of Michigan, and to his influence on other workers. His personal concepts of cellular totipotency and developmental potentials in plants, though not significantly differing from those of Haberlandt and Goebel, were extended to include all aspects of reproduction, and are summed up in a uniquely interesting and challenging paper given before the 6 th Symposium in Biology at the Brookhaven National Laboratory (LaRue 1954). It is noteworthy that in this and in certain other of his works (see Tulecke 1959) he foresaw the employment of pollen and microspore culture as a tool in experimental morphology. However, many other botanists also have been and continue to be active in working with the embryos of gymnosperms, so that the literature dealing with experimental embryology of this group is becoming quite voluminous. While I shall make no attempt at a complete resumé in the following pages, it has been possible to assemble much of the current literature on the subject, and I am especially grateful to Dr C. John Jensen of the Danish Atomic Energy Commission, Roskilde, and Dr Lawson Winton of the Institute of Paper Chemistry, Appleton, Wisconsin, for making their bibliographic materials available to me.

Additionally, the reader may refer to several general reviews for further information on gymnosperm experimental embryology, including those of Rappaport (1954), Narayanaswami and Norstog (1964), and Raghavan (1966). Perhaps, most importantly, valuable lists of references have been assembled by Durzan and Campbell (1974) and Brown and Sommer (1975), citing papers dealing with gymnosperm tissue culture, many of them including embryology and covering, chronologically, the period from 1924 to 1975.

Embryogenic Types

The literature dealing with gymnosperm embryology is extensive (see for example Buchholz 1920; Chamberlain 1935; Maheshwari and Sanwal 1963; Singh 1978) and

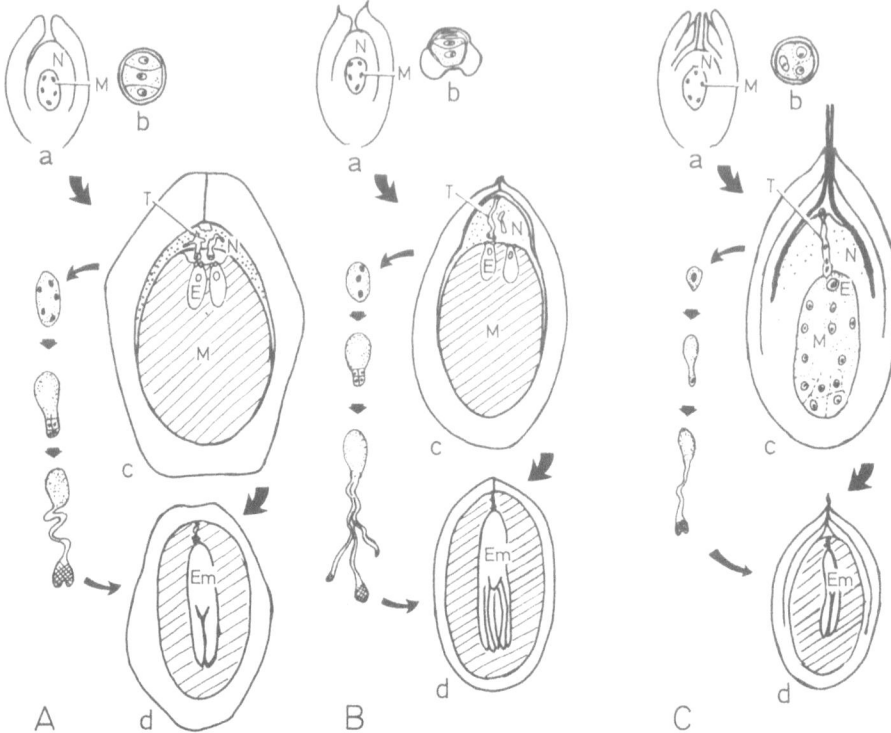

Fig. 1 A–C. Three general embryological types among living gymnosperms. **A** Cycads and *Ginkgo*. **B** Conifers. **C** Gnetopsids. *a*, Ovule with free-nuclear megagametophyte *M*. *b*, Pollen. *c*, Ovule at time of fertilization. *d*, Mature seed. *E*, Egg; *Em*, Embryo; *M*, Megagametophyte; *N*, Nucellus; *T*, Pollen tube in mature megagametophyte. Embryogeny is illustrated to the left of each ovule

it is apparent that considerable variation in embryological development occurs within the group. In many cases these are variations in detail rather than in comprehensive patterns of development, so that one may consider three major embryological types to which most of the living gymnosperms conform (Fig. 1).

One type is found in cycads (Cycadales) and *Ginkgo* (Ginkgoales) in which a major departure from a more common gymnosperm pattern lies in the development of the microgametophyte. The pollen (immature microgametophyte), which is three-celled at the time of pollination, is drawn through the micropyle by the micropylar fluid and comes to lie in a cleft in the nucellus (the pollination chamber), where it slowly develops over a period of about 8–10 months into a mature microgametophyte containing, with one exception, two flagellated and motile male gametes. The exceptional case is *Microcycas*, in which a dozen or more sperms are formed. In the development of the microgametophyte a haustorium is formed but, unlike the situation in other gymnosperms, the proximal part of the microgametophyte, including the pollen wall and the motile sperm cells, forms the pollen tube and grows into the proximity of the archegonium. In other gymnosperms the pollen wall remains where it originally lodged in the nucellus and the pollen tube is formed distally, conveying the nonmotile sperm into the archegonium. Therefore, although a pollen tube is formed in cycads and *Ginkgo*, the microgametophytes are essentially zooidogamous, whereas those of other extant gymnosperms are siphonogamous. During the further growth of the microgametophyte in the nucellus, development of the megagametophyte also occurs. It is at first multinucleate and remains so until about 1,000 nuclei have accumulated, whereupon it becomes cellular. Subsequently a massive megagametophyte, as much as 5 cm in diameter, is formed, having two to many archegonia, depending upon species, but commonly three to five arranged in a cuplike depression (the fertilization chamber) at the apex of the megagametophyte. At the time of fertilization the tissue of the cycad megagametophyte is very dense and its cells are packed with starch; but that of *Ginkgo* is soft and green and exudes a milky substance when cut.

Fertilization occurs when the free-swimming sperm cells penetrate into the archegonium (see Norstog 1972). Both the egg and the zygote tend to be massive (up to 3 mm long) and surrounded by a tough membrane so as to be rather easily excised from the archegonium. Embryo development occurs slowly while the seeds are attached to the parent plant or after the seeds have been shed (some cycads and *Ginkgo*). The embryo is at first free-nuclear but shortly a tier of cells is formed proximally, initiating the suspensor which is composed of several tiers of tubular cells. The suspensor continues to grow and eventually forms a compact button of small cells (the embryo proper) at its distal terminus. (In *Ginkgo* there is no well-defined, elongated suspensor, but a micropylar region of elongated cells is present.) The mature embryo is dicotyledonous. There is no after-ripening requirement, and germination of Cycad and *Ginkgo* seeds takes place whenever a mature embryo and a suitable substrate are present. In both the cycads and *Ginkgo*, the cotyledons remain within the megagametophyte for many months while the seedling becomes established.

A second embryological type, manifested by conifers and taxads (Coniferales) departs from the development described for cycads and *Ginkgo*, principally in the manner of pollen tube formation. The sperms are nonflagellated and are conveyed directly to the egg by the pollen tube (siphonogamy). Development of the megagametophyte resembles that occurring in cycads and *Ginkgo*, although there is greater variation in the structure and size of archegonia among the Coniferales and, in general, the megagametophytes tend to be less massive. Development of microgametophytes and megagametophytes is slow, lasting a number of months (up to 11 months in *Pinus*). Shortly after fertilization the embryo is briefly free-nuclear, but soon an elongated suspensor forms. Its distal end usually separates into several secondary suspensors, each of which terminates in an embryonic bud (cleavage embryony). Ordinarily only one embryo develops further into a mature embryo, which, depending on the genus, may be dicotyledonous or polycotyledonous. Occasionally two or more embryos develop, and haploid plants have been reported among polyembryos of *Picea abies* (Illies 1964).

A third embryological type is represented by *Ephedra*, *Gnetum*, and *Welwitschia* (Ephedrales, Gnetales, and Welwitschiales) departing both from the previous types and from each other. Microgametophytes of all three are siphonogamous, and pollen tube growth may be comparatively rapid (about 10 h required for pollen tube development in *Ephedra*). In *Gnetum* the megagametophyte is free-nuclear at fertilization and later becomes cellular; in *Ephedra* and *Welwitschia* the megagametophytes are cellular at fertilization. Cleavage polyembryony occurs among the gnetopsids, and dicotyledonous embryos are formed by all. A peculiarity in embryos of *Gnetum* is the feeder that develops from the hypocotyl and serves as a haustorium in absorbing nutrients from the endosperm (Vasil 1963).

Culture of Embryos

Most embryological studies with gymnosperms have involved the excision of embryos and their culture upon nutrient media. The earliest of these appears to have been done by A. Schmidt in 1924 (cited in Durzan and Campbell 1974) who cultured embryos of *Pinus sylvestris* in organic nutrient solutions, and Li (1934), and Li and Shen (1934) who reported on the culture of *Ginkgo* embryos. These studies served to demonstrate that excised embryos of gymnosperms could be grown into seedling plants upon relatively simple nutrient media in substitution for their natural substrate (i.e., the megagametophyte). Embryos, representing a number of families of seed plants, were excised and cultured by LaRue (1936). These included *Pinus resinosa*, *Picea canadensis*, *Tsuga canadensis*, and *Pseudotsuga menziesii* *(taxifolia)*. LaRue demonstrated that immature gymnosperm embryos 2–4 mm in length could be grown into seedling plants under aseptic conditions. Subsequently, many investigators cultured excised embryos of gymnosperms, and it has been shown that the mature embryo is capable of seedling growth independent of the megagametophyte. A quite complete survey of these studies is given in the appendix of a review by Durzan and Campbell (1974).

Attention will now be directed to certain investigations dealing with specific problems in experimental embryogenesis.

Developmental Potential of Proembryos

The gymnosperm proembryo may be defined as a stage, or stages, encompassing the early free-nucleate divisions of the zygote, as well as the early cellular stages of embryogenesis. The latter comprise the suspensors and the terminal tier of cells composing the embryonic bud, sometimes called the "embryo proper" or the proembryo. Figure 1 illustrates that all gymnosperms have a basically similar pattern of early embryogenesis, though they differ in the number of embryonic buds formed distally on the suspensor(s).

Among the studies in which proembryos of gymnosperms have been cultured are those of Radforth and his associates (Radforth 1936; Radforth et al. 1958), in which one of the objectives was the assessment of the developmental potentialities residing in the embryo itself. As we have noted, the gymnosperm zygote in its early development forms a more or less elongated suspensor which ends in an embryonic bud. The polarity exhibited by these embryos in their development is parallel to the axis of the archegonium, and a question therefore arises as to whether the orientation, as well as the symmetry of these embryos, is directed from outside the embryo or, conversely, determined by endogenous factors that may be genetically controlled.

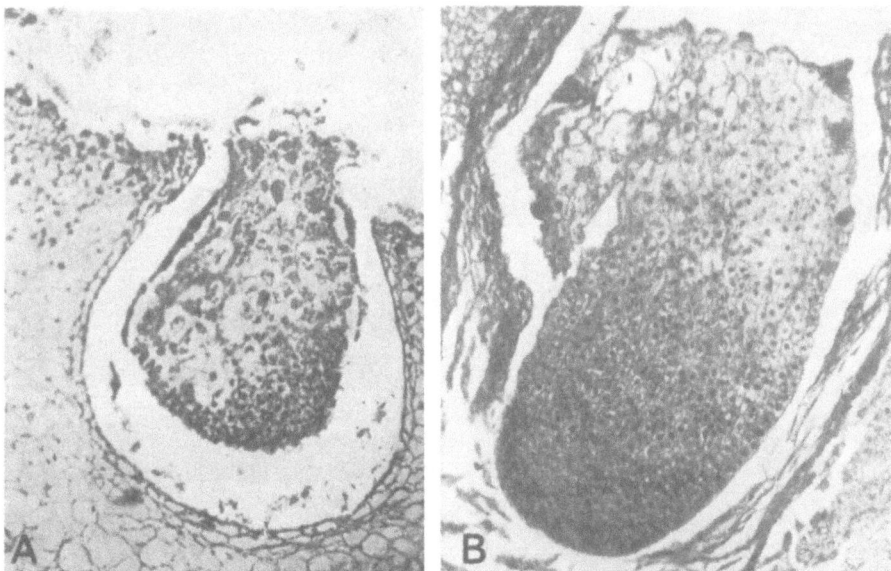

Fig. 2 A, B. *Ginkgo* embryos of equivalent age in vitro and in ovule. **A** Embryo excised with some surrounding gametophytic tissue and cultured in vitro; note lack of polarity. **B** Embryo in ovule; note polarity. (After Radforth et al. 1958)

Radforth and his colleagues (Radforth 1936; Radforth and Bonga 1960; Radforth et al. 1958) cultured proembryos of *Ginkgo* and *Pinus nigra*. They found that in each instance the embryo did not develop suspensorial organization, but rather developed radially, forming callus-like masses. The results in the culture of *Ginkgo* proembryos were particularly interesting (Radforth et al. 1958). Archegonia were excised together with a minimum amount of surrounding gametophyte tissue and cultured on nutrient agar. The archegonia contained free-nuclear embryos, and suspensor formation had not yet begun. Although cell division occurred in these cultured embryos, suspensor formation did not occur in vitro. This failure of the embryo to become polarized was ascribed to the uniform accessibility to nutrients by the embryo (Fig. 2 A, B).

Sterling (1949) attempted to grow immature embryos of larch and observed that proembryonic cells were cleaved in some cultures but none completed normal development. Similar results were reported by others (Loo and Wang 1943; Radforth and Pegoraro 1955; Berlyn 1962; Konar and Oberoi 1965), and were no doubt due partly to suboptimum culture media. More recently, Thomas (1970) cultured undifferentiated proembryos of *Pinus* on the three media of Heller, Halperin, and Murashige and Skoog, which have been used with wide success in culturing both embryos and tissues of many plants. Thomas reported that proembryos survived up to 48 h on all three media, but only the Halperin medium permitted longer survival, up to 8 days. However, neither cell division nor cellular elongation were observed.

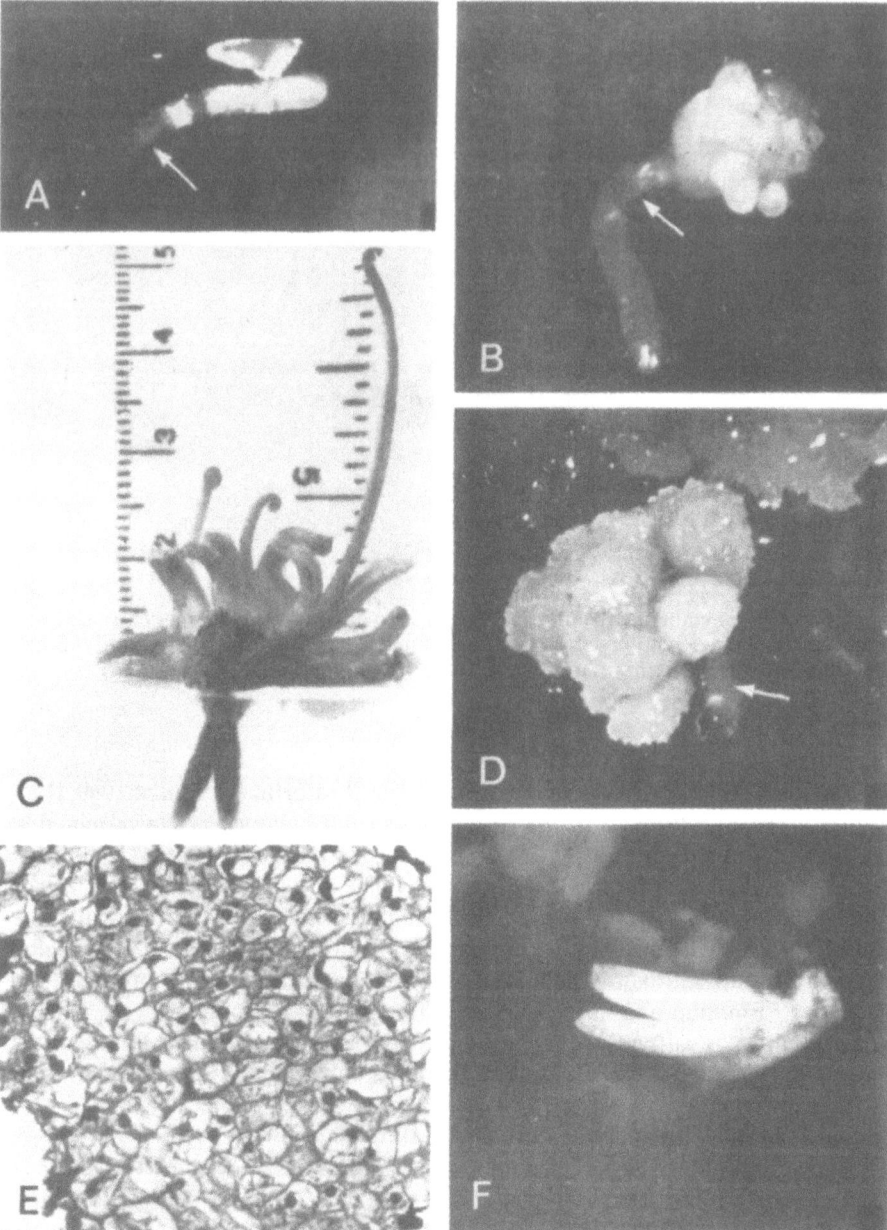

Fig. 3 A–F. Development of *Zamia* proembryos in vitro. **A** Proembryo at time of excision; *arrow* shows suspensor. **B** Proembryo after several weeks of culture; note budding of embryo proper; *arrow* shows suspensor. **C** Adventive embryos reared from callus. **D** Callus derived from proembryos; *arrow* shows suspensor. **E** Section of callus derived from proembryo cultured upon medium containing 2,4-D; note absence of cellular differentiation. **F** Adventive embryo formed from embryo-derived callus grown on a 2,4-D-free medium. (After Norstog and Rhamstine 1967)

On the other hand, Norstog and Rhamstine (1967) cultured proembryos of *Zamia integrifolia* (see Fig. 3 A–F), and found that embryos in which suspensors had differentiated formed essentially normal, dicotyledonous embryos but, in a few cultures, polycotyledony was observed. Normal or near-normal differentiation occurred only on media lacking auxin and kinetin (KN). When relatively low concentrations of 2,4-D were employed (0.01–1.0 mg/l) together with KN (0.5–0.5 mg/l), a callus was formed that subsequently could be sub-cultured. Eventually, upon transfer to an auxin-less medium, embryos were formed.

Vasil (1963) cultured proembryos of *Gnetum ula*, in which the zygote normally divides in vivo to form two daughter cells, each forming a suspensor. These in turn produce other suspensors. A so-called peculiar cell is formed at the tip of each suspensor when the seeds are shed. An embryo is derived from one of the peculiar cells, requiring about 12 months to reach maturity. Vasil, seeking to determine whether embryos might develop more rapidly in vitro than in vivo, cultured embryos at the peculiar cell stage and discovered that growth occurred only on media containing casein hydrolysate (CH) and yeast extract (YE). In addition, she found that in some cases callus formed from the peculiar cell. This callus in turn formed filaments resembling suspensors, from the tips of which proembryos developed. However, differentiation of the proembryos stopped short of the initiation of cotyledons, and an absorptive structure of the natural embryo, the feeder, failed to develop.

Experiments with Full-term (Mature) Embryos

Attempts to culture proembryos of gymnosperms have been disappointing, and even when such efforts have met with partial or complete success, it appears that entirely normal patterns of embryogenesis have not occurred in vitro. Usually the cultured embryos tended to exhibit radial symmetries or asymmetries rather than axiate organization. Excised and cultured full-term or "mature" embryos of gymnosperms, unlike proembryos, have been grown readily and utilized in a number of interesting experiments. Ball (1956a) pointed out an important rationale in plant embryo culture, stating that if one employs excised embryos of suitable maturity, so that there are no difficulties in achieving their growth in vitro, then one may analyze a wide variety of morphological responses by manipulation both of the embryo and of the culture medium. In his cultures of *Ginkgo*, three regions of the embryo were studied: (1) the cotyledons, which in vivo are haustorial structures; (2) the shoot apex, and (3) the tip of the hypocotyl, from which originates the primary root. Using partially matured *Ginkgo* embryos, Ball observed that the primary root originated endogenously between the base of the hypocotyl and the hypocotyl cap. Therefore, he concluded that the hypocotyl meristem is not synonymous with the root meristem [contrary to earlier reports that the primary root arises exogenously from the embryo (Troll 1949, cited in Ball 1956a)]. Further studies by Ball (1956 b) demonstrated the developmental potential of the root initials, which upon being longitudinally split, generated two new and complete apices (Fig. 4 A, B).

To a degree, the capacity for organ regeneration from cultured parts of gymnosperm embryos has also been shown. Hypocotyl segments, excised from em-

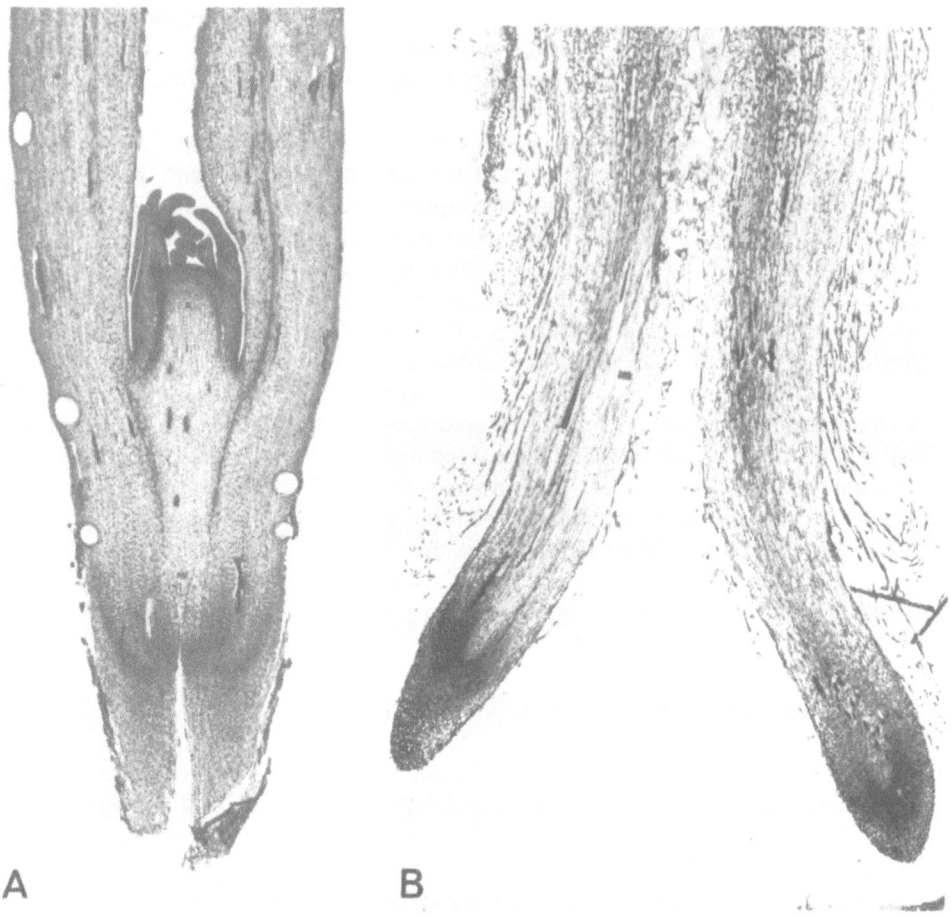

A   B

Fig. 4 A, B. Effect of a longitudinal split in the tip of the hypocotyl of *Ginkgo* embryo on root development in vitro. **A** Longitudinal section of an embryo fixed immediately after the operation. **B** Longitudinal section of an embryo such as shown in **A** after 9 days in vitro; note development of an essentially normal root from each half-hypocotyl tip. (After Ball 1956 b)

bryos of *Pinus gerardiana* and cultured on a medium containing 2,4-D, produced a subculturable callus in 6–8 weeks (Konar 1974). Differentiation of tracheid-like cells at higher sugar concentrations was reported. More recently, rhizogenesis and formation of shoot primordia have also been observed (Konar 1975) in callus cultures of the above species.

Callus cultures derived from excised embryos of Douglas fir *(Pseudotsuga menziesii)* were established on a defined medium supplemented with $5 \mu M$ each of auxin (IAA and IBA) and cytokinin (benzylaminopurine [BAP] and isopentyl-aminopurine [2iP] by Cheng (1975). Organogenesis was achieved by subjecting cultures to a higher concentration of cytokinin (0.5–1.0 mM) prior to transfer to a cytokinin-less medium. Growth of adventitious buds was stimulated by a low salt medium lacking phytohormones. Cotyledon explants obtained from 1 to 3-week-old seedlings also produced adventitious buds when similar procedures were employed. Moreover, tissue subcultured three to four times at about 3-week intervals still retained a potential for bud formation.

Zygotic embryos of Douglas fir placed on a defined medium containing 0.05 or 0.1 mg/l BAP produced shoots from cotyledons (Winton and Verhagen 1977). One vigorous shoot usually developed from the tip of each swollen cotyledon on 0.05 BAP-medium, but small, multiple shoots were produced from cotyledons on 0.1 BAP-medium. After transfer of embryos or excised cotyledons to media without hormones, the vigorous shoots developed normally, but the small multiple shoots did not. Conversely, when cotyledons were grown on a medium containing auxin and cytokinin, callus was produced, which was isolated and subcultured monthly. Shoots were produced from subcultured cotyledon callus in 11 of 35 cultures, a frequency of 31%. Two of 30–50 excised shoots rooted after treatment with 10 mg/l IBA. A few shoots were also produced from subcultured needle callus and from stem explant callus obtained from young seedlings.

Slices of hypocotyl of mature, excised embryos of *Pinus lambertiana* were cultured on nutrient agar by Greenwood and Berlyn (1965), who reported that roots were generated in direct relation to the dimension of the slices and to their relative position in the hypocotyl. Slices taken near the cotyledons did not regenerate as many roots as those taken near the hypocotyl tip. Slices 3 mm in thickness regenerated more roots than those that were 2 mm thick. These findings suggest that growth regulators originally present in the hypocotyl may regulate subsequent cell differentiation (greater amounts presumably being present in the large pieces). Also, they suggested that there is a polar gradient present, either of regulator concentrations or of variations in sensitivity of cells to such regulators.

Bearing somewhat on these assumptions, we may note that Ball (1959) showed that the root and shoot of *Ginkgo* embryos have different requirements with respect to such metabolites as sugars (sucrose, glucose, raffinose, galactose, levulose), indole-3-acetic acid (IAA), glutamine, and coconut milk (CM) (Fig. 5). In all cases, uptake of these substances occurred through the cotyledons alone (Fig. 6), indicating the interesting possibility that the cotyledons may not only take up and distribute growth regulators, but also may mediate the morphogenesis of root and shoot in some unknown way. Interestingly, sole uptake of nutrients through the cotyledons of cultured *Ginkgo* embryos had been reported earlier by Bulard (1952).

In another series of experiments by Greenwood and Goldsmith (1970), pine embryos were ehypocotylized and cultured with only their cotyledons embedded in the media. It was observed that auxin was taken up by the cotyledons and transported basipetally. This resulted in the regeneration of the primary root. This system deviates importantly from the majority of morphogenetic experiments and commercial rooting practice, in which auxins were and are applied directly to callus and/or cuttings. Also tested in this study was the hypothesis that polar transport of auxin requires the presence of a meristem (i.e., root tip) to serve as an auxin sink. Greenwood and Goldsmith concluded that: (1) a growth center was not required, (2) auxin is transported basipetally in the absence of a root apex, (3) auxin accumulates at the wound surface endogenously, and (4) that it stimu-

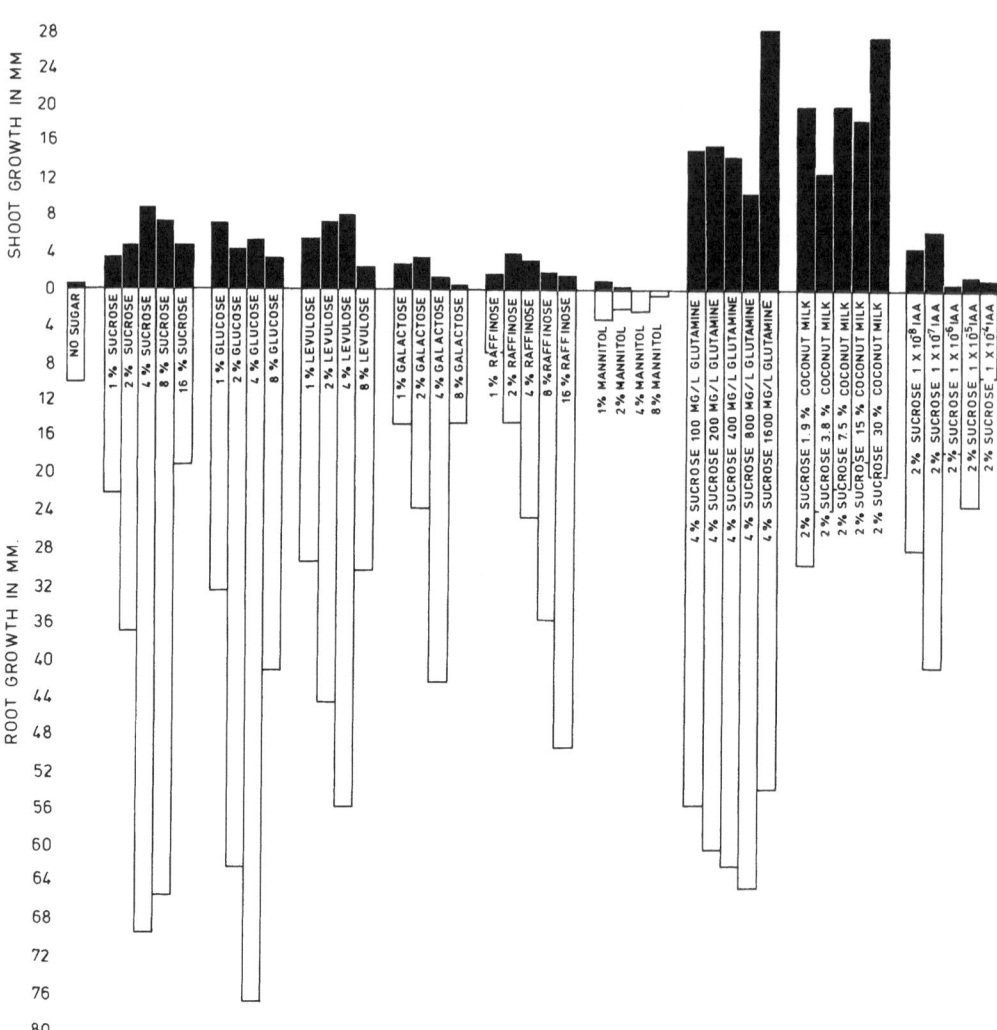

Fig. 5. Histogram of average growth in length of shoots and roots of seedlings of *Ginkgo* grown with the cotyledons in nutrient agar. The naturally-occurring sugars supported better root growth than shoot growth. Glutamine greatly stimulated shoot growth and inhibited root growth. IAA at higher concentrations inhibited both root and shoot growth. Mannitol, a sugar alcohol not occurring normally in these tissues, inhibited growth to a greater extent than did sugar-free agar media. (After Ball 1959)

lates the regeneration of root primoridia in the region of auxin accumulation, i.e., at the wound site. Because of their role in long term uptake, cotyledons of gymnosperm embryos are very appropriate systems for the analysis of absorption, transport, interactions, and developmental effects of growth regulators.

Although substances in the culture medium are taken up only through the cotyledons in *Ginkgo* (Bulard 1952; Ball 1959), this does not appear to be the case universally in gymnosperms. Sacher (1956) cultured excised embryos of *Pinus lam-*

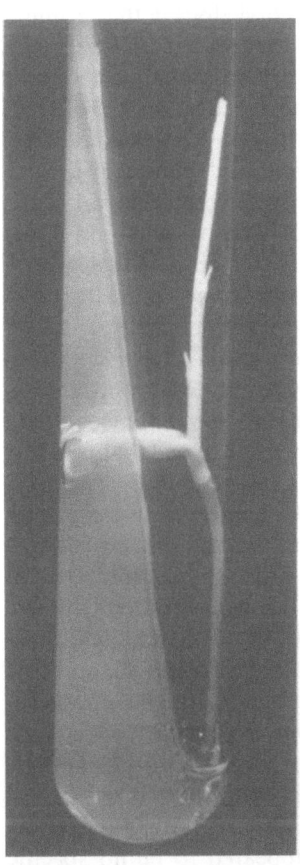

Fig. 6. *Ginkgo* embryo grown in vitro illustrating Ball's method of culturing embryos with only the cotyledons embedded in agar. (After Ball 1959)

bertiana and grew them into seedlings. An important characteristic of these embryo cultures, maintained under continuous light upon media of several types, was that after an initial period of root development, root growth ceased and the root apparently became senescent. However, some measure of root growth could be restored with the application of naphthaleneacetic acid (NAA), which induced the formation of short lateral roots. Interestingly, embryo growth occurred regardless of whether the cotyledons were embedded in the culture medium. However, Bartels (1957a, b) cultured embryos of *Pinus*, *Pseudotsuga*, and *Picea* with only the cotyledons embedded in the media (a technique highly developed by Ball 1956a), and found that root development under these conditions was not inhibited to the extent reported by Sacher. Brown and Gifford (1958) similarly cultured pine embryos, and obtained much the same result as Bartels. They devised a double-medium technique in which the embryos were suspended with their cotyledons embedded in agar, and the hypocotyl tips immersed in a liquid medium contained in an inner vial. Excellent root growth was obtained if the roots were immersed in a sugar-free medium, but root elongation was almost completely inhibited when sugar was present.

Berlyn and Miksche (1965) also studied the role of cotyledons in the germination of excised gymnosperm embryos. In particular, they were concerned with problems of root development which, as noted, Sacher (1956) had reported to be inhibited in cultured pine embryos. Setting out to determine the influence of cotyledons on development, as well as factors affecting root growth, Berlyn and Miksche cultured ecotyledonized embryos of *Pinus lambertiana* and found they grew as well as those with intact cotyledons. They further discovered that removal of the shoot meristem also suppressed root development, and that the orientation of the embryo in relation to gravity and light greatly altered embryo growth. In dark cultures, horizontal embryos grew more rapidly than those oriented vertically (in all cases embryos were cultured with only the cotyledons immersed in the medium). In light, vertical embryos grew more rapidly than horizontal ones and, in addition, the roots were negatively geotropic. These position and light effects appeared to be peculiar to excised embryos, for embryos cultured with gametophytes intact showed no differential effects of light or orientation. They concluded that, in some unknown way, the gametophyte inhibits the tropic responses of the embryo.

Normally germinating embryos of gymnosperms have the ability to develop chlorophyll in the dark (Bogorad 1950; Sacher 1956). This dark synthesis of chlorophyll in intact seeds has been shown to depend on an interaction between megagametophyte and embryo, but excised embryos grown on nutrient media generally do not develop chlorophyll in the dark. However, Bogorad (1950), and Sacher (1956) showed that a germinating *Pinus* embryo may maintain chlorophyll synthesis for a short period in the dark following removal of the megagametophyte. Engvild (1964) attempted to develop a "cotyledon medium" capable of sustaining the dark synthesis of chlorophyll in isolated embryos. He made cultures in which embryos were suspended from an agar medium with the cotyledons embedded in agar media variously containing yeast, malt, and megagametophyte extracts. The results of this study showed that there appeared to be no specific chlorophyll synthesis stimulating molecule involved, but that chlorophyll synthesis could be promoted by sucrose, B vitamins, urea, and an amino acids mixture.

The dormant condition of *Taxus* embryos was investigated and reported in a series of papers by LePage-Degivry (1968, 1970, 1973 a, b, c), and LePage-Degivry and Garello (1973). When immature embryos were cultured on agar media, they grew only slightly and did not exhibit any of the characteristics of germinating embryos in vivo, such as root development or epicotyl elongation. However, when such embryos were first cultured in a liquid medium, then transferred after 8 days to agar media, germination in vitro ensued. They thought that germination inhibitors were leached out of the embryos during the period of liquid culture. Subsequently, methanol extracts of excised, uncultured embryos were analyzed and found to contain a substance having the properties of abscisic acid (ABA). The liquid medium upon which embryos had been cultured for 15–20 days also showed the presence of a substance with the properties of ABA, but extracts of the embryos themselves did not show any ABA-like substance. In addition to leaching, it was shown that there was a chilling requirement involved in breaking embryo dormancy in *Taxus*, and that either chilling at 4 °C or a treatment with gibberellic acid, which substituted for chilling, was required for germination. To a similar extent gibberellic acid was also shown to promote the germination in vitro of excised *Ginkgo* embryos (Bulard and LePage-Degivry 1968).

Adventitious Embryo Development

Recently, several workers have reported that under some conditions excised mature gymnosperm embryos will, in turn, produce adventitious buds and embryos from their cotyledons and hypocotyledonary areas.

The induction of adventitious embryos in angiosperm cultures is by now a well-documented phenomenon (Raghavan 1966; Halperin 1969), and has been reported to occur in cultures of gymnosperms also (Konar 1963 a, b; Konar and Oberoi 1965; Norstog 1967; Norstog and Rhamstine 1967, Sommer et al. 1975).

When mature embryos of *Thuja orientalis (Biota orientalis)*, a member of the Cupressaceae, were cultured on certain media, adventive embryos developed from the cotyledons in 8 weeks (Konar and Oberoi 1965). These were isolated and developed into shoots with spirally arranged leaves, but did not form roots. Sommer et al. (1975) cultured embryos of *Pinus palustris*, and observed that numerous adventive embryos were formed along the elongating cotyledons (Fig. 7 A–D). These could be excised and cultured individually on nutrient agar and subsequently rooted and developed into complete plants. Apart from demonstrating the mor-

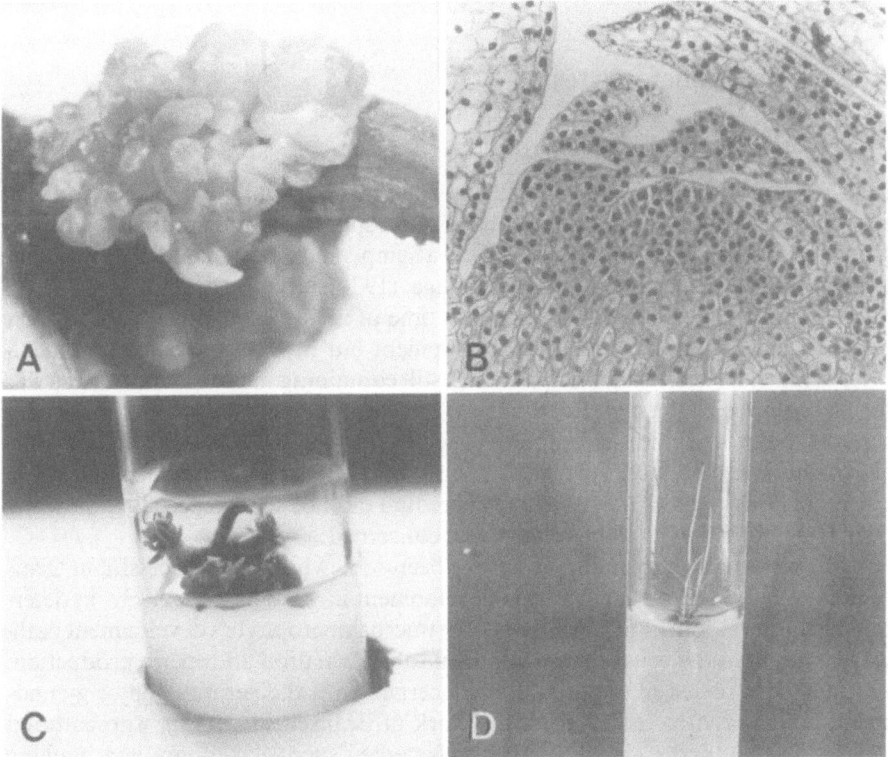

Fig. 7 A–D. Formation of embryos upon cultured embryos of *Pinus*. **A** Group of buds (embryos) upon the cotyledon. **B** Section through apex of embryo showing differentiation of leaves. **C** Explanted cotyledonary buds (embryos). **D** Pine plantlet from explanted embryo. (After Sommer et al. 1975)

phogenetic potential of the cotyledonary tissue of seed embryos, the possibilities of clonal propagation of valuable horticultural and forest species are obvious. However, seed-embryo clones exhibit greater genetic variation than do somatic clones.

Production of adventitious buds in cultures of cotyledon explants derived from young seedlings of Douglas fir *(Pseudotsuga menziesii)* required both cytokinin and auxin (Cheng 1977). A pronounced influence on morphogenetic responses through variations in physiological condition or genetic complement was observed using the natural auxin IAA. However, this effect was nearly eliminated when equivalent concentrations of synthetic auxins, e.g., NAA or 2,4-D, were used. BAP was more effective in stimulation of organ formation than 2iP or KN. A high frequency of organ differentiation was obtained with 5 μM BAP plus either NAA or 2,4-D at 0.5–50 μM.

Not all attempts to obtain adventive embryos in gymnosperm cultures have been successful. Seed embryos of *Welwitschia*, when excised and cultured on nutrient agar without auxin, grew well (Button et al. 1971). When seed embryos were cultured on a medium containing an auxin, a subculturable callus was formed. All attempts to obtain adventive embryos by plating out callus-dervied cells and tissues on auxin-free media failed.

Cultures of Gametophytes

Microspores and Microgametophytes

Microspore cultures of seed plants have been made, usually with excised microsporophylls (anthers) rather than isolated microspores and then primarily with angiosperm material (see Nitsch 1969). Few attempts have been made with such material collected from gymnosperms. Bonga (1974) attempted cultures of *Pinus* microsporophylls explanted at about the time of meiosis, but achieved only very limited results (some post-meiotic development but no callus formation). In cultures using older material, but apparently still containing microspores, a callus was formed composed of a mixture of $1n$ and $2n$ cells. Included in the callus mass were some globular structures resembling proembryos (Fig. 8). Although a subculturable callus was obtained, further differentiation of embryoids in subcultures did not occur. Bonga has stated that the procedure did not seem especially promising insofar as obtaining clonable plants was concerned.

Pollen cultures of gymnosperms have been somewhat more successful in terms of obtaining normal or near-normal development in vitro or, conversely, in determining the developmental potentials of the microgametophyte's development pathways other than the usual course of pollen tube formation and sperm production. LaRue (1954) reviewed the literature concerned with the germination of gymnosperm pollen in vitro, alluding to the work of Branscheidt (1930), who cultured pollen of *Taxus* and *Cupressus* in a supplemented sucrose medium (in a hanging drop culture) and noted that sperm nuclei were formed in about 20 days. LaRue (1954) cultured pollen grains of the cycad *Zamia* on nutrient agar, and recorded their development through several stages ending in the penultimate division of the

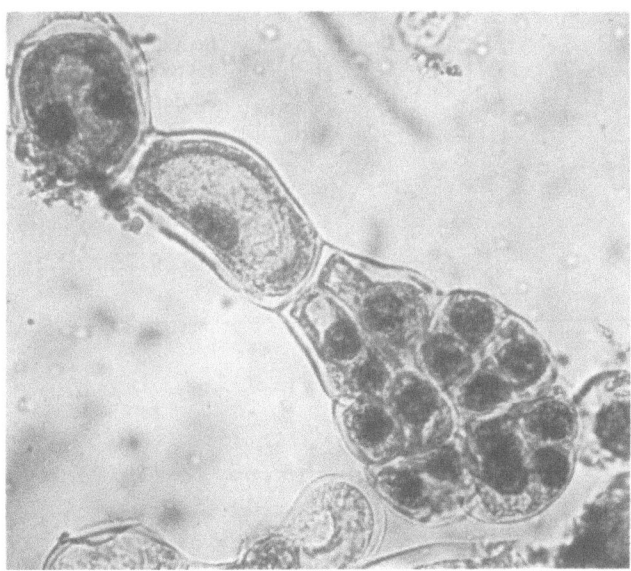

Fig. 8. Embryo-like outgrowth from cultured pollen of *Pinus*. (After Bonga 1974)

generative cell. Blepharoplasts were formed in the sperm mother cells but not spermatozoids.

Konar (1963 b) reported the results of culturing microsporangia of *Pinus* excised at the microspore stage, and stated that microgametophytes were formed, which developed as far as the division of the antheridial cell into a stalk and a body cell. A special case seems to be the germination of pollen of *Ephedra:* fully developed pollen tubes containing apparently normal sperm cells were grown on media containing ovular extracts (Mehra 1938). However, the development of pollen tubes in *Ephedra* is extremely rapid (10 h or so) as compared with other gymnosperms (up to 10 months), and since they are nonhaustorial, their nutritional requirements in vitro may not be so demanding.

The first subculturable callus derived from pollen was from *Ginkgo* (Tulecke 1953). Tulecke (1957) noted that in some cases the cultured pollen of *Ginkgo* developed as far as the immature sperm stage, although maturation of the sperm into motile cells was not observed (Fig. 9). In other cases, *Ginkgo* pollen formed large multinucleated structures, and in still other instances, multicellular bodies were formed which gave rise to a subculturable callus. This tissue culture was maintained in vitro continuously for a number of years without undergoing cell differentiation (Tulecke and Nickell 1960). Similarly, subculturable callus was obtained from the pollen of *Taxus* (Tulecke 1959), and *Torreya* (Tulecke and Sehgal 1963). Rao and Mehta (1969), culturing pollen of *Thuja orientalis*, reported that a subculturable callus was produced in 4–6 weeks. The culture was composed of parenchymatous tissue, but the regeneration of buds, roots, or adventive embryos did not occur. Razmologov (1973) observed that the generative cell of the pollen grain of *Cupressus* produced a callus-like tissue in vitro. Interestingly, none of the above-

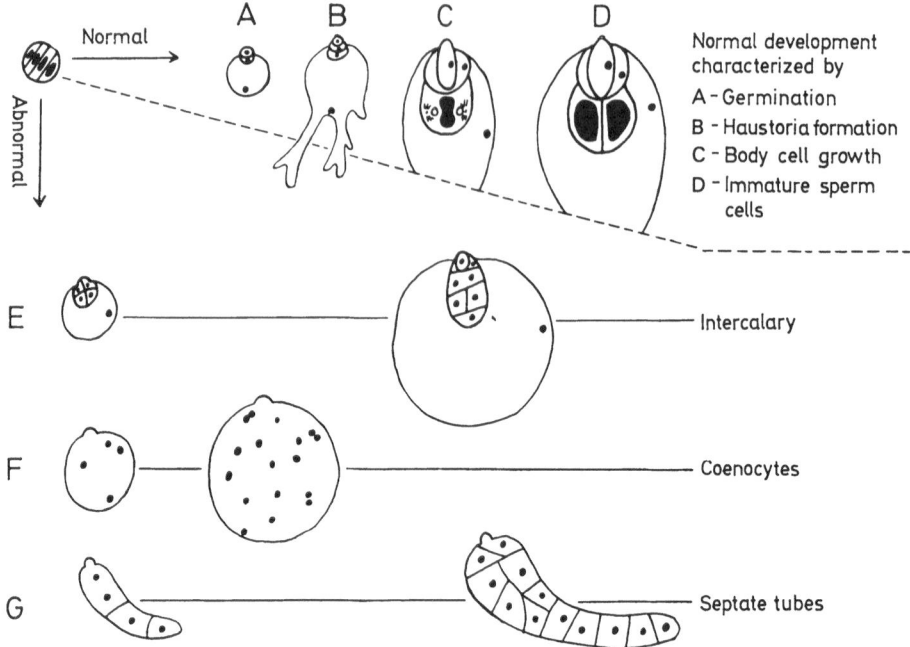

Fig. 9. Normal and abnormal development in vitro of cultured *Ginkgo* pollen. Note embryo-like structures in *E* and *G*. (After Tulecke 1957)

named pollen or microspore cultures or their derived tissue cultures have resulted in the regeneration of buds, roots, or embryos. The reason for this is not clear; it may be related to a greater capacity among gymnosperms for proliferation of undifferentiated vs differentiated cells, or it may simply reflect a failure to select the proper stages for culture. In angiosperms, particularly among the few species of dicotyledons (*Nicotiana, Datura, Petunia*, etc.) which regularly produce androgenous embryos, the developmental pathways leading to embryogenesis are usually quite direct (Nitsch 1969).

Rashid and Street (1974) followed the early stages in adventive embryogenesis in *Nicotiana sylvestris* microspore cultures, and observed that the first division of cultured premitotic microspores producing embryos yielded equal-sized cells. On the other hand, binucleate pollen grains, which have unequal-sized cells, either became swollen and filled with starch or germinated or, if they divided further, fell short of full embryo development. With respect to gymnosperm pollen cultures, it may be that the correct stage for induction of embryos, the premitotic microspore, may not have been utilized. That this may be the case is suggested by recent studies by Bonga and McInnis (1975) concerning development in cultured microsporophylls of *Pinus resinosa*. In these cultures the first division of the immature pollen cells produced equal-sized cells. Cold treatment resulted in production of more pollen with equal-sized cells and an increase in callus initiation. When cold treatment was followed by centrifugation, callusing increased significantly. The authors suggest that centrifugation of pollen may have increased the induction of embryos in

Nicotiana, where centrifugation had been used to separate pollen from other debris (Nitsch 1974; Nitsch and Norreel 1973). However, the example is not clear in the case of gymnosperms, since even the centrifuged pollen failed to form embryos.

In retrospect, the culture of microspores and microgametophytes have not produced the hoped-for clones of haploid tissues and embryos, although, in some instances cited above, the production of subculturable amounts of callus holds promise that future improvements in technique may result in cloning and diploidization productivity comparable with that now possible in some angiosperms. Another promising event is the production of mature microgametophytes in vitro. For example, the microgametophytes of cycads and *Ginkgo* produce flagellated sperm cells. Much additional information regarding spermatogenesis could be learned from the mass culturing of microgametophytes of these plants. Nor would the benefits of such cultures be restricted to those plants having flagellated sperm cells, for very little is known, especially at the ultrastructure level, of the development of microgametophytes and sperm cells of other gymnosperms. Indeed, an ultrastructural study of sperm cells formed within cultured pollen tubes of *Taxus baccata* has been reported recently (Rohr 1973). In certain electron micrographs presented in his report, various ultrastructural aspects of well-defined sperm cells are in evidence.

Megagametophytes

The earliest record of regeneration by megagametophytes of a gymnosperm is that of DuChartre (1888) who described root formation in ovulo by megagametophytes of *Cycas*. A similar observation in *Zamia* was made by Coulter and Chrysler (1904). The production of plants through vegetative growth of prothallial tissues (apogamy) is of considerable scientific interest, and of value also to geneticists, horticulturalists, and foresters as a means of rapid vegetative propagation. It is an important demonstration of the ability of cells of normally restricted ontogeny to manifest almost the entire range of morphogenetic potential of the species (totipotency).

Undoubtedly one of the more vigorous pursuits of this means of expressing the growth potentialities in haploid tissues was that of LaRue. In his 1954 paper on regeneration in gametophytes and sporophytes of gymnosperms, LaRue reported the regeneration of both roots and shoots in cultured megagametophytes of *Zamia* (Fig. 10) and *Cycas*, as well as the production of adventive embryos from megagametophytes of *Zamia*. Cultures of microsporangia of 20 other gymnosperm species were also reported. Subsequently, Tulecke (1964, 1967) reported that *Ginkgo* megagametophytes produced a subculturable callus, but unlike *Zamia*, there was no regeneration of roots, shoots, or embryos.

The female gametophyte of *Ginkgo biloba* was also grown in vitro by Rohr (1977). In gametophytes cultured 4–6 months after fertilization, the superficial tissues turned intensely green and produced a variety of outgrowths. One of them, never observed before, seemed to be derived from the abnormal development of archegonial initials. A tissue has been isolated from the gametophyte and subcultured; it continues to grow after several transfers on the same medium.

Fig. 10. Regenerated roots and shoots on a cultured megagametophyte of *Zamia*. (After LaRue 1948)

   Borchert (1968) reported similar callus cultures to those of LaRue and Tulecke derived from megagametophytes of *Pinus lambertiana*. Megagametophytes of *Pinus resinosa*, *P. nigra*, and *P. mugo* have also been cultured. Only non-subcultur-able proliferations of callus were obtained from the female gametophytes of *P. re-sinosa* (Bonga and Fowler 1970), but after 5–7 weeks of culture, 10% of megagametophytes of *P. nigra*, and about 14% of *P. mugo* began to produce a bright green callus capable of subculture (Bonga 1974). The callus formed by *P. ni-gra* was composed of haploid cells, and that derived from *P. mugo* was predominantly haploid with a few diploid cells. Callus was subcultured a number of times, but at the time of writing none had differentiated except for a few tracheid-like cells.

   Haploid tissues of *Zamia* (LaRue 1948; Norstog 1965) and both haploid and diploid tissues of *Zamia* and *Cycas* have been cultured (LaRue 1950; Norstog and Rhamstine 1967; Norstog 1967) (Fig. 11 A–F). Megagametophytes of *Zamia* re-

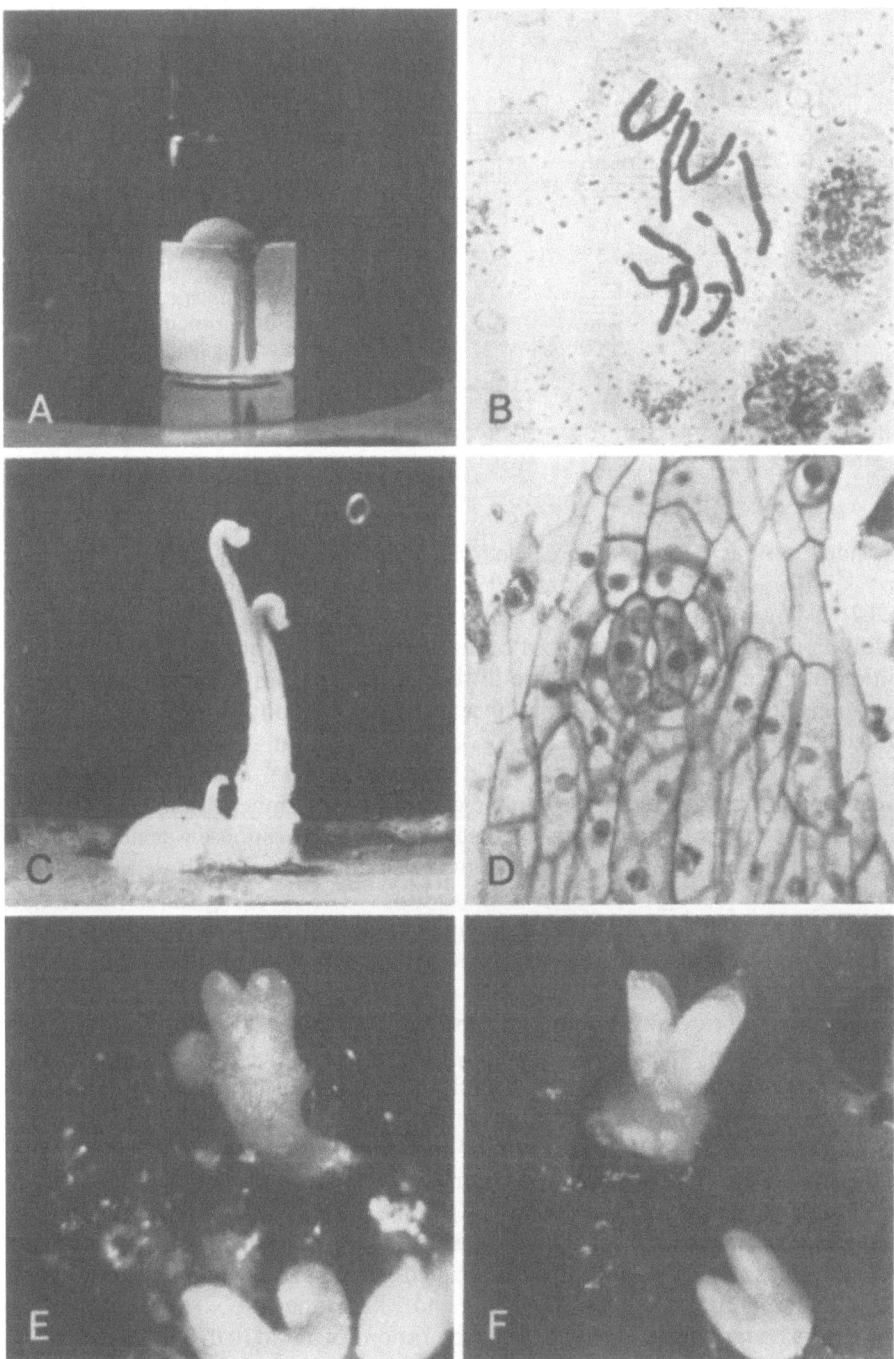

Fig. 11 A–F. Regeneration in cultured megagametophytes of *Zamia*. **A** Root formation. **B** Haploid chromosomes of regenerated roots. **C** Leaves of regenerated buds. **D** Stoma on regenerated leaf. **E** Embryos from $2n$ tissue culture. **F** Embryo from $1n$ tissue culture (cf. **E**). (After Norstog 1965)

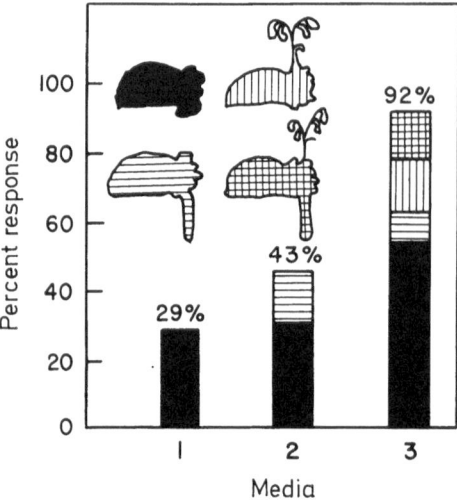

Fig. 12. Growth and development of *Zamia* megagametophytes cultured on three different media. *1* Minimal basal mineral medium. *2* Basal medium with 2,4-D and KN. *3* Basal medium with 2,4-D, KN, glutamine, and asparagine. Approximate proportions of total responses in cultures on the three media for three kinds of development are shown: callusing only *black*, callusing plus rooting *horizontal lines*, callusing plus bud formation *vertical lines*, and callusing, rooting and bud formation *cross-hatched*. (After Norstog 1967)

sponded variously to additions, deletions, and concentration changes in auxins, cytokinin, and amino acids (Norstog 1965; Norstog and Rhamstine 1967; Norstog 1967). Callusing alone occurred in cultures on media lacking all of the above. The addition of 2,4-D and KN induced callusing followed by rooting. The further addition of glutamine and asparagine to the auxin-cytokinin medium resulted in a very considerable increase in the number of megagametophytes exhibiting a morophogenetic response (from 48% to 92%), and, in addition, resulted in the regeneration of shoots as well as roots (Fig. 12; see also Fig. 11 A–F). Regenerated roots were found to have retained the haploid chromosome number. On media with different auxin and cytokinin levels, a somewhat different response was noted (Norstog and Rhamstine 1967). Megagametophytes initially were cultured on a high auxin medium (10 mg/l 2,4-D) in shake cultures. The resulting callus was transferred to a medium with lower levels of auxin and cytokinin (1 mg/l each of 2,4-D and KN), where it was possible to maintain and subculture the tissue without organ regeneration. However, when cells were transferred to a medium containing neither auxin nor cytokinin, embryo production ensued. This sequence of events was repeated many times with essentially the same results. Megagametophyte tissue of *Cycas circinalis* was cultured similarly, and produced subculturable callus which did not form roots, shoots, or embryos.

It was noted earlier that proembryos of *Zamia* also gave rise to subculturable callus tissue. Proembryos cultured prior to cotyledon initiation developed in a rather interesting manner (see Fig. 3 A–F). Some elongation of the suspensor took place, and a few proembryos followed essentially the normal course of in vivo development, except for producing differentiated embryos smaller than those that develop under natural conditions (Norstog and Rhamstine 1967). However, on several media containing relatively low concentrations of auxin (0.01–0.1 mg/l 2,4-D) and cytokinin (0.05–0.5 mg/l KN), proembryos formed a callus which could be subcultured. The tissue culture could be maintained on an auxin-cytokinin medium through repeated subcultures. When cells from such 2*n* tissue cultures were transferred to a medium lacking auxin and cytokinin (the same medium that induced

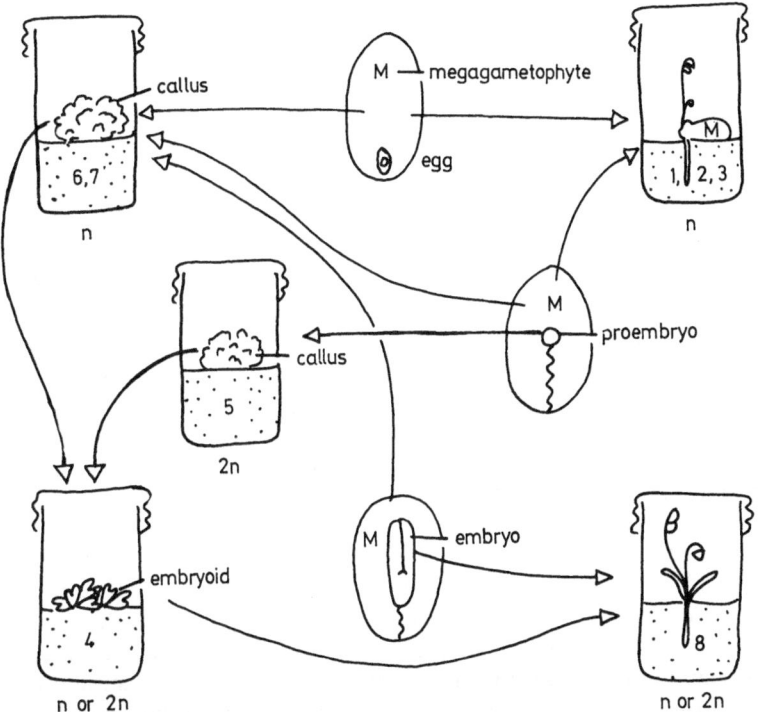

Fig. 13. Responses in cultured embryos and megagametophytes of *Zamia* grown on different media. Media *1,2,3* were basal mineral media, containing either no 2,4-D and KN; 2,4-D and KN (1.0 mg/l each); or 2,4-D and KN (1.0 mg/l each) with asparagine and glutamine (400 mg/l each). Medium *4* contained alanine (100 mg/l), glutamine and asparagine (400 mg/l each), adenine (10 mg/l), and NH_4-malate (100 mg/l). Medium *5* resembled medium *4* but contained 2,4-D and KN (1.0 mg/l each). Medium *6* was a modified Murashige-Skoog medium containing alanine (100 mg/l), glutamin (400 mg/l), 2,4-D and KN (1.0 mg/l each). Medium *7* resembled medium *5* but contained 10 mg/l of 2,4-D. Medium *8* was a basal medium containing sucrose but lacking vitamins (all other media contained both 2% sucrose and vitamins). (After Norstog 1967)

embryoid formation in haploid *Zamia* tissue cultures), embryoid production ensued.

Figure 13 represents a summary of the morphogenetic potentials exhibited by both haploid and diploid cell lines of *Zamia*. Both 1*n* and 2*n* cells of this plant respond similarly to manipulations of the culture environment. Cells and tissues of *Zamia* appear to be very plastic in their morphogenetic responses, apparently much more so than is the case in other gymnosperms. However, this plasticity was only elucidated after the trial of 78 different media in the course of a year's study. From these 78 media, 8 demonstrated marked selective effects on growth and development (Norstog 1967; Norstog and Rhamstine 1967). No doubt, as more is learned about the requirements of haploid and diploid cells of gymnosperms, it will similarly be possible to manipulate microspores, as well as micro- and megagametophytes, and turn these toward commercially useful ends.

Concluding Remarks

Studies of experimental embryology in gymnosperms have dependend upon tissue culture techniques to a considerable extent. Included in such studies have been cultures of embryos and proembryos, microsporangia (microspores and pollen), microgametophytes (pollen), and megagametophytes. Culture of proembryos of the cycad *(Zamia)* have been more successful than those of other gymnosperms (including *Ginkgo, Gnetum, Larix*, and *Pinus*). Although *Zamia* proembryos have been reported to undergo essentially normal development in vitro, those of *Gnetum, Ginkgo, Pinus*, and *Larix* either failed to grow, developed radial symmetry, or were asymmetrical. Even when certain of these were excised together with some surrounding gametophytic tissue, normal bipolar embryos failed to develop. However, it is quite possible that no fundamental reason exists why gymnosperm proembryos have failed to develop normally when cultured, since many recent studies have shown that herbaceous plant cells (notably in flowering plants) have the potential to form polarized embryos in vitro. No doubt improvements in techniques will lead to similar advances in gymnosperm embryo cultures.

Cultures of mature or nearly mature embryos of gymnosperms have been quite successful in elucidating a number of questions related to embryonic nutrition and morphogenesis. For example, excised pine embryos failed to develop chlorophyll when cultured in darkness, although embryos cultured in ovulo are fully competent to form chlorophyll in darkness. In this example it has been demonstrated that competency to form chlorophyll in darkness is due to nutrition of the embryo, and nutrient media have been devised that will substitute for the megagametophyte in supporting the dark-synthesis of chlorophyll.

Another example has shown that dormancy in the *Taxus* embryo is clearly due to soluble factors resembling or perhaps identical to abscisic acid (ABA), and such dormancy may be readily broken using embryo culture techniques. On the other hand, dormancy in *Welwitschia* embryos is due to germination inhibition by the megagametophyte (see Button et al. 1971). Dormancy of this type is immediately broken by excision of the embryo and its culture on nutrient medium. Despite such successes there remains the question as to what degree of control the gymnosperm gametophyte exerts over the growth and development of the embryo. For example, it has been shown that excised pine embryos respond in vitro quite differently to gravity and light than do those grown in ovulo; however, the reasons for this are as yet unclear.

Ginkgo embryos take up nutrients only through the cotyledons, and differential requirements and responses of epicotyl and hypocotyl have been recorded in relation to specific substances fed into the embryo through the cotyledons. Embryo cotyledons have also been shown to possess the capacity for regeneration of roots and, in a few instances, of entire adventive embryos. However, most gymnosperm cell and tissue cultures appear to lack significantly the capacity for either regeneration of buds and roots or production of embryos. This is not only true of cultures derived from embryonic tissues and organs, but appears to be true of a majority of gymnosperm tissue cultures (see Durzan and Campbell 1974).

Possibly, cultures of megagametophytic tissues will turn out to be the most useful means of achieving haploid and doubled haploid trees of gymnosperm species incorporating desirable genetic characters; unlike the situation in herbaceous angiosperms, where much progress has been made to date in the generation of haploids and doubled haploids from microspores and microgametophytes (pollen). Cultures of megagametophytes of cycads and their tissue derivates have demonstrated that such tissues have considerable capacity for regeneration. The prospects for similar or greater control of morphogenesis in embryonic tissues of more advanced gymnosperms are at present somewhat unpredictable. However, the eventual rewards for success in such research seem sufficient at this time to warrant a great deal of further work in this field of study.

Excised embryos from the seeds of *Pinus strobus* were cultured on MS-medium supplemented with different growth hormones (Minocha 1980). On basal medium containing 3%–6% sucrose the embryos developed into seedlings, while on media with 1%–2% sucrose root primordia did not grow satisfactorily. The addition of GA_3 (0.01–20 mg/l) suppressed the growth of root primordia, while auxin (NAA, 2,4-D or IBA – 0.01–2 mg/l) led to callus formation. IBA 1–5 mg/l induced adventitious shoots from the hypocotyl of some embryos. Similar shoots were formed on the tip of cotyledons when TIBA (0.5–1 mg/l) was added to the medium. In no case did the callus show any organogenesis.

Mapes and Zaerr (1981) observed that excised embryos of *Pseudotsuga menziesii* reared on MS-medium, with or without the female gametophyte or its extract, exhibited striking differences in size. The female gametophyte has some substances with a growth-promoting effect on the embryo; this thermostable substances is diffusible in agar. The extracts also have a synergistic effect with CM in promoting morphogenesis in suspension cultures of *Pseudotsuga*.

Callus formation and induction of adventitious buds has been reported in cultures of excised embryos of Norway spruce (*Picea abies*), by Arnold and Eriksson (1978). The embryos were cultured on a medium containing 2iP. Anatomically, different bud primordia were formed at high and low cytokinin concentrations. Transfer of buds to the medium without hormone led to the formation of shoots only (roots were not formed).

Konar and Singh (1980) cultured embryos and hypocotyl segments of *Pinus wallichiana*, on a modified MS-medium supplemented with auxins and cytokinins. With NAA 0.1 ppm, a callus was formed. On subculture on basal medium with BAP 1 ppm, shoots were initiated. The embryos were also induced to form shoots along the surface of cotyledons and hypocotyl; roots were not formed.

Reilly and Washer (1977) reported vegetative propagation of *Pinus radiata* by plantlet formation from embryonic tissues.

In *Zamia pumila*, Webb (1982) studied the effect of light on the in vitro growth of root and nodulation in the embryos and seedlings.

De Luca and Sabato (1979) reported, for the first time, in vitro spermatogenesis. The microsporangia of *Encephalartos altensteinii*, containing microspores at three-celled stage, were cultured on WM-Co-NS (see Tulecke 1957) medium, with or without 2,4-D, for 8 months. Within a week after inoculation masses of pollen tubes emerged from the sporangia. In six months the sperm mother cell was formed at the tip of pollen tube. On division it produced two flagellate sperms. The sperms, however, did not attain maturity.

The megagametophytes of cycads (*Ceratozamia mexicana*, *Cycas revoluta* and *Encephalartos umbeluziensis*) from mature seeds were cultured on modified White's media (according to De Luca et al. 1979) with 2% sucrose, with or without 2,4-D and KN. Callusing was observed in the three species, but regeneration occurred only in *Ceratozamia* and *Cycas*. On the basal medium the callus of *Ceratozamia* differentiated adventive embryos which later developed circinate leaves. On a medium enriched with amino acids, embryos and haploid roots were formed. The megagametophytes of *Cycas* when callussed on an enriched medium, produced a large number of "pseudobulbils" which the authors interpret as root primordia. The megagametophytes, when cultured on a medium containing 2,4-D and KN, formed coralloid roots. Anatomically, these roots were comparable to in vivo roots except that they lacked the endophytic blue-green alga (De Luca and Sabato 1980).

Konar and Singh (1979) cultured the female gametophytes excised from the mature seeds of *Ephedra foliata*. Initiation of callus occurred on BM (MS + 10% CM) + 2,4-D 2 ppm; the cultures were maintained on BM + 2,4-D 1 ppm. Subculture on BM + KN 2 ppm resulted in the formation of shoots and roots. Gametophytes from young ovules (at archegonial stage) and mature seeds also showed callus formation on BM + 2,4-D and, subsequently, roots differentiated (Singh and Konar 1981). Besides 2,4-D, various concentrations of NAA were also tested; best results were obtained on BM + NAA 4 ppm. The addition of BAP led to profuse shoot formation and, eventually, plantlets. A combination of 2,4-D and KN was more effective in inducing roots and shoot buds (Singh et al. 1981). Cytological studies confirmed the haploid chromosome number (n = 7) in roots.

A. David and H. David (1979) isolated protoplasts from the cotyledons of *Pinus pinaster*. These protoplasts regenerated a cell wall, and showed cell division leading to callus formation.

References

Arnold S von, Eriksson T (1978) Induction of adventitious buds on embryos of Norway spruce grown in vitro. Physiol Plant 44:283–287

Ball E (1956a) Growth of the embryo of *Ginkgo biloba* under experimental conditions. I. The apex of the first root of the seedling in vitro. Am J Bot 43:488–495

Ball E (1956b) Growth of the embryo of *Ginkgo biloba* under experimental conditions. II. Effects of a longitudinal split in the tip of the hypocotyl. Am J Bot 43:802–810

Ball E (1959) Growth of the embryo of *Ginkgo biloba* under experimental conditions. III. Growth rates of root and shoot upon media absorbed through the cotyledons. Am J Bot 46:130–139

Bartels H (1957a) Kultur isolierter Koniferenembryonen. Naturwissenschaften 44:290–291

Bartels H (1957b) Kultur isolierter Koniferenembryonen II. Naturwissenschaften 44:595–596

Berlyn GP (1962) Developmental patterns in pine polyembryony. Am J Bot 49:327–333

Berlyn GP, Miksche JP (1965) Growth of excised pine embryos and the role of the cotyledons during germination in vitro. Am J Bot 52:730–736

Bogorad L (1950) Factors associated with the synthesis of chlorophyll in the dark in seedlings of *Pinus jeffreyi*. Bot Gaz 111:221–241

Bonga JM (1974) In vitro culture of microsporophylls and megagametophyte tissue of *Pinus*. In Vitro 9:270–272

Bonga JM, Fowler DP (1970) Growth and differentiation in gametophytes of *Pinus resinosa* cultured in vitro. Can J Bot 48:2205–2207

Bonga JM, McInnis AH (1975) Stimulation of callus development from immature pollen of *Pinus resinosa* by centrifugation. Plant Sci Lett 4:199–203

Borchert R (1968) Spontane Diploidisierung in Gewebekulturen des Megagametophyten von *Pinus lambertiana*. Z Pflanzenphysiol 59:389–392

Branscheidt P (1930) Zur Physiologie der Pollenkeimung und ihrer experimentalen Beeinflussung. Planta 11:368–456

Brown CL, Gifford Jr EM (1958) The relation of the cotyledons to root development of pine embryos grown in vitro. Plant Physiol 33:57–64

Brown CL, Sommer HE (1975) An atlas of gymnosperms cultured in vitro. Georgia Forest Research Council, Macon, Georgia, USA

Buchholz JT (1920) Embryo development and polyembryony in relation to the phylogeny of conifers. Am J Bot 7:125–145

Bulard C (1952) Culture aseptique d'embryons de *Gingko biloba*. Rôle des cotyledons dans l'absorption du sucre et la croissance de la tige. CR Acad Sci Paris 2350:739–741

Bulard C, LePage-Degivry MT (1968) Quelques précisions sur les conditions d'obtention d'une inhibition de la croissance épicotylaire chez *Gingko biloba* L. sous l'éffet de l'acide gibbérellique. CR Acad Sci Paris 2660:356–359

Button J, Bornman CH, Carter M (1971) *Welwitschia mirabilis:* Embryo and cell-free culture. J Exp Bot 22:922–924

Chamberlain CJ (1935) Gymnosperms, structure, and evolution. Univ Chicago Press, Chicago, Illinois

Cheng TY (1975) Adventitious bud formation in culture of Douglas fir (*Pseudotsuga menziessi* [Mirb] Franco). Plant Sci Lett 5:97–102

Cheng TY (1977) Factors effecting adventitious bud formation of cotyledon culture of Douglas fir. Plant Sci Lett 9:179–187

Coulter JM, Chrysler MA (1904) Regeneration in *Zamia*. Bot Gaz 38:130–139

David A, David H (1979) Isolation and callus formation from cotyledon protoplasts of pine (*Pinus pinaster*). Z Pflanzenphysiol 94:173–177

De Luca P, Moretti A, Sabato S (1979) Regeneration in megagametophytes of cycads. Giorn Bot Ital 113:129–143

De Luca P, Sabato S (1979) In vitro spermatogenesis of *Encephalartos* Lehm. Caryologia 32:241–245

De Luca P, Sabato S (1980) Regeneration of corolloid roots on cycad megagametophytes. Plant Sci Lett 18:27–31

DuChartre MP (1888) Note sur l'enracinement de l'albumin d'un *Cycas*. Bull Soc Fr 35:243–256

Durzan DJ, Campbell RA (1974) Prospects for the mass production of improved stock of forest trees by cell and tissue culture. Can J For Res 4:151–174

Engvild KC (1964) Growth and chlorophyll formation of dark-grown pine embryos on different media. Physiol Plant 17:866–874

Greenwood MS, Berlyn GP (1965) The regeneration of active root meristems in vitro by hypocotyl sections from dormant *Pinus lambertiana* embryos. Can J Bot 43:173–175

Greenwood MS, Goldsmith MHM (1970) Polar transport and accumulation of indole-3-acetic acid during root regeneration by *Pinus lambertiana* embryos. Planta 95:297–313

Halperin W (1969) Morphogenesis in cell cultures. A Rev Plant Physiol 20:395–418

Illies AM (1964) Auftreten haploider Keimlinge bei *Picea abies*. Naturwissenschaften 51:442

Konar RN (1963a) A haploid tissue from the pollen of *Ephedra foliata* Boiss. Phytomorphology 13:170–174

Konar RN (1963b) In vitro studies on *Pinus roxburghii* Sarg. In: Maheshwari P, Rangaswamy NS (eds) Plant tissue and organ culture – A symposium. Int Soc Plant Morphologists, Univ Delhi, Delhi, pp 224–229

Konar RN (1974) In vitro studies on *Pinus* I. Establishment and growth of callus. Physiol Plant 32:193–197

Konar RN (1975) In vitro studies on *Pinus* II. The growth and morphogenesis of cell cultures from *Pinus gerardiana*. Phytomorphology 25:55–59

Konar RN, Oberoi YP (1965) In vitro development of embryoids on the cotyledons of *Biota orientalis*. Phytomorphology 15:137–140

Konar RN, Singh MN (1979) Production of plantlets from the female gametophytes of *Ephedra foliata* Boiss. Z Pflanzenphysiol 95:87–90

Konar RN, Singh MN (1980) Induction of shoot buds from tissue cultures of *Pinus wallichiana*. Z Pflanzenphysiol 99:173–177

LaRue CD (1936) The growth of plant embryos in culture. Bull Torrey Bot Club 63:365–382

LaRue CD (1948) Regeneration in the megagametophyte of *Zamia floridana*. Bull Torrey Bot Club 75:597–603

LaRue CD (1950) Regeneration in *Cycas*. Am J Bot 37:664

LaRue CD (1954) Studies on growth and regeneration in gametophytes and sporophytes of gymnosperms. Brookhaven Nat Lab Symp 6:187–208

LePage-Degivry MT (1968) Mise en évidence d'une dormance associée à une immaturité de l'embryon chez *Taxus baccata* L. CR Acad Sci Paris 266 D:1028–1030

LePage-Degivry MT (1970) Acide abscissique et dormance chez les embryons de *Taxus baccata* L. CR Acad Sci Paris 271 D:482–484

LePage-Degivry MT (1973a) Etude en culture in vitro de la dormance embryonnaire chez *Taxus baccata* L. Biol Plant 15:264–269

LePage-Degivry MT (1973b) Intervention d'un inhibiteur lié dans la dormance embryonnaire de *Taxus baccata* L. CR Acad Sci Paris 277 D:177–180

LePage-Degivry MT (1973c) Influence de l'acide abscissique sur le développement des embryons de *Taxus baccata* L. cultivés in vitro. Z Pflanzenphysiol 70:406–413

LePage-Degivry MT, Garello G (1973) La dormance embryonnaire chez les embryons de *Taxus baccata* L.: Influence de la composition du milieu liquide sur l'induction de la germination. Physiol Plant 29:204–207

Li T (1934) The development of *Ginkgo* embryos in vitro. Peiping Nat Tsing Hua Univ Sci Rep B 2:41–52

Li T, Shen T (1934) The effect of "pantothenic" acid on the growth of the radicle of *Ginkgo* embryos in artificial media. Peiping Nat Tsing Hua Univ Sci Rep B 2:53–60

Loo SW, Wang FH (1943) The culture of young conifer embryos in vitro. Science 98:544

Maheshwari P, Sanwal M (1963) The archegonium in gymnosperms: A review. Mem Indian Bot Soc 4:103–119

Mapes MO, Zaerr JB (1981) The effect of the female gametophyte on the growth of cultured Douglas fir embryos. Ann Bot 48:577–583

Mehra PN (1938) The germination of pollen grains in artificial cultures in *Ephedra foliata* Boiss. and *Ephedra gerardiana* Wall. Proc Indian Acad Sci 8:218–230

Minocha SC (1980) Callus and adventitious shoot formation in excised embryos of white pine (*Pinus strobus*). Can J Bot 58:366–370

Narayanaswami S, Norstog K (1964) Plant embryo culture. Bot Rev 30:587–628

Nitsch C (1974) La culture de pollen isolé sur milieu synthétique. CR Acad Sci Paris 278 D:1031–1034

Nitsch C, Norreel B (1973) Effet d'un choc thermique sur le pouvoir embryogène du pollen de *Datura innoxia* cultivé dans l'anthère ou isolé de l'anthère. CR Acad Sci Paris 276 D:303–306

Nitsch JP (1969) Experimental androgenesis in *Nicotiana*. Phytomorphology 19:389–404

Norstog K (1965) Induction of apogamy in megagametophytes of *Zamia integrifolia*. Am J Bot 52:993–999

Norstog K (1967) Some characteristics of growth and development in cycad tissue cultures. 5th Conf Cycad Toxicity. Nat Inst Health Bethesda IV:1–8

Norstog K (1972) Role of archegonial neck cells of *Zamia* and other cycads. Phytomorphology 22:125–130

Norstog K, Rhamstine E (1967) Isolation and culture of haploid and diploid cycad tissues. Phytomorphology 17:1–4

Radforth NW (1936) The development in vitro of the proembryo of *Ginkgo*. Trans R Can
    Inst 21:87–94
Radforth NW, Bonga JM (1960) Differentiation induced as season advances in the embryo-
    gametophyte complex of *Pinus nigra* var. *austriaca* using indole acetic acid. Nature
    185:332
Radforth NW, Pegoraro L (1955) Assessment of early differentiation in *Pinus* proembryos
    transplanted to in vitro conditions. Trans R Soc Can 49:69–82
Radforth NW, Trip P, Bonga JM (1958) Polarity in the early embryogeny of *Ginkgo biloba*
    L. Trans R Soc Can 52:55–58
Raghavan V (1966) Nutrition, growth and morphogenesis of plant embryos. Biol Rev 41:1–
    58
Rao NM, Mehta AR (1969) Callus tissue from the pollen of *Thuja orientalis* L. Indian J Exp
    Biol 7:132
Rappaport J (1954) In vitro culture of plant embryos and factors controlling their growth.
    Bot Rev 20:201–225
Rashid A, Street HE (1974) Segmentations in microspores of *Nicotiana sylvestris* and *Nico-
    tiana tabacum* which lead to embryoid formation in anther cultures. Protoplasma
    80:323–334
Razmologov VP (1973) Tissue culture from the generative cell of the pollen grain of *Cu-
    pressus* spp. Bull Torrey Bot Club 100:18–22
Reilly K, Washer J (1977) Vegetative propagation of radiata pine by tissue culture: Plantlet
    formation from embryonic tissue. New Zealand J Forestry Sci 7:199–206
Rohr R (1973) Ultrastructure des spermatozoïdes de *Taxus baccata* L. obtenus à partir de
    cultures aseptiques de microspores sur un milieu artificiel. CR Acad Sci Paris
    277 D:1869–1871
Rohr R (1977) Evolution en culture in vitro des prothalles femelles âgées chez le *Gingko bi-
    loba* L. Z Pflanzenphysiol 85:61–69
Sacher JA (1956) Observations on pine embryos grown in vitro. Bot Gaz 117:206–214
Singh H (1978) Embryology of gymnosperms. Geb. Bornträger, Berlin
Singh MN, Konar RN (1981) In vitro induction of haploid roots and shoots from female
    gametophyte of *Ephedra foliata* Boiss. Beitr Biol Pflanzen 55:169–177
Singh MN, Konar RN, Bhatnagar SP (1981) Haploid plantlet formation from female game-
    tophytes of *Ephedra foliata* Boiss. in vitro. Ann. Bot 48:215–220
Sommer HE, Brown CL, Kormanik PP (1975) Differentiation of plantlets in longleaf pine
    (*Pinus palustris* Mill.) tissue cultures in vitro. Bot Gaz 136:196–200
Sterling C (1949) Preliminary attempts in larch embryo culture. Bot Gaz 111:90–94
Thomas MJ (1970) Premières recherches sur les besoins nutritifs des embryons isolés du *Pi-
    nus sylvestris* L. Embryons differenciés. CR Acad Sci Paris 270 D:2648–2651
Tulecke W (1953) A tissue derived from the pollen of *Ginkgo biloba*. Science 117:599–600
Tulecke W (1957) The pollen of *Ginkgo biloba*: In vitro culture and tissue formation. Am
    J Bot 44:602–608
Tulecke W (1959) The pollen cultures of C.D. LaRue: A tissue from the pollen of *Taxus*. Bull
    Torrey Bot Club 86:283–289
Tulecke W (1964) A haploid tissue culture from the female gametophyte of *Ginkgo
    biloba* L. Nature (London) 203:94–95
Tulecke W (1967) Studies on tissue cultures derived from *Ginkgo biloba* L. Phytomorpho-
    logy 17:381–386
Tulecke W, Nickell LG (1960) Methods, problems, and results of growing plant cells under
    submerged conditions. Trans NY Acad Sci Ser III 22:196–206
Tulecke W, Sehgal N (1963) Cell proliferation from the pollen of *Torreya nucifera*. Contr
    Boyce Thompson Inst 23:33–46
Vasil V (1963) In vitro culture of embryos of *Gnetum ula* Brongn. In: Maheshwari P,
    Rangaswamy NS (eds) Plant tissue and organ culture – A Symposium. Int Soc Plant
    Morphologists, Univ Delhi, Delhi, pp 278–280
Webb D (1982) Effects of light on root elongation, apogeotropism and nodulation of *Zamia
    pumila* embryos and seedlings in sterile culture. Am J Bot 69: 298–305
Winton LL, Verhagen SA (1977) Shoots from douglas-fir cultures. Can J Bot 55:1246–1250

4. Flower Culture

R.N. KONAR and S. KITCHLUE

The onset of the reproductive phase in the life cycle of a plant is heralded by flower initiation, a very important morphogenetic event. Intense morphogenetic activities occur in this short life-span of an angiosperm flower. Floral initiation and development have been delineated as two distinct processes by Nitsch (1965), and Evans (1969). Although a large volume of data has accumulated on flower initiation, most of the information on developmental studies is available only from experiments carried out with intact plants. Growth and development of floral buds in culture and their nutritional requirements are also not sufficiently well understood. The technique of culture of flowers and ovaries, introduced as early as 1942 by LaRue, opened up a promising avenue for study of growth interactions and floral morphogenesis. In recent years, with the refined in vitro culture technique, experiments with isolated floral buds have led to a greater understanding of the orderly development of floral organs and their morphogenetic potentialities.

Work on floral morphogenesis has been basically two-pronged, using (1) vegetative explants, and (2) floral apices. Both kinds of explant are grown on a variety of culture media and under different physical conditions to fully exploit the advantages of a controlled regimen. Current investigations can be arbitrarily classified as follows:

1. Culture of vegetative parts with a view to induction of flowering.
2. Culture of induced explants to study persistence of stimulus and further initiation.
3. Culture of floral buds and organs to study their capacity for further growth.

Attention is invited to reviews covering various aspects of vegetative and floral morphogenesis (see Wadhi 1967; Johri and Ganapathy 1967; Mohan Ram and Jaiswal 1974).

The technique of culturing young excised floral primordia affords a potentially useful tool for any study concerned with the control of flower morphogenesis. It permits assessment of the influence on buds of growth regulators and nutrients, without the presence of intervening vegetative tissues. Secondly, one can test whether the floral apex behaves autonomously, i.e., whether the sequence of organ initiation and subsequent growth is regulated from within the isolated immature bud. Lastly, it facilitates easy handling and allows a variety of surgical manipulations of the excised buds in culture.

Differentiation in Floral Bud Cultures: Herbaceous Species

The young floral primordia of a number of taxa have proved rather difficult to grow to maturity in culture. Cultured immature flower buds produce viable pollen grains and ovules in only a few instances. The developmental details of micro- and megasporogenesis in cultured buds are not well documented.

Though tissue culture of floral buds was first attempted by LaRue in 1942, it was only two decades later that Galun et al. (1962, 1963) achieved success in growing detached floral buds of *Cucumis sativus*. Sex expression has been experimentally modified by manipulation of growth substances in the medium. An attempt was made to culture young floral buds at a stage when they are morphologically and physiologically bisexual. Floral buds of the same developmental stage taken from a pure monoecious line of *C. sativus* were selected. All the floral buds, if left in situ, would have developed into male flowers, depending on the nodes on which they were borne. However, the excised buds were induced to develop into female flowers on modified White's medium (WM; White 1943) supplemented with indole acetic acid (IAA). Galun et al. (1963) cultured three types of buds: potentially male, potentially female, and potentially hermaphrodite, though all these were at the bisexual stage at the time of inoculation. In plants under greenhouse controlled environmental conditions the first 20 nodes of the main shoot would bear only staminate flowers. Flower buds of the same developmental stage (0.5–0.7 mm diam) were used for culture. They were at the bisexual stage, and borne on about the eighth node. IAA in the medium promoted ovary development in the potentially male buds, while gibberellic acid (GA_3) counteracted this effect but had no stimulatory effect of its own. Isolated female and potentially hermaphrodite buds continued their normal development and were little affected either by IAA or GA_3.

Hermaphrodite buds (0.7–1.1 mm diam) from one line of *Cucumis melo* (uniform with respect to sex expression) at the archesporium to pre-leptotene stage of microspore mother cell were cultured by Porath and Galun (1967). After 14–16 days almost all the buds isolated at 0.9–1.1 mm diam and a few of the 0.7 mm diam buds showed post-meiotic stages. In some buds pollen grains developed. The development in vitro followed, in general, the in vivo pattern. Nevertheless, there were some exceptions – in the intact buds (in vivo) there is a certain uniformity with respect to differentiation of microspore mother cells, but it is quite common for individual cultured buds to be in various stages. Moreover, while in intact buds there always exists a correlation between the size of the anther and its diffferentiation, the cultured buds frequently had large anthers in which the microsporangial tissue was arrested at or before meiosis (meiotic arrest).

It is typical of melon floral buds that differentiation of sporogenous tissue precedes the organisation of ovules. Thus, while the exine of the maturing pollen grains was already visible, the ovules had not been initiated. The invagination of ovary and differentiation of carpels proceeded normally under in vitro conditions, but floral buds 0.7–1.1 mm diam rarely reached the ovule stage.

In *Cucurbita pepo* staminate flowers are initiated monosexually, and pistillate bisexually (they are physiologically bisexual from a stage below 0.3 mm until they are 0.8 mm in diam). Rodrigues Pereira (1968) raised both male and female buds on WM. In excised female buds development of stamens was better than in vivo floral buds, but the development of ovary was very poor. Smaller pistillate buds showed greater tendency to maleness as compared with larger buds. With IAA and an anti-auxin p-chlorophenoxy isobutyric acid (PCIB) added together, the development of buds was somewhat better than when each was added alone. However, the percentage of female buds among explants originating from identical nodes was not altered by IAA or PCIB, i.e., there was no qualitative change in sex.

Masculinisation of female inflorescence of a dioecious species, *Mercurialis annua*, was reported by Champault (1969). By raising cultures of female inflorescences on Heller's medium (1953) containing IAA (1 ppm), normal male flowers were obtained.

Subsequent to the report of Galun et al. (1962), Tepfer et al. (1963, 1966) grew the floral buds of *Aquilegia* to maturity. Floral buds varying from early stages before sepal initiation to late stages with young carpellary primordia were grown on different culture media. Their critical studies on *A. formosa* indicate that for a satisfactory growth of buds the medium should contain White's minerals, Nitsch's trace elements, coconut milk (CM), sucrose, and assorted water-soluble vitamins.

The addition of IAA, GA_3, and kinetin (KN) to the above medium extended the developmental limits of buds at nearly all stages, and decidedly improved the continued development of carpels. The removal of sepals was essential for the floral parts to develop normally. The buds from early stages initiated organ primordia in the normal sequence and grew to about the size of flowers at anthesis. However, staminal development was consistently poor, and the primordia failed to form even sporogenous tissue, despite the addition of hormones. The abortion of carpels promoted the development of stamens, but sporogenous tissue failed to differentiate. In the medium lacking IAA, organs other than carpels developed normally. Addition of tri-iodo benzoic acid (TIBA) led to a continued growth of floral apex without organ formation.

Thus, on a complex medium supplemented with KN, GA_3, and IAA, organ primordia in the buds were initiated in the normal sequence and early development of the different organs was realised. However, the stamen primordia later aborted, the petals did not grow to maturity, and the carpels were sterile.

Cultures of *Aquilegia* have also been used by Bilderback (1971, 1972) to study the effect of amino acids and hormones on floral bud development. Though the addition of hormones IAA, GA_3, and KN enhanced bud development, complete differentiation of floral primordia into mature organs was not achieved. Jensen (1971, 1972) conducted studies on growth, regeneration, and cytological changes that occur in the cultured buds following bisection.

Blake (1966) first reported complete development of the excised buds into flowers (Fig. 1 A–D). Excised floral apices of two species of *Viscaria*, *V. candida*, and *V. cardinalis*, were raised on MS medium (Murashige and Skoog 1962) supplemented with CM, casein hydrolysate (CH), inositol, amino acids, GA_3, and IAA. *Viscaria* is a long-day plant and vegetative apices cultured under such conditions subsequently flowered after production of the same number of pairs of leaves as pot-grown plants.

In apices excised with sepal primordia only or with petal and sepal primordia, the succeeding floral parts were initiated normally, and none of the adjuvants disturbed the sequence. With such small apices, however, development of the flower to maturity occurred rarely unless the explant had a pair of leaves also. When the buds were taken without leaves, but with all the organs just initiated, development of flower was quite normal. In buds of less than 2 mm diam, sepals were well-developed, petals were small, the ovary was at an early stage of initiation, and the

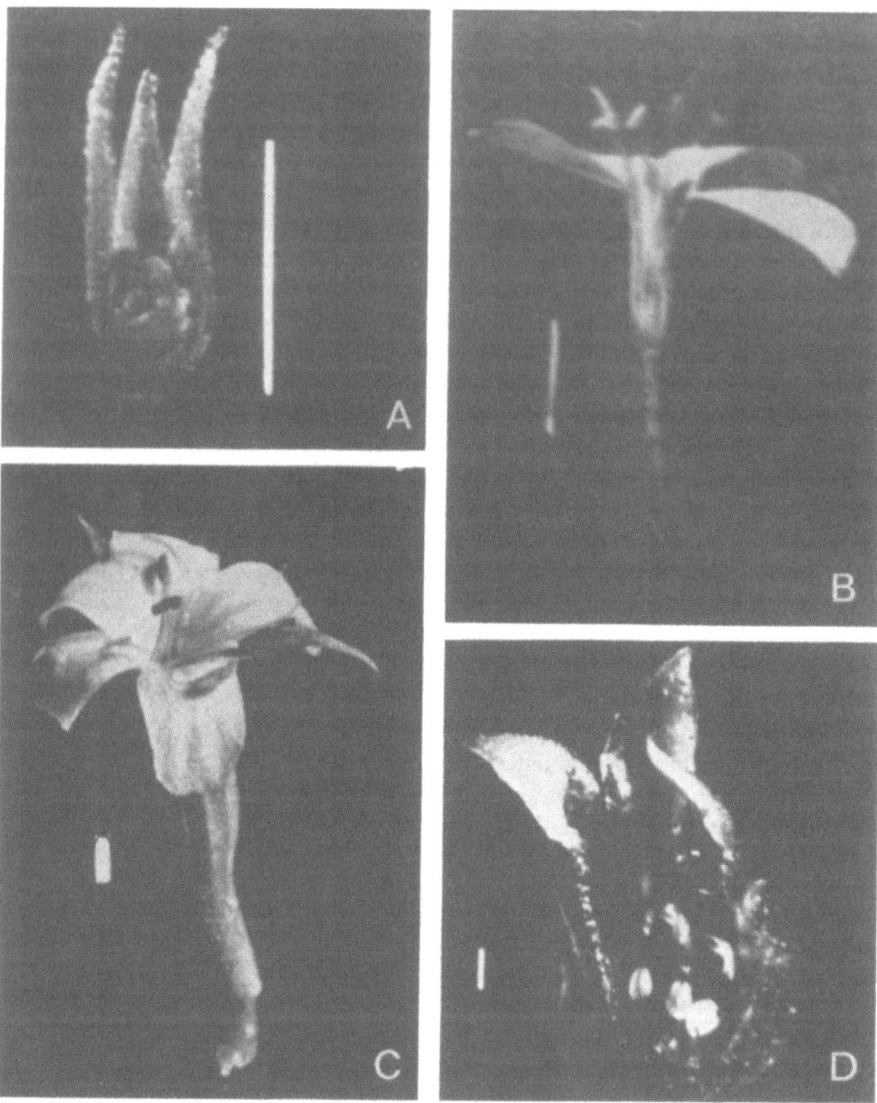

Fig. 1 A–D. *Viscaria candida.* **A** Flower bud (1–2 mm) at culture (scale equals 1 mm). **B** Normal flower (calyx, two petals and one stamen removed) grown on plant (in vivo) under the same conditions as the flower cultured in **C** (scale equals 5 mm). **C** Flower bud grown in culture with GA_3 present in the nutrient medium; four anthers exserted, and styles visible. **D** Bud grown without GA_3 (**C** and **D** scale equals 1 mm). Note stamens and ovary; petals (two sepals removed) are just visible. (After Blake 1966, 1969)

archesporial tissue had not yet differentiated in the stamens. On the complex medium used, normal growth of sepals, petals, ovary, and ovules was obtained. The development of stamens was variable; in many cases normal development occurred and up to 50% of the cultures produced one or more stamens with pollen. Larger buds (4–5 mm diam) containing pre-meiotic archesporial tissue tended to produce pollen more easily.

Even though the flower buds were grown successfully, meiosis did not occur in all the anthers and ovules. Blake (1969) made a detailed study of the factors affecting the development of floral parts in *V. candida*. Floral buds of 1–2 mm diam were cultured especially to study the production of pollen. In assessing the results of treatments on normal development of stamen, the criterion employed was the formation of recognisable pollen grains with sculptured wall which stained with aceto-orcein. The number of pollen grains varied from a few to numerous. Unlike other parts, production of pollen-bearing anthers was greatly affected by constituents of the medium, and environmental conditions like temperature and light. Pollen production was promoted by temperature at 15 °C as opposed to 25 °C, by the addition of KN or CH, and by the growth in continuous darkness as compared with that in 8 or 16 h daily light. Calyx and corolla developed under a wide range of conditions provided GA_3 was present. Development of ovary and ovules tended to be irregular, and consistent effects of treatment could not be obtained. In most cases an ovary (of measurable size) with styles was formed.

Brulfert and Fontaine (1967) grew normal flowers of *Anagallis arvensis* from floral apices on which sepal primordia were present. This is the only report of complete development of flowers from buds grown at the primordial stage, on a relatively simple medium consisting (according to Brulfert 1965) of mineral salts, vitamins, and sucrose, supplemented with 10^{-6} M IAA.

Floral primordia of *Nicotiana tabacum* "Wisconsin 38" have been successfully cultured by Hicks and Sussex (1970). Buds with only the sepal primordia were cultured on Linsmaier and Skoog's (1965) medium with 0.01–10 ppm KN. The petal, stamen, and carpel primordia formed in normal acropetal sequence on the apex that initially bore only sepal primordia (Fig. 2 A–C). Relatively normal morphogenesis of the organs ensued on optimal concentration (0.1 or 1 ppm) of KN (Fig. 2 A, C). Generally, the floral organs compared favourably with those at various stages of development on the plant, though they were slightly shorter in length. At termination of experiments after 4 weeks, anthers in control buds were only at the archesporial or microspore mother cell stage (Fig. 2 F), while buds grown on 0.01 and 0.1 ppm KN showed one- or two-nucleate pollen grains (Fig. 2 G, H). At higher concentrations of KN, pollen did not develop. The pollen from cultured buds did not germinate successfully. In the buds grown on 0.1 KN the ovaries were well-developed, containing ovules at 7-nucleate embryo sac stage comparable to those in Nature (Fig. 2 D, E). At other concentrations of KN, there were no mature ovules. Fertilization with normal pollen grains was not successful.

When KN was omitted from the medium, all the primordia were initiated but, subsequently, remained minute through 4 weeks of culture. Callus formation occurred at the base of the pedicel with 0.1 or 1 ppm KN during the second or third week (Fig. 2 A). The following week vegetative buds appeared on the callus, and quickly gave rise to young leaves and leaf-like organs.

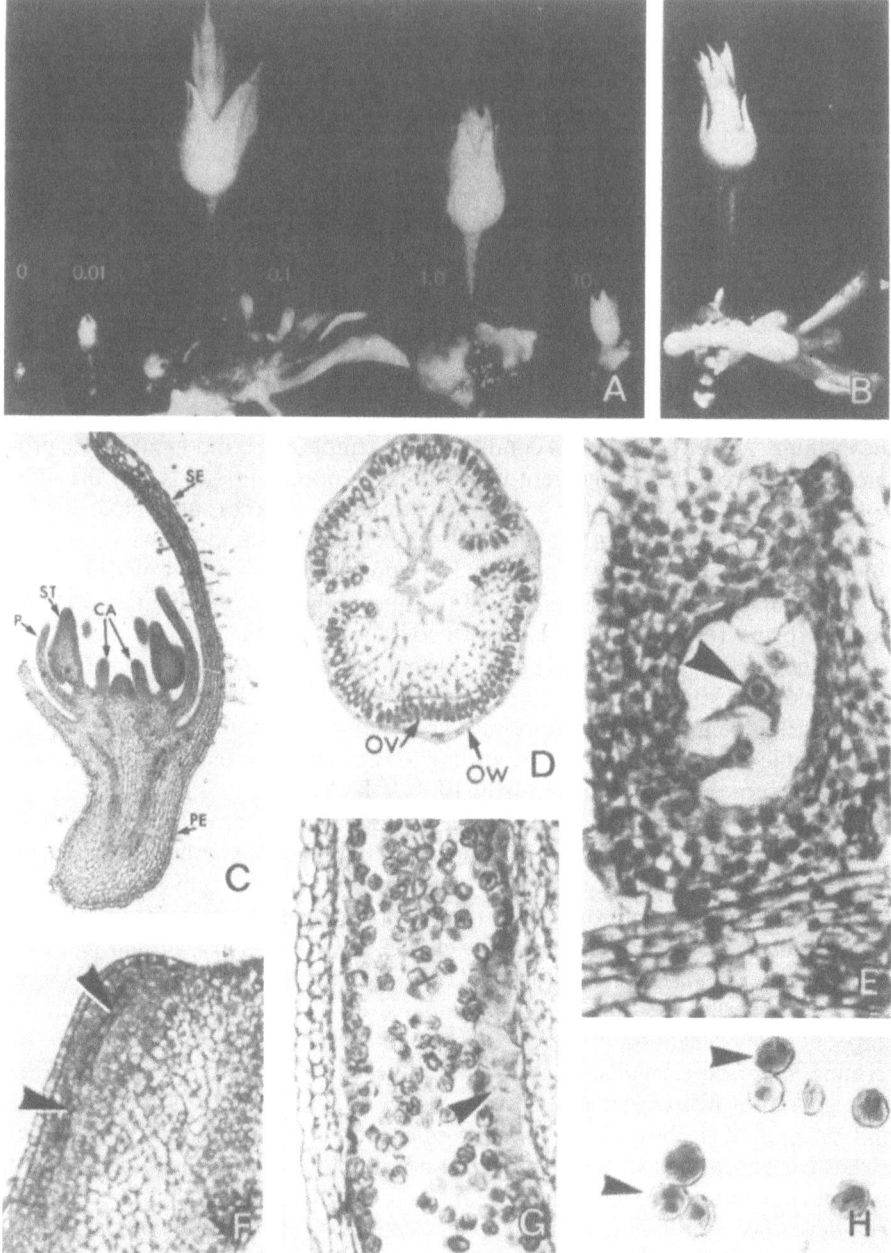

Fig. 2 A–H. *Nicotiana tabacum.* **A** Differentiation of buds; the concentration of KN (ppm) is given along with each bud. Note the well-developed pedicel, calyx, corolla, and extensive basal callus with vegetative buds at 0.1 ppm; 27-day-old culture. **B** Flower, on WM with 0.1 ppm KN from undifferentiated floral apex; 28-day-old culture. **C** Longisection through a bud grown for 27 days on WM, showing a short pedicel *PE*, and primordia of sepals *SE*, petals *P*, stamens *ST*, and carpels *CA*. **D** Transection through mid-portion of ovary from a bud, on WM + 0.1 ppm KN, showing ovary wall *OW*, and placental tissue bearing numer-

Organ primordia of the floral buds of *N. tabacum* have also been subjected to surgical manipulations, and cultured in vitro (Hicks and Sussex 1971; Hicks 1972, 1973).

Inflorescences of *Begonia franconis* comprising two male flowers (1.0–2.5 mm long) and one female (0.4–0.7 mm) were reared on a chemically-defined nutrient medium to analyse the requirements for optimum growth (Berghoef and Bruinsma 1979a). On liquid medium the buds attained normal and complete development. For satisfactory growth, both nitrate and ammonium along with a cytokinin were essential. The optimum concentration of cytokinin required for female buds was 10–30-times higher than that for male buds. IAA and Ethephon [(2-chloroethyl)-phosphonic acid] had no effect on bud-size, but abscisic acid (ABA) decreased growth if applied together with cytokinin. Although GA_3 had no effect, GA_{4+7} promoted the length of male perianth (see also pp 72, 73).

Konar and Nataraja (1969) cultured flower buds of *Ranunculus sceleratus* at three stages of development. Buds of Stage I had a meristematic dome and primordia of sepals and stamens (Fig. 3 A). In slightly older buds (Stage II) the anthers contained microspore mother cells, and carpels were at different stages (ovular primordia to ovules with massive nucelli). The oldest buds (Stage III) had anthers with two-celled pollen grains, and ovules with mature embryo sac. Buds were grown on modified WM supplemented with CM, CH, 2,4-dichlorophenoxyacetic acid (2,4-D), GA_3, IAA, KN, and yeast extract (YE) either alone or in various combinations. In addition, supplements of H_1 medium of Tepfer et al. (1963) were also tried. Petals or stamens failed to develop in buds of Stage I, but carpels developed even on WM. They were small (ca. 0.5 mm), spherical, with a minute stigma, and the number varied from 10 to 15 per flower bud as compared to about 100 in vivo (Fig. 3 B). Addition of CM alone or with IAA improved carpel development, stimulating them to attain normal size. On the other hand, the buds with anthers at pollen mother cell stage (Stage II) completed sporo- and gametogenesis and two-celled pollen grains were formed. The torus elongated considerably and emerged through folded sepals, carpels enlarged slightly but later shrivelled. On WM + CM + KN, the torus enlarged rapidly and branched to form several spherical structures on which a fresh crop of carpels developed. In the cultures of buds of Stage III, sepals, petals, and stamens turned brown; the torus merely elongated and emerged through the sepals. Regeneration of roots, and shoot buds (especially in buds of Stages II and III) was common.

There are various other reports dealing with the study of in vitro morphogenesis of floral tissues. Ganapathy (1969), working with *Browallia demissa*, cultured buds at three stages of development, on the medium used by J.P. Nitsch and C. Nitsch (1967) together with KN, benzyl aminopurine (BA), 6-benzyl-9-tetrahy-

---

ous ovules *OV*; 28-day-old culture. **E** Enlarged view *l.s.* of an ovule from **D**; the seven nuclei in the embryo sac comprise the fusion nucleus *arrow*, egg apparatus (towards micropyle), and two antipodals. **F** Longisection through distal portion of anther from a bud grown on WM, showing pollen mother cells *arrow*. **G** Longisection through anther from a bud, on WM + 0.01 ppm KN, showing numerous pollen grains and tapetal cells *arrow*; 27-day-old culture. **H** Pollen (propionic acid-orcein stained) from an anther cultured on WM + 0.01 ppm KN for 27 days. Note binucleate grains *arrow*. (After Hicks and Sussex 1970)

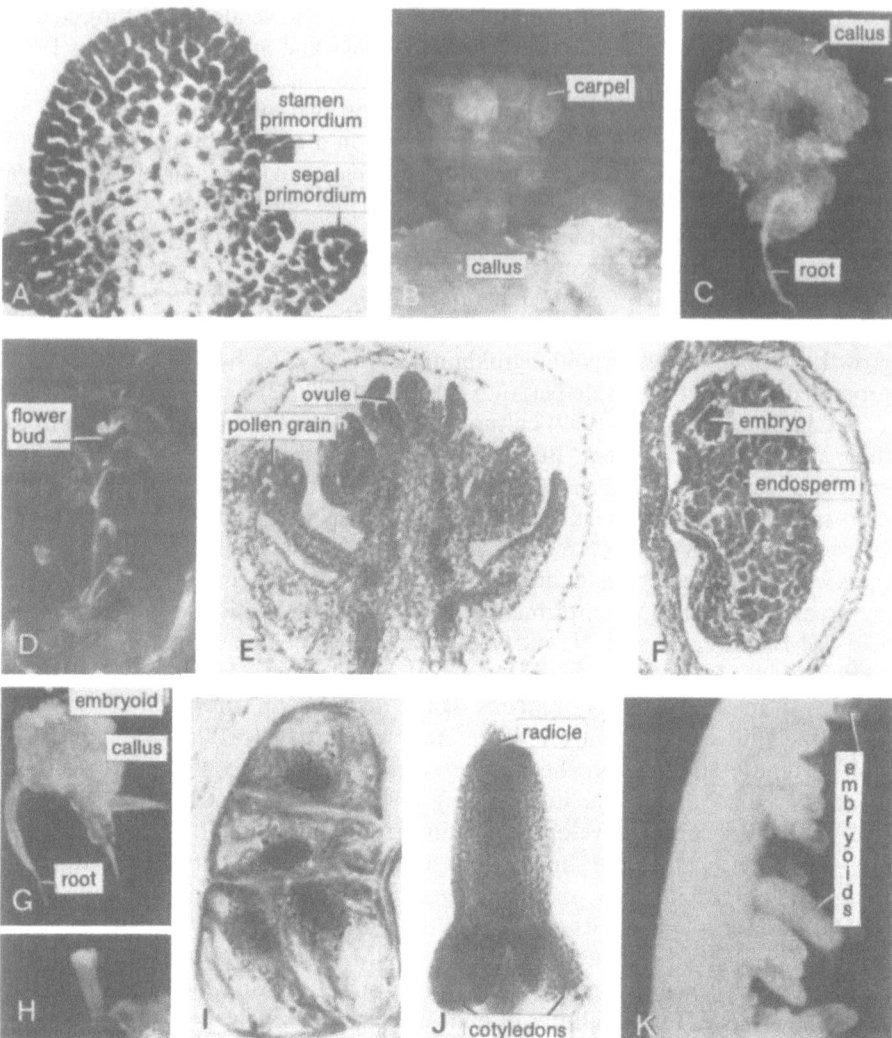

Fig. 3 A–K. *Ranunculus sceleratus*. **A** Longisection of flower bud; Stage I at culture. **B** Young spirally-arranged carpels and profuse callusing from the cut end of pedicel on WM; 3-week-old culture, **C** 3-week-old culture, on WM + CM (10% v/v) + IAA (1 ppm); note the callus mass and roots. **D** Plantlet with flower bud, on WM + CM (10% v/v) + IAA (1 ppm). **E** Flower bud at anthesis, the anther sacs contain pollen grains. **F** Seed, longisection. **G** 6-week-old culture, on WM; note the rooted callus with an aggregate of embryoids. **H** One embryoid on callus. **I** Acetocarmine squash preparation of callus to show four-celled embryoid. **J** Whole mount of mature embryoid. **K** Portion of stem of in vitro plantlet enlarged to show embryoids. (After Konar and Nataraja 1969)

dropyrane adenine (SD 8339), and GA_3. In addition, the effects of TIBA and l-arginine and l-cysteine were also examined. Buds at Stage I were 0.1–0.2 cm long, anthers contained microspore mother cells, but ovules had not been initiated; sepals were well-formed while petals were represented only by their primordia. After 3 weeks of culture, the calyx and corolla did not show any visible change, and further development did not occur in anthers or ovules; only partial development of stigma and style was observed. Slightly older (Stage II) buds (0.3 cm long) in which petals had differentiated and anthers contained microspores, also failed to give rise to well-developed flowers. In the oldest (Stage III) buds (0.5 cm), where the pollen grains had already been formed and ovules just initiated, the anthers became shrivelled and pollen grains became deformed. The differentiation of ovules progressed beyond the primordial stage but only a few ovules developed. Finally, the ovules showed endothelium but failed to develop embryo sacs irrespective of the treatment. With the addition of TIBA to the medium, even the integument did not differentiate. Normal anthesis was reported from flowers in which, at culture, the corolla tube had already grown beyond the calyx tube. All three stages failed to undergo complete development.

Floral morphogenesis has also been studied by Mohan Ram and Batra (1970) in a monocot, *Cyperus rotundus*. They cultured immature spikes with terminal meristematic dome followed by six floral primordia, and nine or ten immature buds. On WM the flowers already present turned brown, and in 42% of spikes two or three primordia matured into new flowers with stamens and pistils, but the anthers contained only degenerated sporogenous tissue and the ovary a shrunken ovule. The addition of CM enhanced the percentage of spikes developing new flowers, though the number of flowers per spike did not increase. Moreover, the anthers showed normal microsporogenesis with a characteristic configuration of one large and three degenerated microspores in a tetrad. The stigma and style emerged from the flowers, and the ovary showed fleshy ovules.

Flowers produced in response to treatment with 1 ppm KN and 1 ppm BA were sectioned. The anther wall comprised the epidermis, two middle layers, and tapetum, but the endothecial layer (characteristic of the plant) was absent. The microspore mother cells developed into tetrads with three small nuclei and one large nucleus, but the embryo sacs reached only the four-nucleate stage. When WM was fortified with 10 ppm KN, the total number of flowers borne on each spike was nearly the same as that in Nature.

The in vivo inflorescences of *Cleome* show alternating zones of pistil development and suppression in the flowers (de Jong and Bruinsma 1974a). Flower buds of *C. iberidella*, about 2.5 mm length, were reared in vitro (de Jong et al. 1974) on a synthetic medium enriched with 10^{-6} M zeatin and 10^{-7} M naphthalene acetic acid (NAA). Within 14 days the buds grew into perfect flowers of 8 mm length, and by 21 days they attained the size of buds in vivo (13 mm). While development of the pistil was stimulated by the cytokinins, zeatin and BA, GAs caused pistil abortion. This effect was counteracted by ABA, Ethephon, and (2-chloroethyl)-trimethyl ammonium chloride (CCC). However, in the absence of GA, the latter substances reduced pistil growth, ABA being particularly inhibitory. Thus, without the addition of growth regulators, predominantly male flowers developed in vitro, while cytokinins enhanced pistil development (de Jong and Bruinsma 1974b).

Kandeler and Hügel (1974), Kandeler et al. (1974), and Hügel (1976 a) cultured the inflorescence primordia (on Pirson-Seidel medium with sucrose, KN, GA_3, IAA, some vitamins, nucleotides, and CH) of the long-day plant *Lemna gibba* and the short-day plant *L. paucicostata*, and observed that the pattern of development in both was similar. In both species young, fully meristematic primordia failed to develop or showed insignificant growth response, whereas primordia with pistil length more than 0.08 mm, whose spathes were beginning to enclose them, always showed some growth. With pistil length more than 0.15 mm, there was complete development of floral organs on an appropriate medium. Primordia with pistil length of more than 0.35 mm invariably produced not only mature flowers, but also fructification with ripe seeds. Pirson-Seidel medium (Pirson and Seidel 1950) together with 2% sucrose and 1 ppm KN gave better results with younger stages. However, on this simple medium, relative pistil development in *L. gibba* was enhanced, resulting in femaleness, whereas in *L. paucicostata* enhanced stamen development led to masculinisation of explants.

Addition of CCC and Ethephon led to a normalisation of development in *Lemna paucicostata* by promotion of pistil growth. In *L. gibba* both substances enhance feminisation by a further promotion of pistil growth. The same effect is brought about by IAA. GA_3, however, causes normalisation of the explants of *L. gibba* due to promotion of stamen growth (Hügel 1976 b). Addition of ABA also resulted in differing effects on sex expression: increasing concentrations of ABA reduced stamen development in *L. gibba*, whereas pistil development in *L. paucicostata* was strongly inhibited. This caused an increase in the one-sidedness of sex expression of explants (Hügel 1976 c).

Greyson and Raman (1975) studied *Nigella damascena* using genotypically dissimilar "single" and "double" flowers, the double being a single gene recessive inherited trait. Floral apices of both forms were cultured on modified MS medium from pre-sepal to carpellary stages. The immature floral buds, after removal of leaves and bracts, were planted on the surface of the medium. A differential nutritional requirement for organ initiation was detected. MS medium supplemented with KN supported organ (stamen and carpel) initiation in both genotypes, and singles could initiate organs on MS alone. In the absence of KN, GA_3 was necessary for organ initiation in doubles, but it completely inhibited stamen initiation in singles (Fig. 4 A, B). Raman and Greyson (1978) further extended these studies and found that all concentrations of GA_3 inhibited the initiation and growth of stamens and nectaries in singles. The use of IAA or KN in any combination failed to overcome the inhibitory action of GA_3 on singles. These workers also succeeded in obtaining graft unions between bisected floral meristems of the two forms (Raman and Greyson 1977).

Flowers and inflorescences of the diplosporous apomict *Aerva javanica* ($= A. tomentosa$) were cultured on J.P. Nitsch's (1951) medium (NM) supplemented with White's vitamins, and CH, CM, YE, IAA, and KN (Puri 1964). The plant is dioecious, but male flowers are extremely rare. The egg and endosperm of this species develop autonomously without fertilization. When the ovaries isolated at megaspore to mature embryo sac stage were cultivated, there was no or only scanty seed-set. If, however, more of the maternal tissue remained attached to the explant, i.e., when flowers or inflorescences were cultured at the time of anthesis,

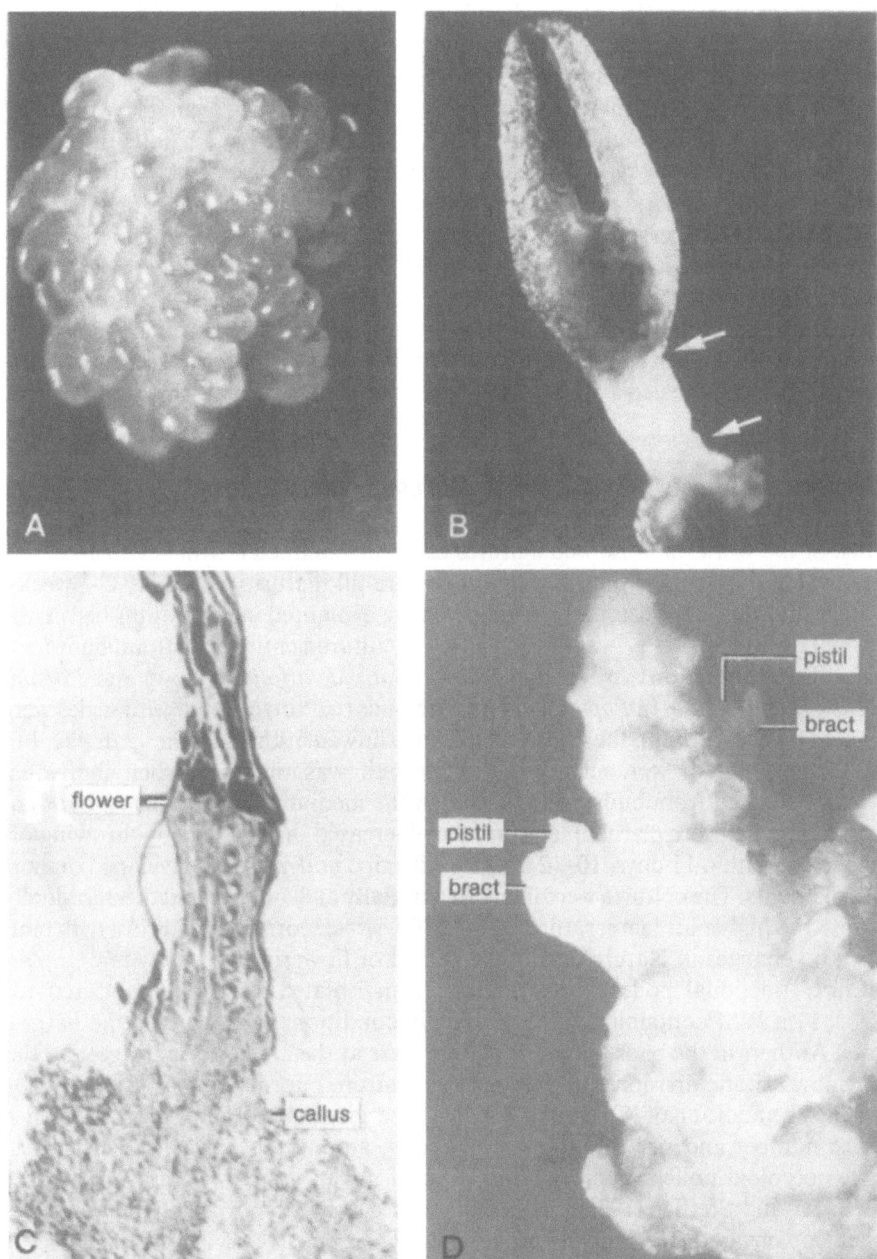

Fig. 4A, B *Nigella damascena;* **C** *Haworthia maughanii;* **D** *Populus trichocarpa.* **A** "Double" flower bud (21-day-old culture) inoculated at sepal stage, on WM + GA₃. **B** "Single" bud (21-day-old culture) inoculated at early sepal stage, on WM + GA₃. *Arrows* indicate the site of initiation of stamens; sepals removed. **C** Longisection of flower showing callused pedicel; 55-day-old culture. **D** Developmental stage of primordia of catkin; 70-day-old culture. Note initiation of differentiation of pistil in the axil of bract. (**A** and **B** after Greyson and Raman 1975; **C** after Kaul and Sabharwal 1972; **D** after Bawa and Stettler 1972)

the seed-set was greatly enhanced and, in about 3 weeks, seeds with mature embryos developed. On NM containing 1% CM and 1 ppm KN, 50% of ovules had mature dicot embryos, which was well over even the in vivo percentage.

Very few reports about the culture of monocot flowers are available. Majumdar and Sabharwal (1968) and Majumdar (1970a) studied the growth potential of various species of *Haworthia* (Liliaceae). The inflorescence segments bearing flower buds were inoculated, and optimal growth recorded on WM supplemented with CM, IAA, and KN. In some cases the flower buds shrivelled, while in others mature flowers developed. Kaul and Sabharwal (1972) reported the formation of callus from inflorescence segments (see Fig. 4 C). A male-sterile species of *Haworthia* was also cultured. In Nature the microspore mother cells undergo normal meiosis and tetrad formation, but the mature pollen grains are sterile. Surprisingly, in vitro-formed flowers revealed a higher percentage of pollen fertility (Majumdar 1970b).

Differentiation in Floral Bud Cultures: Woody Species

Most of the work on floral bud cultures has been conducted with herbaceous species, where the period from the stage at culture till maturity is usually 2–3 weeks. But, in the woody species, catkin primordia are explanted many months before the normal time of anthesis. Leshkovsteva (1967) cultured entire male floral buds (catkin inflorescences) of two woody species – *Salix acutifolia* (willow) and *Populus balsamifera* (poplar). In *Populus* when entire buds (catkin) covered with scales were planted on the medium, the buds opened and showed anthers by the 15th day, but no further growth was achieved. Best growth was obtained when individual flowers, isolated from buds, were placed on the medium. In *Salix*, when buds enclosed in scales were placed in culture, they increased in size, thereby throwing off the scales. Within 11 days, 10–12 anthers with ripe pollen grains developed on normal filaments. The cultures were incubated initially at 8°–10 °C, and then gradually exposed to higher air temperature (22°–25 °C), which corresponded closely to temperature changes in Nature during the period of flowering of willow and poplar. Mature functional pollen was obtained from isolated male buds cultured for 25 days on WM containing KN and IAA in conditions analogous to the natural state. Anthers in the buds taken for culture were at the archesporial stage, and did not show mitotic divisions. In cultures they matured normally and produced ripe pollen. The addition of KN and IAA had a synergistic effect, and led to an increase in the number and size of anthers containing, among the normal pollen grains, large polyploid pollen. In situ germination of pollen grains was also observed.

Bawa and Stettler (1972) cultured female catkin primordia of the dioecious black cottonwood tree *Populus trichocarpa*. The objective of the experiment was to induce growth and differentiation of female catkin in vitro up to the stage where they could be morphologically recognised as female flowers. Catkin primordia, with bud scales removed, were cultured for 70 days on modified MS medium. Many of them developed floral structures characteristic of the female sex, the most advanced stage showing the dome shape of the pistil primordia, distinct from the cup-shaped staminal primordia. Therefore, the most successful cultures reached a stage where the floral primordia could be morphologically distinguished as female (Fig. 4D).

Callusing of Floral Explants and Differentiation

Buds placed in culture may not follow the sequential development into flowers and, instead, form a callus. Occasionally, the floral primordia may exhibit a dual response: some primordia may continue to develop uninterruptedly and form flowers, while others may proliferate into a callus as in *Ranunculus sceleratus* (Konar and Nataraja 1969), *Browallia demissa* (Ganapathy 1969), *Nigella damascena* (Raman and Greyson 1974), *Populus trichocarpa* (Bawa and Stettler 1972), and *Haworthia* (Kaul and Sabharwal 1970, 1972). In *Ranunculus* and *Browallia* the undifferentiated floral buds developed progressively, and produced pollen grains and carpels. Later they callused and differentiated plantlets that even flowered.

In *Ranunculus sceleratus* (Konar and Nataraja 1964, 1969) the buds comprising the primordia of various organs failed to complete normal development on any of the nutrient media tested, but they all formed callus which subsequently differentiated roots, shoots, and embryoids leading to the formation of plantlets (Fig. 3 C, D). The cut-end of undifferentiated flower buds proliferated rapidly, resulting in a mass of white, friable callus. In a few weeks the shape of the original explant was completely distorted due to proliferation all over, including the apex. Optimal callus formation was observed on WM containing 2% sucrose, 10% CM, 500 ppm CH, and 1 ppm 2,4-D.

On WM, WM+CM (10%), or WM+CM+IAA (1 ppm), roots and shoot buds appeared within 3–4 weeks (Fig. 3 C). The latter developed into a shoot terminating in a flower bud. Flowers were also borne directly on the callus (Konar and Nataraja 1964). The flower bud was miniature and the number of stamens and carpels was much reduced, but microsporogenesis in anthers was normal and produced two-celled viable pollen grains (Fig. 3 D, E). After anthesis all floral parts abscised except the carpels, which produced empty achenes (devoid of embryo) due to lack of pollination. Rarely, mature seeds were formed (Fig. 3 F).

Besides roots and shoot buds, within 4–6 weeks, on 2,4-D-free medium, numerous embryoids invariably differentiated on the callus and developed into mature plantlets through stages resembling germination of mature dicotyledonous embryo in Nature (Fig. 3 G–J). Though embryoids also developed on the medium with 2,4-D (1, 2 ppm), in 10–12 weeks their growth was arrested and they formed fasciated plantlets. Nevertheless, normal plantlets were obtained when the embryoids were isolated and transferred to a 2,4-D-free medium. In about 12 weeks these plants flowered in vitro. At 5 ppm 2,4-D, embryoids failed to differentiate.

The subculture of embryoids in an auxin-rich medium resulted in proliferation and formation of a fresh mound of callus from which numerous embryoids differentiated. Thus, the cycle of dedifferentiation and redifferentiation could be repeated.

The in vitro plantlets developing from the embryoids showed an interesting feature of differentiation of embryoids all along the stem surface. The number of embryoids on each plantlet was variable (5–50), and they were irregularly distributed. Histologically, the embryoids originated from the epidermal cells of the stem (Fig. 3 K).

In *Browallia demissa* Ganapathy (1969) cultured three stages of buds on J.P. Nitsch's and C. Nitsch's (1967) medium. All of these failed to undergo complete development, but some buds proliferated and roots developed from the cut end of pedicel. The callus developed "meristemoids" that differentiated into roots and/or shoots. Cytokinins promoted bud formation, but IAA and GA_3 depressed it. The vegetative shoots differentiated from callus flowered after they bore as few as four leaves (in Nature flowering occurs only after the formation of nine leaves), and set normal fruits and seeds. Rarely, flowers also developed directly from the callus (Fig. 5 A–C). These flowers followed normal embryo sac development as in vivo.

The differentiation of both roots, and shoot buds in the callus of floral origin has also been reported in *Phlox drummondii* (R.N. Konar and A. Konar 1966). Floral buds with primordia of sepals and petals with a central meristematic dome grown on WM supplemented with CM or CM + IAA produced a fast-growing callus. Subsequently the callus differentiated roots and shoots and established new plantlets. The plantlets, finally, bore solitary flowers, in contrast to the in vivo control where they are borne in involucrate cymose heads.

Internodal segments of inflorescence of *Mazus pumilus* (Raste and Ganapathy 1970) were grown on MS medium with various supplements (IAA, KN, GA_3, adenine sulphate). These segments callused and either showed direct differentiation of flower buds, or differentiation of roots and leafy shoots which subsequently flowered.

Embryologically, the flowers which differentiated on leafy shoots revealed normal development of pollen grains and ovules, and viable seeds were formed. However, buds that arose directly on the callus differed from the former in floral morphology, produced only a few deformed pollen grains and scant initiation of ovules (Fig. 5 D, E).

Raman and Greyson (1974) reported induction of embryoids in *Nigella damascena* starting with the meristem bearing the first few whorls of floral primordia (along with a short pedicel). They were able to initiate callus formation on MS medium together with sucrose, CM, and 2,4-D. After repeated subcultures, the callus consisted mostly of parenchymatous cells with numerous tracheary elements. However, 6–7 days after subculture on a liquid medium devoid of CM, compact, globular masses became organised. In 4 weeks the entire medium was full of such organised structures, some clearly visible as small plantlets. Numerous embryonic structures resembling sequential development of dicotyledonous embryoid were also observed; these are presumed to have arisen from single cells. According to Raman and Greyson two important points emerge: embryogenesis occurs on liquid medium, and also in the absence of CM.

In *Populus trichocarpa* (Bawa and Stettler 1972) the number of primordia that lost all organisation as a result of extensive callus production was low. After 70 days, however, callus had overgrown most primordia. Pieces of callus isolated and subcultured on medium containing BA but lacking KN and auxin resulted in the differentiation of many shoots with leaf primordia; these later developed into plantlets lacking roots.

Nataraja (1971) reported his findings on the culture of floral buds of a few other members of Ranunculaceae: *Consolida orientalis*, *Delphinium brunonianum*, and *Clematis gouriana*. Floral buds, ranging from those with primordia of sepals surrounding a meristematic zone to older buds with anthers at pollen mother cells or pollen grain stage, were placed on modified WM supplemented with CM, IAA, 2,4-D, CH, and YE alone or in different combinations. Callus was usually formed

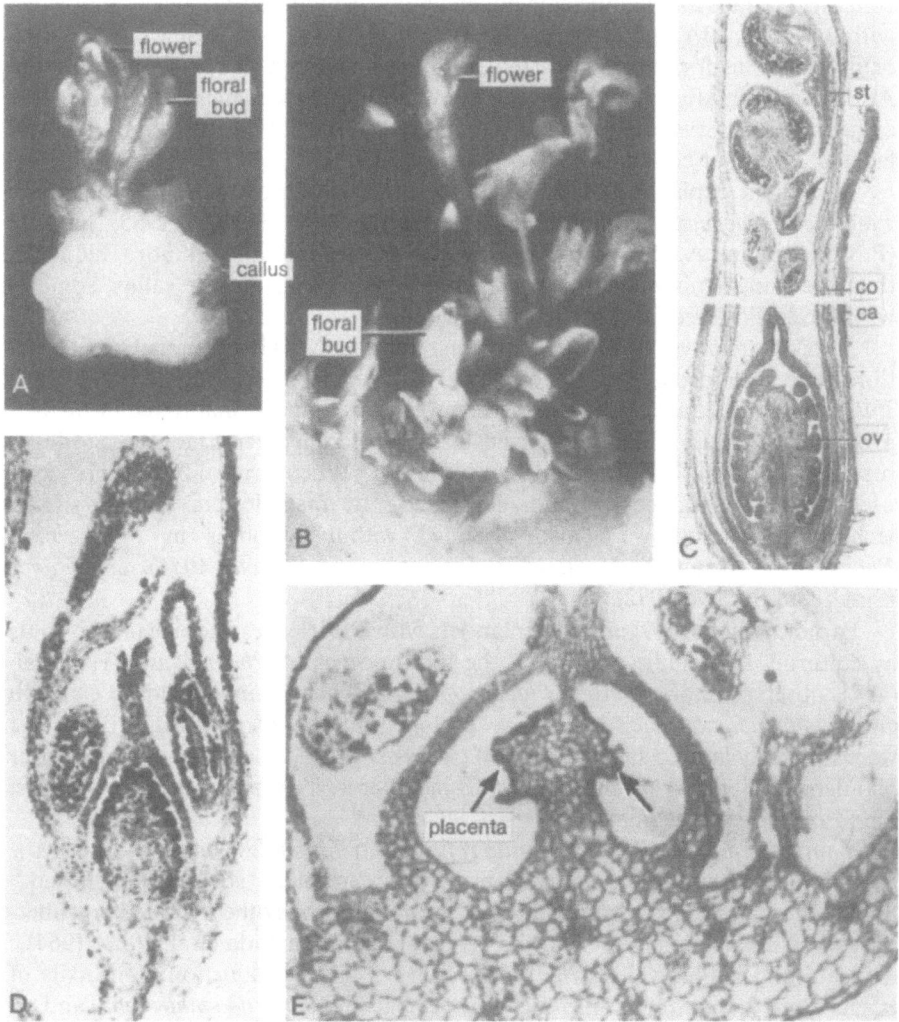

Fig. 5 A–C. *Browallia demissa;* **D, E** *Mazus pumilus.* **A** Floral bud and flowers developing directly from the callus, on WM + IAA + KN + GA$_3$. **B** Differentiation of a large number of floral buds, flowers, and shoots on WM + cytokinins (KN/SD 8339). **C** Differentiation of normal flower on WM + IAA + KN; *ca* calyx, *co* corolla, *ov* ovary, *st* stamen. **D** Longisection of flower from inflorescence developed on in vitro differentiated shoots; development of floral parts is normal. **E** Longisection of flower differentiated directly from the callus; the development of placentae *arrow* and pollen is inhibited. (**A–C** after Ganapathy 1969; **D** and **E** after Raste and Ganapathy 1970)

in all explants. *C. orientalis* differentiated both shoot buds and roots resulting in plantlets, whereas tissues of *D. brunonianum* developed only roots. *C. gouriana* did not show any differentiation.

Iizuka et al. (1973) cultured floral tissues from three cultivars of *Chrysanthemum* and one cultivar of *Senecio cruentus*. The inflorescence is a capitulum, and only tubular florets were used in the study. At the time of inoculation the anthers contained tetrads and uninucleate pollen grains. A modified MS medium together with BA, IAA, IBA (indole butyric acid) and 2,4-D was used. A few days after inoculation callus formation occurred from the lower end of corolla and ovaries. About a month after inoculation, differentiation of leaves, stems and roots from calluses was observed. Plantlets differentiated from calluses grown on a medium lacking 2,4-D but containing IAA, BA, and inosine.

Capitulum explants and floral parts (petals, ovary, style, and stigma) of *Pterotheca falconeri* (Compositae) were cultured on Earle's (1965) medium (P.N. Mehra and A. Mehra 1971). While mature flowers callused more frequently than the buds, embryoids and plantlets differentiated from the callus, some of which even flowered in vitro.

After the removal of bracts, Intuwong and Sagawa (1973) inoculated the inflorescence primordia (0.5 cm long) of the orchids *Ascofinetia*, *Neostylis*, and *Vascostylis* in a liquid medium. In a few weeks the hypodermal tissue of the axis proliferated, and this was transferred to agar medium after 4 months. Seedlings ready for transplantation developed in 15 months. When larger inflorescences were used, the axis elongated and buds developed into small flowers. Using a similar technique, plantlets were also obtained within 5 months in *Dendrobium*, *Phalaenopsis* (Intuwong et al. 1972; Intuwong and Sagawa 1974), and *Vanda* (Singh and Sagawa 1972).

Earlier, Rotor (1949) obtained plantlets with a few leaves and one or two roots by culturing the middle portion of the inflorescence of *Phalaenopsis*. Tse et al. (1971) observed poor growth in *Phalaenopsis*, but when the intact buds of *Cattleya* and *Cymbidium* were cultured, they developed into shoot primordia. Buds that were surgically injured showed signs of callus growth. Subsequently, three or four plantlets developed from the callus. MS medium + NAA gave better results than Knudson's (1951) medium.

Young floral buds of *Tradescantia reflexa* and *T. paludosa* cultured on WM + IAA formed subculturable callus of three types: "friable," "sticky," and "liquid." Some calluses developed "root hair-like" structures, while others formed protuberances which had some vessels in the central region (Yamada et al. 1963, 1964).

Kaul and Sabharwal (1970, 1972) reported callus production from axils of bracts when they cultured inflorescence segments of *Haworthia maughanii*, and also in extra-axillary positions and on cut-end of pedicel on inflorescence axis (Fig. 4C). The callus differentiated shoot buds which, when subcultured, produced roots.

Preil et al. (1977) claimed that, in Compositae, haploids can be obtained from in vitro culture of explants of capitulum, bypassing the method of anther or pollen culture. Medianly-cut explants from closed capitulum, when cultured, produced haploid callus from which shoots differentiated within 61 days.

Reversal of Floral Explants to Vegetative Condition

Apart from the normal development of floral organs, a variety of morphogenic responses have been observed in cultured flower buds. Reports relating to a changed morphogenic pattern of development mention either a reversion to the vegetative condition or disdifferentiation. The development of leafy shoots from excised flower buds is extremely rare. As early as 1942, LaRue observed the occurrence of leafy shoots in cultured floral buds in two (*Kalanchoe globulifera* and *Nemesia strumosa*) out of 92 species of angiosperms he investigated. He suspected that in *N. strumosa* the shoot buds might have been formed prior to culture. In *K. globulifera* shoots originated from the cut-end of the pedicel and not from the flower per se. In LaRue's experiments shoot buds organised on the enlarged but non-callused torus.

An interesting observation on the reversion from floral to vegetative condition is in *Kalanchoe pinnata* (Mohan Ram and Wadhi 1966, 1968). This investigation was undertaken to test whether immature floral buds could attain full development in vitro and to what extent morphogenetic pattern could be altered by manipulation of the nutrient medium. Flower buds of two different stages were used. The young buds had well-developed sepals, but only the primordia of other floral parts. The older buds contained all four floral whorls and the anthers were at the archesporial stage. On WM + NAA the cultures showed profuse rooting from the torus region. In the younger buds even the cut-end rooted. On WM + CM + 2,4-D, the buds formed either roots only or roots with a little callus, or both roots and shoots. As many as 10–12 shoots emerged from a single floral bud; the shoot meristem also arose from the torus (Fig. 6 A–C). None of the media tried supported the normal development of buds into flowers. The authors attribute the phenomenon of reversion to vegetative growth in *Kalanchoe* to early excision, which apparently hinders the normal sequence of floral morphogenesis.

Roest and Bokelmann (1973) cultivated [medium: Knop's (1865) mineral salts with Heller's microelements] the young capitulum of pyrethrum *(Chrysanthemum cinerariaefolium)*. It was divided into two segments, each with a portion of floral stalk. Before inoculation the disc and ray florets along with upper involucral bracts were "nearly completely" cut off, and placed on nutrient medium containing BA at 1 ppm. Two weeks after culture, a large number of adventitious shoots developed from receptacles, remnants of disc florets, axils of disc florets, or involucral bracts. Shoots that were treated with IAA rooted and formed plantlets.

In conformity with these results, shoot development was also observed following the above procedure from capitulum explants of other Compositae (Roest and Bokelmann 1973): *Chrysanthemum leucanthemum, C. parthenium, C. segetum, Anthemis arvensis, Calendula officinalis, Helenium autumnale, Hypochaeris radiata, Leontodon autumnalis,* and *Matricaria maritima;* and floral explants of *Hypericum perforatum* (Hypericaceae).

In the above reports no mention has been made of the fate of anthers and carpels, or any other embryological features.

Peduncle segments of many cultivars of *Chrysanthemum morifolium* rapidly regenerated numerous adventitious shoots in vitro (Roest and Bokelmann 1975). When the young shoots were excised and cultured, they formed roots and plantlets.

Pierik et al. (1973, 1975) developed a new technique for vegetative propagation using capitulum of *Gerbera jamesonii*. Fully-developed inflorescences were denuded of florets, and segments of receptacles implanted on nutrient medium. Shoot buds developed from the base of involucral bracts. The shoots were excised and rooted, leading to plantlet formation.

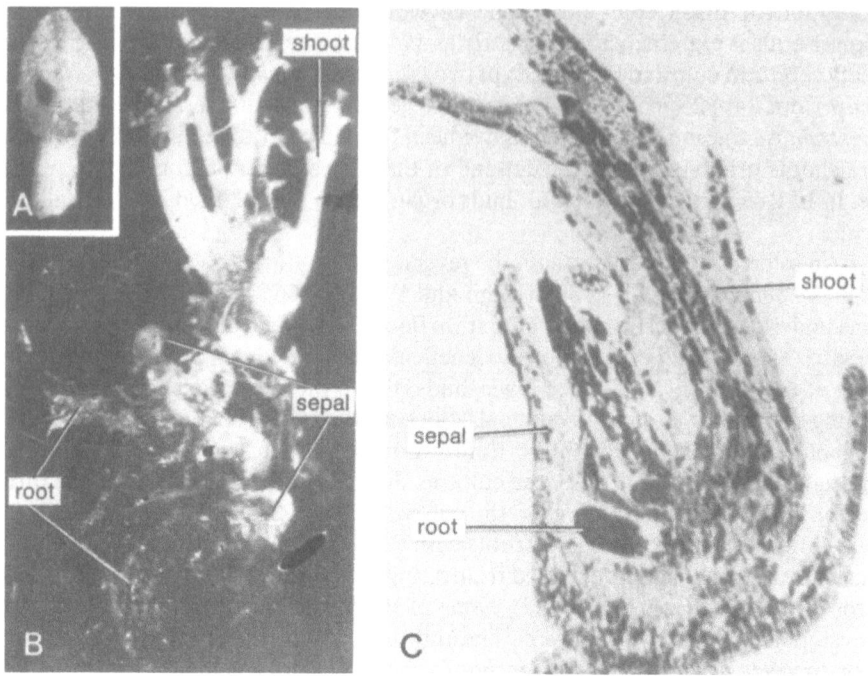

Fig. 6 A–C. *Kalanchoe pinnata.* **A** Flower bud (0.5 mm long at time of inoculation) with sepals, and primordia of other floral parts. **B** Development of shoots and roots from flower bud, in 28 weeks, on WM + CM (10%) + 2,4-D (1 ppm); remnants of sepals seen as bulges. **C** Longisection of flower bud, 20-week-old culture, on WM + NAA (1 ppm); note roots and shoots from the torus. (After Mohan Ram and Wadhi 1968)

Inflorescence segments of *Haworthia* spp were grown on WM supplemented with IAA, CM, and KN. There was either formation of flowers or, sometimes, the inflorescence segments shrivelled up; in 20–25 days vegetative buds simulating plantlets appeared on the floral axis. Simultaneously, formation and further differentiation into vegetative buds and roots also occurred from the floral axes (Majumdar 1970a; Kaul and Sabharwal 1970).

The floral stalks (bearing buds) of *Phalaenopsis* were cut into one-node sections and cultured (Tanaka and Sakanishi 1978). The buds showed three modes of growth: remaining dormant, or responding like vegetative or reproductive structures. The position of bud, temperature of culture, and BA influenced the mode of growth.

Bawa and Stettler (1972) placed the female catkin primordia of *Populus trichocarpa* in culture, and though they were not successful in growing them to maturity, they found a general tendency to callus formation with increasing age of culture, occasionally followed by reversal to vegetative growth. Catkin primordia raised on Wolter's medium (see Winton 1968) without auxin or KN but with BA proliferated axillary shoots in place of pistils.

Concluding Remarks

The credit for the early work on flower bud culture goes to LaRue (1942). In most of the cultures, only callusing and root formation were observed. The development of excised young buds into normal flowers with mature anthers and ovules has been achieved only in a few species: *Viscaria candida* (Blake 1966), *Anagallis arvensis* (Brulfert and Fontaine 1967), *Cucumis melo* (Porath and Galun 1967), *Nicotiana tabacum* (Hicks and Sussex 1970), and *Cyperus rotundus* (Mohan Ram and Batra 1970).

In *Cucumis sativus* (Galun et al. 1962, 1963) only potential female or hermaphrodite floral buds continued normal differentiation, while potential male floral buds expected to form male flowers tended to develop ovaries or did not grow at all.

Buds of *Aquilegia* (Tepfer et al. 1963, 1966) and *Ranunculus* (Konar and Nataraja 1964, 1969) at the sepal-primordia stage, grown on IAA-containing medium, showed normal development of carpels but microsporogenesis failed to occur in anthers. Ovules also differentiated in *Ranunculus*, but not in *Aquilegia* (Tepfer et al. 1963).

In *Aquilegia* (Tepfer et al. 1963) and *Cucumis sativus* (Galun et al. 1963) buds excised at pre-meiotic stage failed to complete microsporogenesis. But in *Viscaria* (Blake 1966), *C. melo* (Porath and Galun 1967), *Ranunculus* (Konar and Nataraja 1969), *Nicotiana* (Hicks and Sussex 1970), and *Cyperus* (Mohan Ram and Batra 1970), buds cultured at the pre-meiotic stage did complete microsporogenesis and pollen grains were formed. Obviously, "meiotic stimulus" was already present in the explants before excision. In this context, let us examine the studies of Leshkovsteva (1967) on *Populus balsamifera* and *Salix acutifolia*, and of Bawa and Stettler (1972) on *Populus trichocarpa*. In these plants, in Nature, the floral primordia are exposed to a variety of sequential stimuli induced by prolonged seasonal and environmental changes before maturity. Leshkovsteva (1967) obtained mature pollen grains in culture by simulating conditions analogous to the natural state, i.e. by gradually transferring them from cooler to warmer temperature. The failure to simulate these changes in vitro was perhaps a major reason for growth not proceeding beyond a certain stage in *P. trichocarpa* (Bawa and Stettler 1972).

The nutritional requirements for normal development of flowers appear to be complex, and it seems the buds depend on the vegetative parts of the plant body for some of their requirements. The culture of buds of *Aquilegia* (Tepfer et al. 1963) and *Viscaria* (Blake 1966), without leaves attached to the explants, did not succeed.

The development of ovary is never accompanied by the development of stamens in flower buds of *Cucumis sativus* (Galun et al. 1963), irrespective of using any growth hormones. Likewise, abortion of carpels led to better development of stamens in *Aquilegia* (Tepfer et al. 1966).

In an unusual and interesting report Majumdar (1970 a, b) records the production of viable pollen grains from the culture of a male sterile line of *Haworthia*.

Sex reversal has been achieved in many cases by the use of sex hormones using flowers attached to the plant body; only a few instances are known of the modification of sex in excised flowers. While Galun et al. (1962) reported ovary development in male buds of *Cucumis sativus*, in *Cucurbita pepo* (Rodrigues Pereira 1968) there was enhancement of stamen development in cultured female buds. In *Mercurialis* (Champault 1969) female inflorescences showed a tendency towards maleness.

Callusing is a common phenomenon in tissues or organs cultured in vitro. LaRue (1942) observed callus formation in isolated flower buds of *Hibiscus* and *Kalanchoe* on a simple medium. Similarly, the buds of *Ranunculus sceleratus* (Konar and Nataraja 1964, 1969) and *Nigella damascena* (Raman and Greyson 1974) developed callus capable of producing embryoids, while another member of the same family (*Aquilegia formosa*, Ranunculaceae) showed a completely different response. However, embryogenesis in *Nigella* occurred on liquid medium in the absence of CM, in *Ranunculus* embryogenesis was completed on agar medium containing 10% CM.

The production of flowers in vitro has been reported in *Ranunculus* (Konar and Nataraja 1969), *Phlox* (R.N. Konar and A. Konar 1966), *Browallia* (Ganapathy 1969), and *Mazus* (Raste and Ganapathy 1970). Flowers formed both on the plantlets and directly on the callus in *Ranunculus* were similar to those in Nature, except that the former had fewer stamens and carpels. On the other hand, in *Phlox* solitary flowers were obtained in cultures, instead of the cymose heads that are formed in vivo. Similarly, in *Browallia* flowers produced in vitro on leafy shoots and directly on the callus are comparable to those in Nature. This is not always so, and flowers characteristic of the species developed in vitro on the leafy shoots of *Mazus* (Raste and Ganapathy 1970) revealed normal structural and embryological features. Flowers arising directly on the callus differed not only in morphology but also produced only a few ovules and deformed pollen grains.

The potentiality of the flower buds for growth, development, and differentiation in vitro varies from species to species, as is also true of many other tissues. Consideration of the developmental stage of the explant, presence of leaves or absence thereof at the time of inoculation, temperature conditions during in vitro growth, and an appropriate nutrient medium are essential for normal development. Many investigators have been concerned with merely the morphological differentiation either in the explanted flower buds, or those differentiated from callus; our knowledge of the in vitro development of pollen grains in anthers, and embryo sac, endosperm and embryo in ovules is very meager.

Hicks et al. (1981) investigated cytoplasmic male sterility using the technique of floral meristem culture. Young meristems (with visible floral primordia, or mere sepal primordia) of a male sterile tobacco hybrid were cultured on Linsmaier and Skoog's (1965) medium, supplemented with KN and GA$_3$, singly or in combination, in continuous light. In the medium containing KN all the floral organs were initiated in the normal sequence, and pattern of male sterile phenotype was expressed. GA$_3$ suppressed the formation of floral organs.

In *Begonia franconis*, a monoecious species, when young floral primordia were cultured on a medium (suggested by de Jong et al. 1974) containing BA, normal flower buds developed (Berghoef and Bruinsma 1979 a). Gibberellins promoted the initiation of floral organs. With the addition of IAA, ABA, or ethephon, to the medium, the primordia failed to develop into flower buds. The growth regulators did not affect the sex of the bud. While a low-level of sucrose in the medium increased the number of male buds, a high level did not induce female buds. It appears that the regulation of sex expression is controlled endogenously by the carbohydrate-level in the central region of inflorescence-primordium, which is a limiting factor for female differentiation.

The pedicels of inflorescences (flowers at anthesis) cultured on the same medium (Berghoef and Bruinsma 1979b) with IAA, NAA, BA, CCC, ethephon, GA_3, at first showed callus formation followed by differentiation of flower buds (Berghoef and Bruinsma 1979 c). Both auxins and cytokinin are required for the explant to callus. The floral buds invariably originate from this callus, and the male buds differentiate first. On the bracts of these male buds differentiate female buds. None of the growth regulators tested affected the ratio of male and female buds. The authors suggest that the precedence of male buds does not specifically regulate the differentiation of female buds, but they merely enhance the nutritional status of the tissue and, hence, promote female bud formation.

The floral parts (sepals, petals, and ovaries) of African violet, *Saintpaulia ionantha* cv Blue Rhapsody, produced callus on MS + BA (0.2 ppm) + NAA (2 ppm) (Vazquez and Short 1978). Later, the callus produced roots and shoots. From the epidermal cells of floral parts, roots differentiated on MS+KN (1 ppm) + NAA (0.5 ppm), and shoot buds on MS + BA (1 ppm) + NAA (1 ppm), without any callus formation. On transfer to MS alone, the shoots developed into plantlets which were transferred to pots. The morphology and cytology of the regenerated plants was comparable to that of the parents.

Morphactin [EMD 7301 W, methyl 2-chloro-(9)-hydroxyfluorene-(9)-carboxylate] sprayed on ten-week-old plants (just after floral evocation) of *Tagetes erecta* induced "barren capitula" with a bulged receptacle and suppressed rim (Mohan Ram and Mehta 1982). The barren capitula (diam 3–6 mm) cultured on B_5-medium (see Gamborg et al. 1968) $+ 10^{-5}$ M and 10^{-6} M BA produced callusing from the cut-end, involucral bracts, and surface of capitulum. Callus differentiated small, green buds two weeks after culture. On $B_5 + 10^{-7}$ M BA, callusing was less-marked, but the explants produced numerous buds. On $B_5 +$ NAA (10^{-5} and 10^{-6} M) there was profuse callusing followed by differentiation of roots. With BA (10^{-5} or 5×10^{-6} M) + NAA (5×10^{-6} M) both callus and shoot buds were formed; sometimes roots were also formed from the callus. The excised shoot buds grown on 10^{-5} or 10^{-6} M BA produced green vegetative shoots, or strap-shaped yellow to orange structures mid-way between the leaves and ray florets. The shoots, when subcultured on $B_5 +$ BA and/or NAA, developed into normal plantlets. After four weeks in culture, the plantlets attained a height of 3–4 cm; miniature terminal capitula developed within two months of culture. Over 115 shoots were obtained from one capitulum in three months. Thus, the morphactin-induced inhibition of capitula was partially overcome, and entire flowering plants appeared in culture.

Wilson (1977) and Wilson et al. (1978) cultured intact spikes of barley (*Hordeum vulgare* var Dissa) on agitated Linsmaier and Skoog's (1965) medium with 10% sucrose and 1 mg/l each of IAA, 2,4-D and BA. A high percentage of pollen in the anthers (from the florets in the middle of spike) which was at the two-nucleate stage, developed into callus masses within two weeks. The response was maximal when the tillers were clipped off at the ground-level, and allowed to stand in water for one or two days prior to culture. In 60 days, large, white, callus masses emerged from some of the florets. The pollen callus comprised haploid, diploid and polyploid cells. When the pollen callus was transferred to semi-solid medium devoid of 2,4-D, and with lower levels of sucrose (3%), IAA and BA (0.4 mg/l each), approximately 40 plantlets regenerated; two were green and short-lived and the rest albino.

Following the above technique, Kasperbauer et al. (1980) raised haploid plants of *Festuca arundinacea* var Kentucky 31. Field-grown plants cut at soil-surface were dipped in tap water and placed in darkness at 5 °C. The segments of panicle, 2.5 cm long, were cultured on liquid MS-medium with 2 mg/l 2,4-D. In a five-week-old culture, several anthers showed loose proliferation of cells, and plantlets appeared in another two weeks. Over 30 plantlets were formed – all green, and these grew vigorously on transfer to soil.

Wilson (1977) states that it is possible to stimulate callus formation in pollen (in anthers intact in the florets on the spike) to a much higher percentage as compared to culture of isolated anthers. Another advantage of culturing the whole spike, or parts thereof, is that a very large number of anthers can be induced simultaneously. The enhanced pollen response is probably due to culturing on a liquid medium, that allows efficient removal of inhibitors and a more rapid transport of nutrients into the anthers. It is also important that the anthers remain connected to the floret. Thus, the effects of culture are at least partly mediated through endogenous substances.

In our opinion, the culture of entire inflorescence, or its segments, is an attractive procedure for the production of androgenic haploids from recalcitrant systems, including cereals (see also Preil et al. 1977; Wilson 1977; Kasperbauer et al. 1980).

References

Bawa KS, Stettler RF (1972) Organ culture with black cottonwood: Morphogenetic response of female catkin primordia. Can J Bot 50:1627–1631

Berghoef J, Bruinsma J (1979a) Flower development of *Begonia franconis* Liebm. II. Effects of nutrition and growth regulation substances on the growth of floral buds in vitro. Z Pflanzenphysiol 93:345–347

Berghoef J, Bruinsma J (1979b) Flower development of *Begonia franconis* Liebm. III. Effects of growth regulating substances on organ initiation in flower buds in vitro. Z Pflanzenphysiol 93:377–386

Berghoef J, Bruinsma J (1979c) Flower development of *Begonia franconis* Liebm. IV. Adventitious flower bud formation on excised inflorescence pedicels in vitro. Z Pflanzenphysiol 94:407–416

Bilderback DE (1971) The effects of amino acids upon the development of excised floral buds of *Aquilegia*. Am J Bot 58:203–208

Bilderback DE (1972) The effects of hormones upon the development of excised floral buds of *Aquilegia*. Am J Bot 59:525–529

Blake J (1966) Flower apices cultured in vitro. Nature (London) 211:990–991

Blake J (1969) The effect of environmental and nutritional factors upon the development of flower apices cultured in vitro. J Exp Bot 20:113–123

Brulfert J (1965) Etude expérimentale du développement végétatif et floral chez *Anagallis arvensis* L. ssp, *phoenicea* Scop. Formation de fleurs prolifères chez cette même espèce. Rev Gen Bot 72:641–694

Brulfert J, Fontaine D (1967) Utilisation de la culture in vitro pour une étude de développement floral chez *Anagallis arvensis* ssp. *phoenicea* Scop. Biol Plant 9:439–446

Champault A (1969) Masculinisation d'inflorescences femelles de *Mercurialis annua* L. ($2n = 16$) par culture in vitro, de noeuds isolés en présence d'auxines. CR Acad Sci Paris 269:1948–1950

Earle E (1965) Cell colony formation from isolated plant cells. In: White PR, Grove AR (eds) Proc Int Conf Plant Tissue Culture, Berkeley, pp 401–409

Evans LT (1969) The nature of flower induction. In: Evans LT (ed) The induction of flowering: Some case histories. Cornell Univ Press, Ithaca, pp 457–480

Galun E, Jung Y, Lang A (1962) Culture and sex modification of male cucumber buds in vitro. Nature (London) 194:596–598

Galun E, Jung Y, Lang A (1963) Morphogenesis of floral buds of cucumber cultured in vitro. Dev Biol 6:370–387

Gamborg OL, Miller RA, Ojima K (1968) Nutrient requirments of suspension cultures of soybean root cells. Exptl Cell Res 50:151–158

Ganapathy PS (1969) Floral morphogenesis and flowering in aseptic cultures of *Browallia demissa* L. Biol Plant 11:165–174

Greyson RI, Raman K (1975) Differential sensitivity of "double" and "single" flowers of *Nigella damascena* (Ranunculaceae) to emasculation and to GA_3. Am J Bot 62:531–536

Heller R (1953) Recherches sur la nutrition minérale des tissus végétaux cultivés in vitro. Ann Sci Nat Bot Biol Veg 14:1–223

Hicks GS (1972) Development of surgically-halved stamen primordia of tobacco. Can J Bot 50:2396–2400

Hicks GS (1973) Studies on tobacco petal primordia in culture: Petal duplication induced by surgery. Bot Gaz 134:154–160

Hicks GS, Sussex IM (1970) Development in vitro of excised flower primordia of *Nicotiana tabacum*. Can J Bot 48:133–139

Hicks GS, Sussex IM (1971) Organ regeneration in sterile culture after median bisections of the flower primordia of *Nicotiana tabacum*. Bot Gaz 132:350–363

Hicks GS, Browne R, Sands SA (1981) Organogenesis from cultured floral meristems of a male sterile tobacco hybrid. Can J Bot 59:1665–1670

Hügel B (1976a) Gegensätzliche Geschlechtsausprägung von Blütenstandsanlagen der Langtagpflanze *Lemna gibba* und der Kurztagpflanze *Lemna paucicostata* in vitro. Z Pflanzenphysiol 77:395–405

Hügel B (1976b) Wirkung von GA_3, CCC, Ethrel und Indolessigsäure auf die Geschlechtsausprägung isolierter Blütenstandsanlagen von Lemnaceen. Z Pflanzenphysiol 80:283–297

Hügel B (1976c) Wirkung von Kinetin und Abscicinsäure auf die Entwicklung von Lemnaceen – Blütenstandsanlagen in vitro. Z Pflanzenphysiol 80:298–305

Iizuka M, Matsumoto E, Doi A, Madrigal R, Fukushima A (1973) Tubular floret culture of chrysanthemum and cineraria cultured in vitro. Jpn J Genet 48:79–87

Intuwong O, Sagawa Y (1973) Clonal propagation of Sarcanthine orchids by aseptic culture of inflorescence. Am Orchid Soc Bull 42:209–215

Intuwong O, Sagawa Y (1974) Clonal propagation of *Phalaenopsis* by shoot tip culture. Am Orchid Soc Bull 43:893–895

Intuwong O, Kunisaki JT, Sagawa Y (1972) Vegetative propagation of *Phalaenopsis* by flower-stalk cuttings. Na Okika Hawaii 1:13–18

Jensen LCW (1971) Experimental bisection of *Aquilegia* floral buds cultured in vitro. I. The effect on growth, primordia initiation and apical regeneration. Can J Bot 49:487–493

Jensen LCW (1972) Experimental bisection of *Aquilegia* floral buds cultured in vitro. II. Cytological changes following bisection. Can J Bot 50:1611–1615

Johri BM, Ganapathy PS (1967) Floral differentiation and morphogenesis in vitro. J Indian Bot Soc 46:374–388

Jong AW de, Bruinsma J (1974a) Pistil development in *Cleome* flowers. I. Effects of mineral nutrition and of the presence of leaves and fruits on female abortion in *Cleome spinosa* Jacq. Z Pflanzenphysiol 72:220–226

Jong AW de, Bruinsma J (1974b) Pistil development in *Cleome* flowers. III. Effects of growth-regulating substances on flower buds of *Cleome iberidella* Welw. ex Oliv. grown in vitro. Z Pflanzenphysiol 73:142–151

Jong AW de, Smit AL, Bruinsma J (1974) Pistil development in *Cleome* flowers. II. Effects of nutrients on flower buds of *Cleome iberidella* Welw. ex Oliv. grown in vitro. Z Pflanzenphysiol 72:227–236

Kandeler R, Hügel B (1974) Development in vitro of flower primordia of Lemnaceae (Abstr). 3rd Int Congr Plant Tissue and Cell Culture. Univ Leicester, Leicester, p 160

Kandeler R, Hügel B, Rottenburg Th (1974) Gegensätzliche Wirkung der Sproßalterung auf die Blütenbildung bei *Lemna paucicostata* und *Lemna gibba*. Biochem Physiol Pflanz 165:331–336

Kasperbauer MJ, Buckner RC, Springer WD (1980) Haploid plants by anther-panicle culture of tall fescue. Crop Sci 20:103–106

Kaul K, Sabharwal PS (1970) In vitro induction of vegetative buds on inflorescence segments of *Haworthia*. Experientia 26:433–434

Kaul K, Sabharwal PS (1972) Morphogenetic studies on *Haworthia*: Establishment of tissue culture and control of differentiation. Am J Bot 59:377–385

Knop W (1865) Quantitative Untersuchungen über den Ernährungsprozeß der Pflanzen. Landwirtsch Vers Stu 7:93–107

Knudson L (1951) Nutrient solution for orchids. Bot Gaz 112:528–532

Konar RN, Konar A (1966) Plantlet and flower formation in callus cultures from *Phlox drummondii*. Phytomorphology 16:379–382

Konar RN, Nataraja K (1964) In vitro control of floral morphogenesis in *Ranunculus sceleratus* L. Phytomorphology 14:558–563

Konar RN, Nataraja K (1969) Morphogenesis of isolated floral buds of *Ranunculus sceleratus* L. in vitro. Acta Bot Neerl 18:680–699

LaRue CD (1942) The rooting of flowers in sterile culture. Bull Torrey Bot Club 69:332–341

Leshkovsteva II (1967) Formation of sporogenous cells in the culture of isolated male flower buds of woody plants (effect of heteroauxin and kinetin). Sov Plant Physiol 14:599–605

Linsmaier EM, Skoog F (1965) Organic growth factor requirements of tobacco tissue cultures. Physiol Plant 18:100–127

Majumdar SK (1970a) Culture of *Haworthia* inflorescences in vitro. J S Afr Bot 36:63–68

Majumdar SK (1970b) In vitro culture of flower buds of *Haworthia* and *Astroloba*. Φyton (Argentina) 27:31–34

Majumdar SK, Sabharwal PS (1968) Induction of vegetative buds of inflorescence of *Haworthia*. Abstr Am J Bot 55:705

Mehra PN, Mehra A (1971) Morphogenetic studies in *Pterotheca falconeri*. Phytomorphology 21:174–191

Mohan Ram HY, Batra M (1970) Stimulation of flower formation by cytokinins in the excised immature inflorescences of *Cyperus rotundus*. Phytomorphology 20:22–29

Mohan Ram HY, Jaiswal VS (1974) Some aspects of flower morphogenesis in vitro. In: Mohan Ram HY, Shah JJ, Shah CK (eds) Form, structure, and function in plants. Sarita Prakashan, Meerut, pp 236–241

Mohan Ram HY, Mehta G (1982) Regeneration of plantlets from cultured morphactin-induced barren capitula of African marigold (*Tagetes erecta* L.). Plant Sci Lett (In Press)

Mohan Ram HY, Wadhi M (1966) Reversion of floral buds of *Kalanchoe pinnata* to vegetative state in culture. Naturwissenschaften 53:387–388

Mohan Ram HY, Wadhi M (1968) Morphogenic potentialities of flower buds of *Kalanchoe pinnata* grown in vitro. Ann Bot 32:825–832

Murashige T, Skoog F (1962) A revised medium for rapid growth and bioassays with tobacco tissue cultures. Physiologia Plant 15:473–497

Nataraja K (1971) Morphogenic variations in callus cultures derived from floral buds and anthers of some members of Ranunculaceae. Phytomorphology 21:290–296

Nitsch JP (1951) Growth and development in vitro of excised ovaries. Am J Bot 38:566–577

Nitsch JP (1965) Physiology of flowering and fruit development. Handb Pflanzenphysiol 15:1537–1648

Nitsch JP, Nitsch C (1967) The induction of flowering in vitro in stem segments of *Plumbago indica* L. I. The production of vegetative buds. Planta 72:355–370

Pierik RLM, Steegmans HHM, Marelis JJ (1973) *Gerbera* plantlets from in vitro cultivated capitulum explants. Scientia Hortic 1:117–119

Pierik RLM, Jansen JLM, Maasdam A, Binnendjik CM (1975) Optimalization of *Gerbera* plantlets production from excised capitulum explants. Scientia Hortic 3:351–357

Pirson A, Seidel F (1950) Zell- und Stoffwechselphysiologische Untersuchungen an der Wurzel von *Lemna minor* unter besonderer Berücksichtigung von Kalium- und Calciummangel. Planta 38:431–473

Porath D, Galun E (1967) In vitro culture of hermaphrodite floral buds of *Cucumis melo* L.: Microsporogenesis and ovary formation. Ann Bot 31:283–290

Preil W, Huhnke W, Engelhardt M, Hoffman M (1977) Haploide bei *Gerbera jamesonii* aus in vitro-Kulturen von Blütenköpfchen. Z Pflanzenzüchtg 79:167–171

Puri P (1964) In vitro culture of floral organs of an apomict *Aerva javanica*. Phytomorphology 14:564–573

Raman K, Greyson RI (1974) In vitro induction of embryoids in tissue cultures of *Nigella damascena*. Can J Bot 52:1988–1989

Raman K, Greyson RI (1977) Graft unions between floral half-meristems of differing genotypes of *Nigella damascena*. Plant Sci Lett 8:367–373

Raman K, Greyson RI (1978) Further observations on the differential sensitivities to plant growth regulators by cultured "single" and "double" flower buds of *Nigella damascena* L. (Ranunculaceae). Am J Bot 65:180–191

Raste AP, Ganapathy PS (1970) In vitro behaviour of inflorescence segments of *Mazus pumilus*. Phytomorphology 20:367–373

Rodrigues Pereira AS (1968) Sex expression in floral buds of acorn squash. Planta 80:349–358

Roest S, Bokelmann GS (1973) Vegetative propagation of *Chrysanthemum cinerariaefolium* in vitro. Scientia Hortic 1:120–122

Roest S, Bokelmann GS (1975) Vegetative propagation of *Chrysanthemum morifolium* Ram. in vitro. Scientia Hortic 3:317–330

Rotor G (1949) A method of vegetative propagation of *Phalaenopsis* species and hybrids. Am Orchid Soc Bull 18:738–739

Singh H, Sagawa Y (1972) Vegetative propagation of *Dendrobium* by flower-stalk cuttings. Na Okika Hawaii 1:19

Tanaka M, Sakanishi Y (1978) Factors affecting the growth of in vitro cultured lateral buds from *Phalaeonopsis* flower stalks. Scientia Hortic 8:169–178

Tepfer SS, Greyson RI, Craig WR, Hindman JL (1963) In vitro culture of floral buds of *Aquilegia*. Am J Bot 50:1035–1045

Tepfer SS, Karpoff AJ, Greyson RI (1966) Effects of growth substances on excised floral buds of *Aquilegia*. Am J Bot 53:148–157

Tse AT, Smith RJ, Hackett WP (1971) Adventitious shoot formation on *Phalaenopsis* nodes. Am Orchid Soc Bull 40:807–810

Vazques AM, Short KC (1978) Morphogenesis in cultured floral parts of African violet. J exp bot 29:1265–1271

Wadhi M (1967) Floral morphogenesis in vitro. In: Johri BM (ed) Seminar on "Plant cell, tissue, and organ cultures." Univ Delhi, Delhi, pp 85–87

White PR (1943) A handbook of plant tissue culture. The Jacques Cattell Press, Lancaster

Wilson HM (1977) Culture of whole barley spikes stimulates high frequency of pollen cal-
    luses in individual anthers. Plant Sci Lett 9:223–238
Wilson HM, Mix G, Foroughi-Wehr B (1978) Early microspore divisions and subsequent
    formation of microspore calluses at high frequency in anthers of *Hordeum vulgare* L.
    J exp bot 29:227–238
Winton LW (1968) Plantlets from aspen tissue culture. Science 160:1234–1235
Yamada T, Shoji T, Sinoto Y (1963) Formation of calli and free cells in the tissue culture
    of *Tradescantia reflexa*. Bot Mag Tokyo 76:332–339
Yamada T, Shoji T, Sinoto Y (1964) Cytological studies on cultured cells. II. Formation of
    calli and general behaviour of their cells in the tissue cultures of *Tradescantia paludosa*.
    Bot Mag Tokyo 77:436–446

5. Anther Culture

S. Narayanaswamy and L. George

More than two decades ago, C.D. LaRue (1954) of the University of Michigan, Ann Arbor, USA, conceived the idea of culturing pollen (gametophytes) of higher plants in order to explore their growth potential under in vitro conditions. He carried out extensive investigations with the male and female gametophytes of a number of gymnosperms in aseptic media and showed that these were capable of unusual growth by repeated mitotic divisions of the pollen or the ovular nucellus. The nucellar tissue regenerated roots and shoots. In spite of their sporophytic appearance, these structures were haploid. Similar studies with angiosperms, however, did not succeed.

In subsequent years continuous growth in vitro was achieved in cultures of pollen grains of several gymnosperms (see Vasil and Nitsch 1975). The vegetative cell of the pollen could be stimulated to divide and form haploid tissue extensively, but organogenesis or embryogenesis failed to occur. The chance observation of the development of embryo-like structures in anther culture of *Datura innoxia* by Guha and Maheshwari (1964, 1966, 1967), and later, confirmation of their origin from pollen grains, gave a new impulse to studies on the potentialities of a number of other angiosperm pollen for cell division and growth in vitro in several laboratories of the world.

Starting from a single cell (microspore) and ending in a whole organism passing through a series of cell divisions and differentiation, the pollen grain has now been shown to be totipotent. Pollen of monocotyledons, on the other hand, forms a subculturable callus mass and retains potentiality for organogenesis.

The totipotent nature of pollen grains has, in recent times, been exploited to advantage in many plants of economic and ornamental value with varying success. The problems of experimental androgenesis in angiosperms have been earlier reviewed by several authors (Melchers and Labib 1970; Sunderland 1971, 1973; Melchers 1972; Pandey 1973), and the significance of production of haploids in studies of induction of mutations and as a tool in various biological disciplines has been stressed (Vasil and Nitsch 1975; Clapham 1977; Jensen 1977; Nitzsche and Wenzel 1977; Reinert and Bajaj 1977; Sink and Padmanabhan 1977).

Literature pertaining to anther culture to date shows that this technique has been especially successful in species belonging to the family Solanaceae and Gramineae. Most solanaceous plants have proved versatile in that embryoids developed directly from the pollen, unlike the monocotyledonous group which is characterised by the intervention of a callus before organogenesis sets in. A list of species in which haploid embryoids/calli/plantlets have been obtained from cultured anther or pollen is given in Table 1.

Table 1. List of species in which successful production of androgenic embryoids, calli and plantlets has been achieved by anther/pollen culture

Species	Family	Nature of regeneration	Reference
Dicotyledons			
Aesculus hippocastanum	Sapindaceae	Embryoids and plantlets	Radojevic (1978)
Arabidopsis thaliana	Cruciferae	Callus and plantlets	Gresshoff and Doy (1972a)
Atropa belladonna	Solanaceae	Embryoids and plantlets	Zenkteler (1971), Narayanaswamy and George (1972), Rashid and Street (1973)
Brassica campestris	Cruciferae	Embryoids and plantlets	Keller et al. (1975)
B. napus	—	Embryoids and plantlets	Thomas and Wenzel (1975), Keller and Armstrong (1977)
B. oleracea	—	Callus and plantlets	Kameya and Hinata (1970)
B. oleracea x *B. alboglabra*	—	Callus and plantlets	Kameya and Hinata (1970)
Capsicum annuum	Solanaceae	Embryoids and plantlets	George and Narayanaswamy (1973), Wang et al. (1973), Kuo et al. (1973)
Coffea arabica	Rubiaceae	Callus and embryoids	Sharp et al. (1973)
Datura innoxia	Solanaceae	Embryoids and plantlets	Guha and Maheshwari (1964, 1966, 1967), Engvild et al. (1972), Geier and Kohlenbach (1973), Nitsch and Norreel (1973), Sunderland et al. (1974), Collins et al. (1974a)
D. metel	—	Embryoids and plantlets	Narayanaswamy and Chandy (1971), Chandy and Narayanaswamy (1971), Iyer and Raina (1972), Narayanaswamy and George (1974)
D. meteloides	—	Embryoids and plantlets	Kohlenbach and Geier (1972), Geier and Kohlenbach (1973)
Digitalis purpurea	Scrophulariaceae	Callus and plantlets	Corduan and Spix (1975)
Fragaria virginiana	Rosaceae	Callus and plantlets	Rosati et al. (1975)
Gerbera jamesonii	Compositae	Callus and plantlets	Preil et al. (1977)
Hevea brasiliensis	Euphorbiaceae	Embryoids and plantlets	Cheng-hua et al. (1978), Chen et al. (1979)
Hyoscyamus niger	Solanaceae	Embryoids and plantlets	Corduan (1975), Raghavan (1975)
Iberis amara	Cruciferae	Embryoids	Babbar et al. (1980)
Lycopersicon esculentum	Solanaceae	Callus, embryoids and plantlets	Debergh and Nitsch (1973), Gresshoff and Doy (1972b)
Nicotiana tabacum	—	Embryoids and plantlets	Bourgin and Nitsch (1967), Nakata and Tanaka (1968), Nitsch et al. (1968), Nitsch (1969), Sunderland and Wicks (1969), Anagnostakis (1974), Reinert et al. (1975), Bajaj et al. (1977), Horner et al. (1977), Niizeki (1977)
Paeonia hybrida	Ranunculaceae	Embryoids and plantlets	Sunderland (1974)
P. lutea	—	Embryoids	Zenkteler et al. (1975)
Pelargonium hortorum	Geraniaceae	Callus and plantlets	Abo El-Nil and Hildebrandt (1971)

Species	Family	Type of response	References
Petunia axillaris	Solanaceae	Callus and/or embryoids and plantlets	Doreswamy and Chacko (1973), Engvild (1973)
P. hybrida x *P. axillaris*	–	Callus and/or embryoids and plantlets	Bernhard (1971), Raquin and Pilet (1972), Wagner and Hess (1974), Sangwan and Norreel (1975)
Physalis minima	Solanaceae	Embryoids and plantlets	George and Rao (1979)
Populus	Salicaceae	Callus and plantlets	Li-kung (1978)
Primula obconica	Primulaceae	Callus and plantlets	Bajaj (1976)
Primula obconica	Primulaceae	Callus and/or embryoids and plantlets	Bajaj (1981)
Saintpaulia ionantha	Gesneriaceae	Embryoids and plantlets	Hughes et al. (1975)
Solanum dulcamara	Solanaceae	Embryoids and plantlets	Zenkteler (1973)
S. melongena	–	Callus and plantlets	Raina and Iyer (1973)
S. nigrum	–	Callus and plantlets	Harn (1971)
S. tuberosum	–	Embryoids and/or plantlets	Kohlenbach and Geier (1972), Dunwell and Sunderland (1973), Foroughi-Wehr et al. (1977)
Monocotyledons			
Aegilops caudata x *A. umbellulata*	Gramineae	Callus and plantlets	Kimato and Sakamoto (1972)
Asparagus officinalis	Liliaceae	Callus and plantlets	Pelletier et al. (1972)
Festuca arundinacea	Gramineae	Callus and plantlets	Kasperbauer et al. (1980)
Freesia sps	Iridaceae	Callus and plantlets	Bajaj and Pierik (1974)
Hordeum vulgare	Gramineae	Callus and plantlets	Clapham (1971, 1973), Malepszy and Grunewaldt (1974), Grunewaldt and Malepszy (1975)
Lolium multiflorum	–	Callus and plantlets	Clapham (1971)
Lolium multiflorum	–	Callus and/or embryoids and plantlets	Pagniez and Demarly (1979)
L. multiflorum x *Festuca arundinacea*	–	Callus and plantlets	Nitzsche (1970)
Oryza glaberrima	–	Callus and plantlets	Woo and Huang (1980)
O. glaberrima x *O. sativa* cultivar Taichung 65	–	Callus and plantlets	Woo and Huang (1980)
Oryza sativa	–	Callus and plantlets	Niizeki and Oono (1968), Nishi and Mitsuoka (1969), Guha et al. (1970), Woo and Tung (1972), Guha-Mukherjee (1973)
Secale cereale	–	Embryoids and plantlets	Wenzel and Thomas (1974), Thomas et al. (1975)
Setaria italica	–	Callus and plantlets	Ban et al. (1971)
Triticum aestivum	–	Callus and plantlets	Ouyang et al. (1973), Wang et al. (1973), Craig (1974), Picard and de Buyser (1975)
Triticum aestivum (haploid)	–	Embryos and plantlets	DeBuyser and Henry (1980)
Triticale	–	Callus and plantlets	Wang et al. (1973), Sun et al. (1973)
Zea mays	–	Callus and plantlets	Murakami et al. (1973)

Methods of Culture

It has been the general practice to culture the microspores when they are confined within the anther sac. Young flower buds at the appropriate stage of pollen development are surface-sterilized and the intact, uninjured anthers inoculated on nutrient media. For most species, a nutrient medium solidified with agar has been used to induce pollen divisions. However, in species of *Nicotiana* and *Hyoscyamus* the yield of pollen plantlets could be increased considerably by using a liquid medium (Wernicke and Kohlenbach 1976; Sunderland and Roberts 1977; Sunderland 1978).

Success in obtaining division of pollen grains outside the anther sac has been reported in a number of plants. Mature pollen grains of *Brassica* could be induced to divide to form cell clusters in hanging drop cultures (Kameya and Hinata 1970). In *Lycopersicon esculentum* pollen could be triggered into active cell division by the nurse culture technique (Sharp et al. 1972). Induction of plantlets from microspores isolated from anthers of *Petunia* (Sangwan and Norreel 1975) and tobacco (C. Nitsch 1974; Reinert et al. 1975) could be achieved by plating them on nutrient agar. In *Nicotiana* and *Hyoscyamus* microspores isolated from precultured anthers when grown on liquid medium gave rise to plantlets (Wernicke and Kohlenbach, 1977). In several species anthers dehisce soon after culture, and if they are floated on liquid medium the microspores will be liberated into the medium, where they will continue to grow. By exploiting this phenomenon several batches of pollen plants could be obtained from the same anthers in tobacco (Sunderland and Roberts 1977).

Nutritional and Hormonal Conditions for Androgenesis

Success in the production of embryoids depends to a large extent upon the constituents of the media used for culturing the anthers. Since the anthers contain both somatic and gametophytic (pollen) tissues, it is essential to induce division in the microspore alone. Usually, this is achieved by the judicious applications of auxins and cytokinins in the medium. Plant extracts, coconut milk (CM) (Guha and Maheshwari 1964), plum juice (Guha and Maheshwari 1967), potato extract (Wenzel et al. 1977), and yeast extract (YE) (Guha et al. 1970) and casein hydrolysate (CH) (Zenkteler and Misiura 1974) are other growth adjuvants used to augment cell division and growth. The composition of basal medium containing mineral elements does not appear to be very critical. J.P. Nitsch's (1951), P.R. White's (1963), and T. Murashige and F. Skoog's (MS, 1962) media have all been tried for one species or another. Generally an induction-medium contains mineral elements, vitamins, sucrose, and growth regulators. In the initial stages of anther culture, media similar to those used in somatic tissue culture were used. But with refinements in technique, the significance of pollen-age in embryoid induction came to be realised and the growth media could be simplified. J.P. Nitsch (1969) demonstrated that for tobacco, only sucrose is essential for the initiation of cell division in pollen. By floating anthers on liquid medium containing major salts (half) and iron-EDTA of

Murashige and Skoog (1962) plus 2% sucrose, embryoids and plantlets could be obtained in *Nicotiana* and several other species (Sunderland 1978).

For most species 2%–3% sucrose in the media is most favourable, although in special cases higher levels have been used (Clapham 1971; Ouyang et al. 1973; Thomas and Wenzel 1975; Keller and Armstrong 1978). However, continuous culture in a high sucrose medium proved deleterious to the normal differentiation of embryoids into plantlets unless the embryoids are transferred to a low sucrose medium (Keller et al. 1975).

The nutritional requirements vary greatly from species to species. For instance, Kinetin (KN) added alone to the medium is sufficient to bring about initiation of pollen division in *Datura* (Guha and Maheshwari 1967) and *Petunia* (Engvild 1973). Incorporation of YE, CH, or YE and 2,4-dichlorophenoxy-acetic acid (2,4-D) triggers callus growth from anther wall tissues.

In a number of instances, however, auxins have promoted microspore division. The monocots, in general, respond to a medium containing 2,4-D (Clapham 1977). Other auxins like indoleacetic acid (IAA) and naphthaleneacetic acid (NAA) are also effective for certain species (Ban et al. 1971; Dunwell and Sunderland 1973).

For most species both an auxin and a cytokinin in certain specific concentrations are essential for induction of division as in *Atropa* (Zenkteler 1971), *Solanum dulcamara* (Zenkteler 1973), *Petunia* (Doreswamy and Chacko 1973), *Brassica* (Kameya and Hinata 1970), *Arabidopsis* (Gresshoff and Doy 1972 a), and *Triticum* (Wang et al. 1973).

Several species of *Datura* respond to a medium containing CM (Guha and Maheshwari 1967; Narayanaswamy and Chandy 1971; Geier and Kohlenbach 1973). For *Geranium*, NAA and KN are essential in addition to CM (Abo El-Nil and Hildebrandt 1971). A highly complex medium containing CM or YE, in addition to auxins such as 2,4-D and IAA, is necessary for successful pollen division in *Oryza* (Guha et al. 1970, Guha-Mukherjee 1973).

Activated charcoal in the medium has been reported to have stimulatory effect on embryoid formation and plantlet development in tobacco (Anagnostakis 1974; Bajaj et al. 1977; Horner et al. 1977). Incorporation of serine (Nitsch and Norreel 1973) and glutamine (Keller et al. 1975) into the medium also favoured pollen embryogenesis in certain species.

Developmental Stage of Microspore and Other Factors for Pollen Division

The significance of age of pollen in anther culture has been widely discussed (Sunderland 1971, 1973, 1974; Pandey 1973). The developmental stage of microspores at culture is critical for the induction of division. Different stages of meiosis in pollen mother cells (Gresshoff and Doy 1972 a), tetrads (Nakata and Tanaka 1968), uninucleate microspores (J.P. Nitsch and C. Nitsch 1969), first pollen mitosis (Sunderland and Wicks 1969), and mature pollen (Kameya and Hinata 1970) have all shown a bearing on the initiation of pollen division. The critical stage varied from species to species. Generally, pollen cultured just prior to or after the first mitosis gave a positive response by producing typical embryoids or callus masses.

In tobacco the yield of embryoids could be significantly increased by culturing anthers at the free microspore stage (J.P. Nitsch and C. Nitsch 1969). Young anthers showing meiotic stages in pollen mother cells and older ones with starch-filled grains failed to produce embryoids. Using different cultivars of *N. tabacum*, however, Nakata and Tanaka (1968) obtained plantlets from anthers cultured at the tetrad stage of microsporogenesis. *Petunia* and *Brassica* were exceptional in that pollen grains cultured at the binucleate stage, i.e., with the tube nucleus and generative cell, responded better (Kameya and Hinata 1970; Raquin and Pilet 1972; Engvild 1973).

The correlation between the developmental stage of pollen and embryoid formation has been critically examined in *Nicotiana tabacum* (Sunderland and Wicks 1971) and *Datura innoxia* (Engvild et al. 1972). The most responsive stage coincided with the first pollen mitosis and, by restricting the culture of anthers to this critical stage, the yield of embryoids could be improved considerably. Once the pollen grains were filled with starch, embryoid formation was precluded (J.P. Nitsch et al. 1968). Either the hormone level in the anther tissue became unfavourable for induction, or some essential component for cell division was depleted, resulting in the failure of pollen embryogeny (Sunderland 1973).

Besides culture media and developmental stage of microspore, certain other factors are also known to influence pollen division. For instance, age of the plant at anther culture has shown a bearing on microspore division. Buds developed towards the beginning and peak of flowering season responded better than those produced later in the season, indicating that the physiological status of donor plants was an important factor. Pollen fertility declined with seasonal variations, possibly due to low endogenous hormone levels. Zenkteler (1972) observed that in *Lycium* only those anthers inoculated in the month of June gave rise to embryoids in culture.

Growth conditions of the donor plants also affect the embryogenic potential of the cultured anthers. By subjecting the anther donor plants to nitrogen starvation, an increase in anther response and embryo-yield could be obtained in *Nicotiana* (Sunderland 1978).

Information regarding the effect of physical factors like temperature and light on anther productivity is limited to a few species. In general, a temperature regime ranging from 23°–28 °C is suitable for pollen divisions. Alternating night and day temperatures are favourable to some extent (J.P. Nitsch and C. Nitsch 1969), although not especially advantageous.

In *Brassica napus* temperature treatments had significant effect on the formation of pollen embryoids from anthers (Keller and Armstrong 1978). Maximum number of embryoids were obtained when anthers were cultured for 14 or 21 days at 30 °C before transfer to 25 °C.

Similarly, chilling of anthers prior to isolation and culture of microspores stimulated embryoid formation in species of *Nicotiana* (Nitsch 1974; Reinert et al. 1975; Bajaj 1978) and rye (Wenzel 1978). Cold pretreatment of flower buds was an essential prerequisite for float cultures of *Nicotiana* (Sunderland and Roberts 1977). The optimum temperature and duration of pretreatment varied from species to species. Recent studies on *Nicotiana tabacum* have shown that the stimulatory effect of cold pretreatment is an indirect one; it helps in preserving pollen viability, thus enhancing embryoid formation (Sunderland 1978).

The effect of light on pollen embryogeny has not been critically examined for all species. Initial incubation of anthers in darkness was essential for embryoid formation in float cultures of *Nicotiana* (Sunderland and Roberts 1977). In callus-forming species such as *Lycopersicon* (Gresshoff and Doy 1972 b) and *Digitalis* (Corduan and Spix 1975), and also in most cereals (Clapham 1977), incubation of cultures in darkness favoured callus growth from microspores. However, regeneration of plantlets occurred only when cultures were transferred to light.

The role of anther tissues in the ontogeny of embryoids is not fully understood. Occasionally, the embryoids developed suspensor-like structures attached to the anther wall, which indicated a nutritive role for the diploid tissue (Nitsch 1969; Guha et al. 1970; Rashid and Street 1974). More often, the embryoids lay free inside the anther sac. C. Nitsch and Norreel (1973) were able to grow plantlets from isolated pollen grains of *Nicotiana* after pretreating them with a water extract of the anther wall. The active substance responsible for this was reported to be the amino acid serine. The degeneration of anther tissue has been observed in all successful cultures, and release of hormones by senescing cells is known (Sheldrake 1973). Many of the multicellular structures from pollen of rye proved abortive, and it is possible that wall factors (inducers and inhibitors) are involved in the induction and/or growth of proembryoids (Thomas et al. 1975).

Ontogeny of Pollen Embryoids

Under normal conditions the microspores mature into pollen grains in situ, but in culture their morphogenesis is altered. Depending upon the composition of medium, the development may lead either to the formation of embryoids and/or plantlets, or a mass of parenchymatous callus.

Initial divisions in the microspores have been critically examined in *Nicotiana tabacum* (Sunderland and Wicks 1969, 1971) and *Datura innoxia* (Sunderland et al. 1974), and three routes to embryogenesis have been distinguished. In route A the first division of microspore is asymmetric, giving rise to a small generative cell and a large vegetative cell. The haploid embryoid is formed by repeated divisions in the vegetative cell. The generative cell may or may not divide further, but degenerates eventually. In route B the microspore divides into two cells of equal size with diffuse nuclei. The embryoid is formed by divisions in one or both the cells. When there is nuclear fusion, a diploid embryoid is obtained by this route. In route C the first division is asymmetric, but the generative cell also participates in embryogenesis. Variations of route C can lead to the production of triploid and other types of polyploid embryoids (Sunderland 1974).

Irrespective of the nature of first pollen mitosis, further development may lead to the formation of embryoids and plantlets, or a callus mass. In plants like *Nicotiana* (Bourgin and Nitsch 1967), *Datura* (Guha and Maheshwari 1967), *Atropa* (Zenkteler 1971), *Capsicum* (George and Narayanaswamy 1973; Fig. 1 A–I), *Petunia* (Engvild 1973), *Brassica* (Keller et al. 1975), and species of *Solanum* (Zenkteler 1973; Dunwell and Sunderland 1973) the pollen grains produce embryoids and plantlets directly without the intervention of a callus phase. Development of embryoids and embryogenic callus from pollen grains later regenerating into plant-

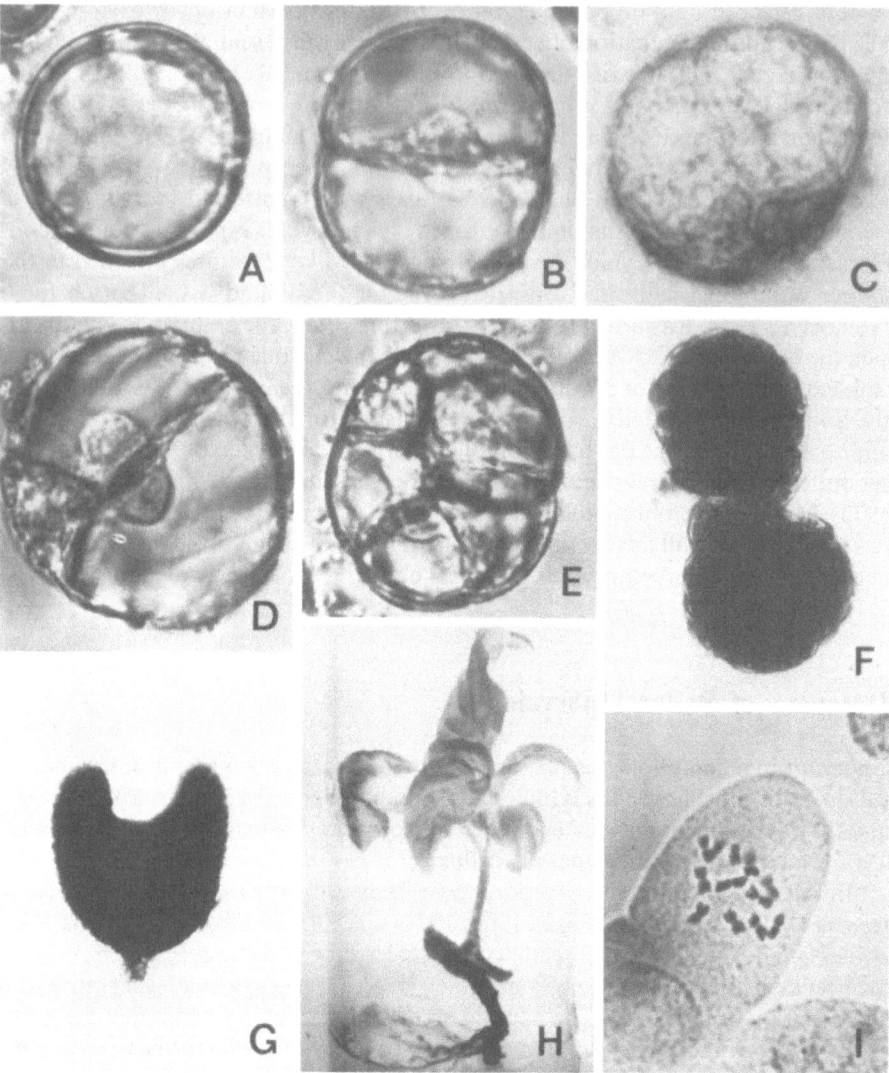

Fig. 1 A–I. Stages in the differentiation of pollen embryoids in *Capsicum*. **A** Normal pollen grain. **B** Pollen grain divided into equal halves. **C** Pollen grain showing degeneration of generative cell (germ cells), and division of tube cell. **D** Oblique divisions in two-celled pollen. **E** Multicellular pollen grain. **F** Twin embryoids. **G** Heart-shaped embryoid. **H** A well-developed pollen plantlet. **I** Root tip cell of a plantlet showing haploid complement of chromosomes ($2n = 12$). (After George and Narayanaswamy 1973)

lets may occur as in *Datura meteloides* (Geier and Kohlenbach 1973; Fig. 2 A–F). In many of the monocotyledonous plants like *Oryza* (Niizeki and Oono 1968), *Triticum* (Ouyang et al. 1973), *Hordeum* and *Lolium* (Clapham 1971) no embryoid formation ever occurred; instead, the pollen grain developed into an unorganized mass of callus which could be manipulated by transfer to a suitable regenerating medium to differentiate into plantlets (Fig. 3 A–D).

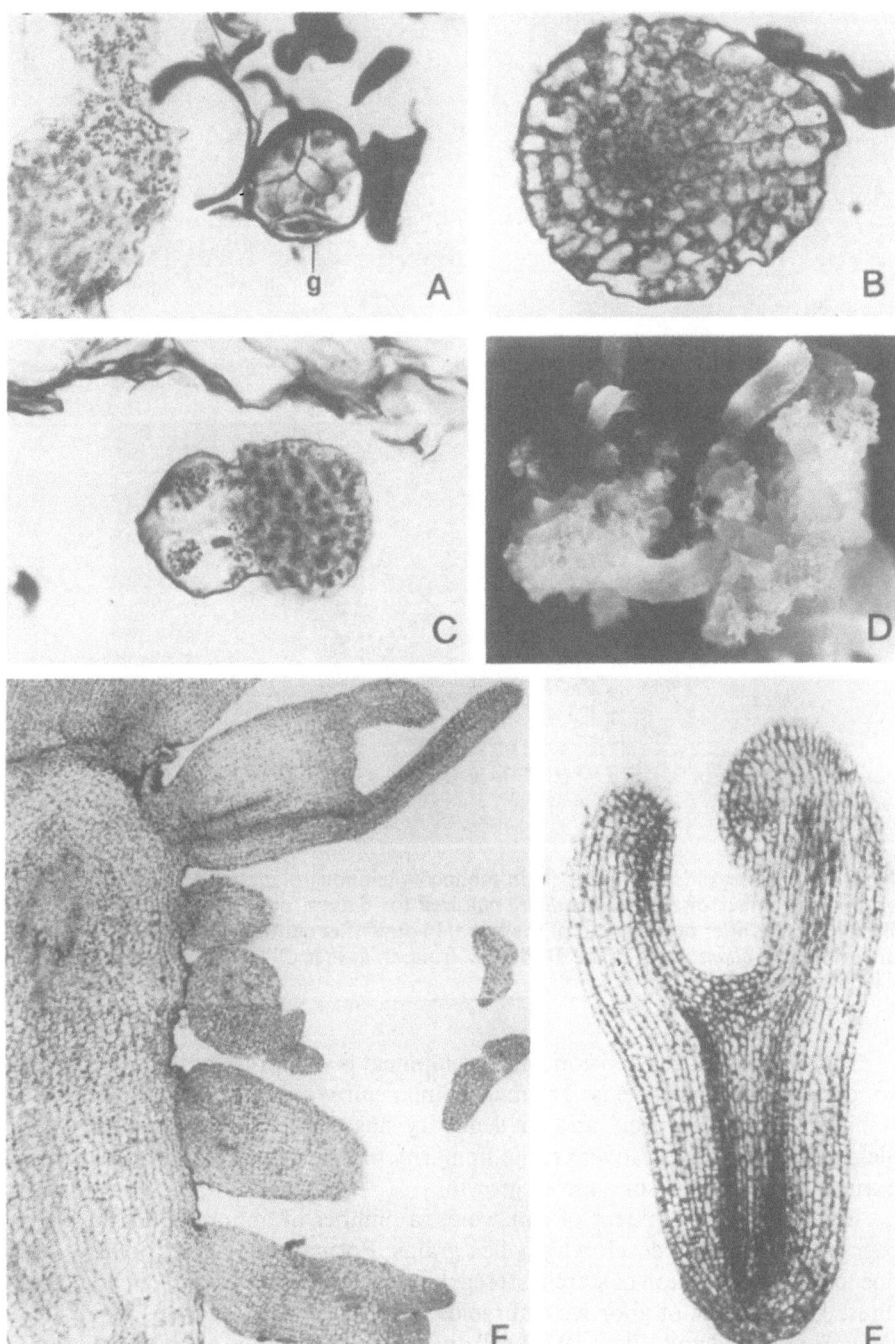

Fig. 2 A–F. Embryoid formation in pollen of *Datura meteloides*. **A** Division in vegetative cell of pollen. Note degeneration of the generative cell, *g*. **B** Young, globular pollen embryoid. **C** Small, globular embryoid with callused basal end. **D** Production of embryoids from pollen callus. **E** Adventive embryoids developed from the hypocotyl of pollen embryoids in vitro. **F** An embryoid with cotyledons. (After Geier and Kohlenbach 1973)

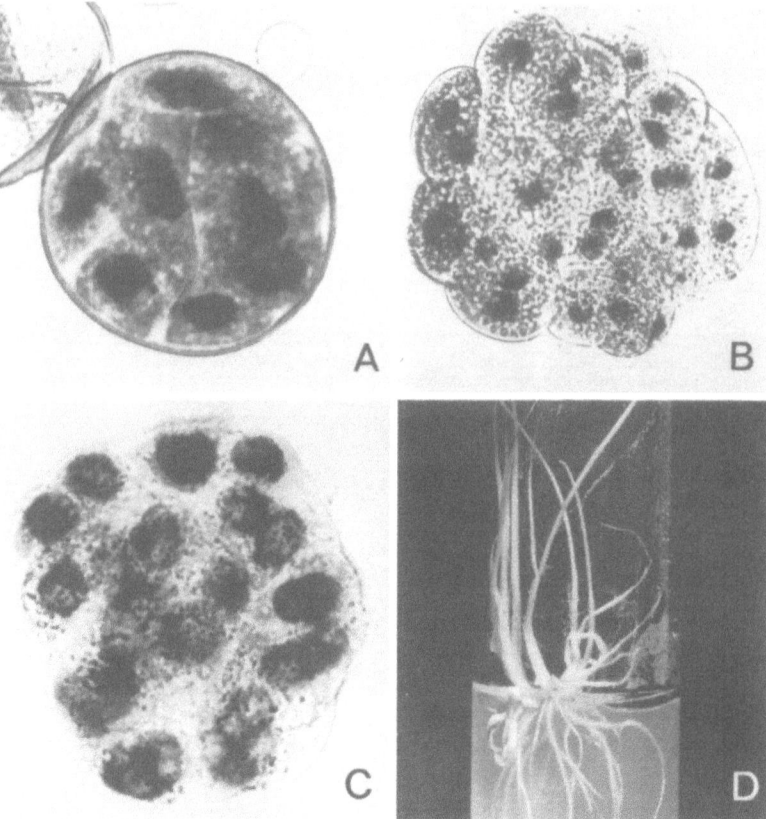

Fig. 3 A–D. Embryoids and plantlets in monocotyledonous plants. **A** Multicellular pollen grain from anther of *Hordeum vulgare* cultured for 6 days; note intact exine and intine. **B** Pollen callus after bursting out of the exine, 14 days after culture. **C** Same as above, from an anther of *Lolium multiflorum*. **D** Plantlet from an anther of *Lolium multiflorum*. (After Clapham 1971)

The sequence of cell division and development is similar to that observed in zygotic or somatic embryogeny. The multicellular embryoids pass through the globular, heart-shaped, torpedo, and cotyledonary stages before developing into complete plantlets (Fig. 1). However, the similarity to zygotic embryos is evident only during the later stages of embryo growth.

Besides the development of embryoids, a number of abnormalities have also been observed in the developing pollen grains. Formation of giant pollen grains due to the accumulation of starch is a regular phenomenon, especially in liquid cultures. The presence of abortive embryoids is not an unusual feature. As in *Nicotiana* (Sunderland and Wicks 1971), cell division might occur in hundreds of grains, but only a few attain maturity. Quiescent proembryoids in various stages of differentiation and degeneration were present in large numbers in *Nicotiana tabacum* (Sunderland and Wicks 1971), and *Solanum tuberosum* (Dunwell and Sunderland 1973). Divisions in the pollen might occur, as in *Prunus* (Harn and Kim 1972), but further differentiation has not been observed. Rupture of the exine caused the vacuolation of the cells of the embryoids, resulting in their early degeneration. Confinement of embryoids within exine, and their incubation inside the anther cavity up to a certain critical stage, appears to be essential for normal plantlet formation.

Embryoid formation is a continuous process in several species extending up to 12 weeks, and a competition among the embryoids for the available nutrients cannot be ruled out. As suggested by Sunderland (1973), those formed first may utilise the essential components of the medium for differentiation, whereas the later ones may remain quiescent and then degenerate as a consequence of depletion of essential elements. Damage to the anther at culture could disturb the gaseous exchange within the anther wall or result in the production of toxic substances (Sunderland and Wicks 1971). Any error in the orderly sequence of divisions during the initial stages may also upset the entire process of development culminating in the degeneration of potential embryoids.

Emergence of differentiating embryoids and plantlets occurs about 6 weeks after culture, and in *Nicotiana* it lasted up to 3 months (Sunderland and Wicks 1971). The plantlets were mostly normal with well-developed shoots and roots.

The plantlets removed from the culture and freed of agar by washing are grown as potted plants. They eventually grow to maturity and flower.

The production of partial or complete albinos in anther culture is not an uncommon phenomenon. Albinos have been observed in *Nicotiana tabacum* (Devreux 1970) and *Datura metel* (Narayanaswamy and Chandy 1971) in very low frequency. But in cereals like *Oryza sativa* (Niizeki and Oono 1971), *Hordeum* (Clapham 1971), *Secale cereale* (Wenzel et al. 1977), and *Aegilops* (Kimato and Sakamoto 1972) large numbers of albinos have been reported. Since the generative cell is considered to be devoid of green plastids, it has been suggested that the albinos are derived from such cells (Devreux 1970), or they might have originated from microspores carrying chlorophyll mutations (Clapham 1977). The regression of plastids during meiosis and their reorganisation during the inductive period of anther culture may also have some correlation with the development of albinos (Sunderland 1973). However, in cereals not only the pollen but also the cultured somatic cells regenerate high proportions of albinos (Gamborg et al. 1970; Wenzel and Thomas 1974).

Ploidy Status of Androgenic Plants

The occurrence of non-haploids among the pollen progeny was first reported in *Datura metel*, by Narayanaswamy and Chandy (1971), and Chandy and Narayanaswamy (1971). Of the 100 plantlets grown to maturity on soil, haploids, diploids, triploids, and tetraploids were noticed (Fig. 4 A–I). *D. innoxia* behaved in a similar fashion under culture conditions (Engvild et al. 1972; Sunderland et al. 1974). Non-haploids have also been reported to occur in anther cultures of other species like *Solanum nigrum* (Harn 1971), *S. melongena* (Raina and Iyer 1973), *Atropa belladonna* (Zenkteler 1971), *Petunia* (Raquin and Pilet 1972; Engvild

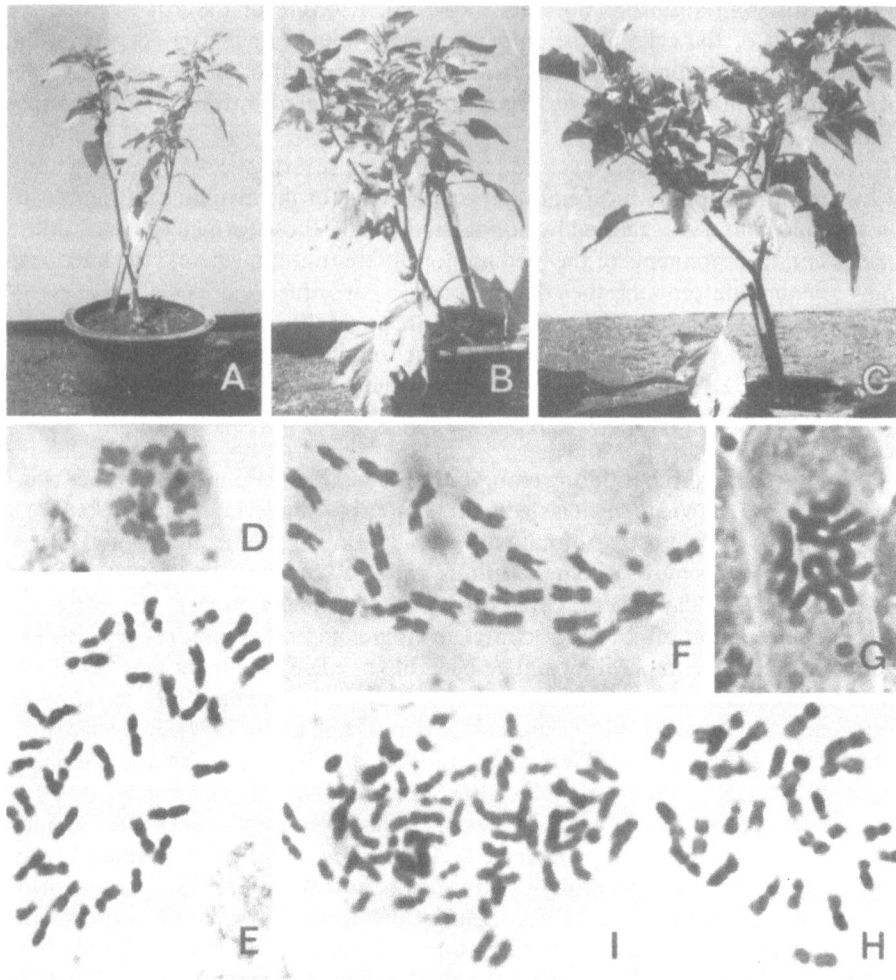

Fig. 4 A–I. Ploidy plants of *Datura metel* L. obtained through pollen embryogenesis in vitro. **A–C** Haploid, diploid, and triploid plants reared in vitro, and grown as potted plants. **D–F** Chromosome numbers from root tip cells: $n = 12$ (haploid), $2n = 24$ (diploid), and $3n = 36$ (triploid). **G–I** Somatic tissue squashes from a triploid plant showing $2n = 12$, 27, and 62 chromosomes. (After Narayanaswamy and Chandy 1971)

1973), *Brassica* (Keller et al. 1975; Keller and Armstrong 1977), and *Oryza sativa* (Nishi and Mitsuoka 1969).

Endoduplication of chromosomes, fusion between nuclei occurring within the pollen (Narayanaswamy and Chandy 1971; Engvild et al. 1972), or the functioning of unreduced microspores formed during microsporogenesis (Collins et al. 1974a) may lead to the production of non-haploids. When haploids are obtained by the functioning of unreduced pollen they are heterozygous, whereas those originating from haploid pollen are truly homozygous.

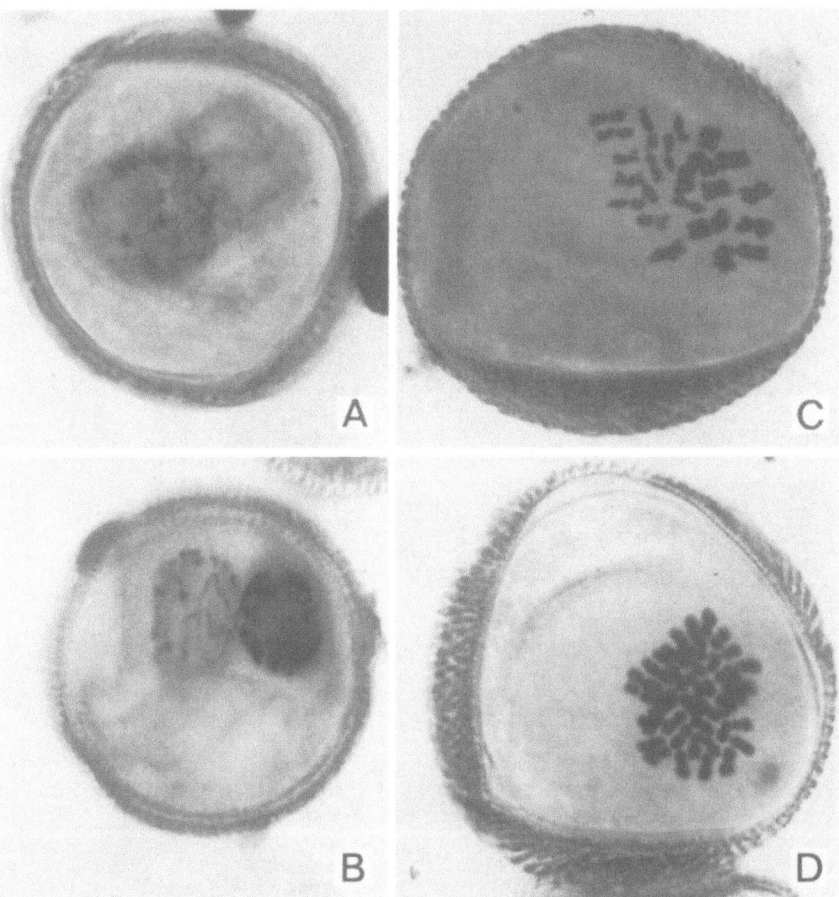

Fig. 5 A–D. Role of nuclear fusion in pollen embryogenesis of *Datura innoxia* Mill. **A** Embryogenic type C grain of *Datura innoxia*. The generative nucleus is assumed to have undergone two cycles of DNA synthesis without intervention of mitosis, whereas the vegetative nucleus has undergone only one cycle. **B** Later stage of **A**; both nuclei are entering prophase. **C** Later stage of **B**; both nuclei are at metaphase. Note the group of 12 diplochromosomes from the endoreduplicated generative nucleus, and the group of 12 ordinary chromosomes from the vegetative nucleus. **D** Compound metaphase figure in a type C grain. Note diplochromosomes from the generative nucleus (mostly out of focus), and two groups of ordinary chromosomes from two vegetative daughter nuclei. All are from the same anther. (After Sunderland et al. 1974)

In *D. innoxia* the role of nuclear fusion in pollen embryogenesis has been critically examined by Sunderland et al. (1974). In type C grains divisions may be simultaneous in both generative and vegetative cells, as in A grains, following two successive cycles of DNA synthesis in the generative nucleus, but limited to one in the vegetative nucleus. The diplochromosomes formed by the endoreduplicated generative nucleus in conjunction with the vegetative chromosomes give rise to a $3n$ proembryoid following division on a common spindle (Fig. 5 A–D). The origin

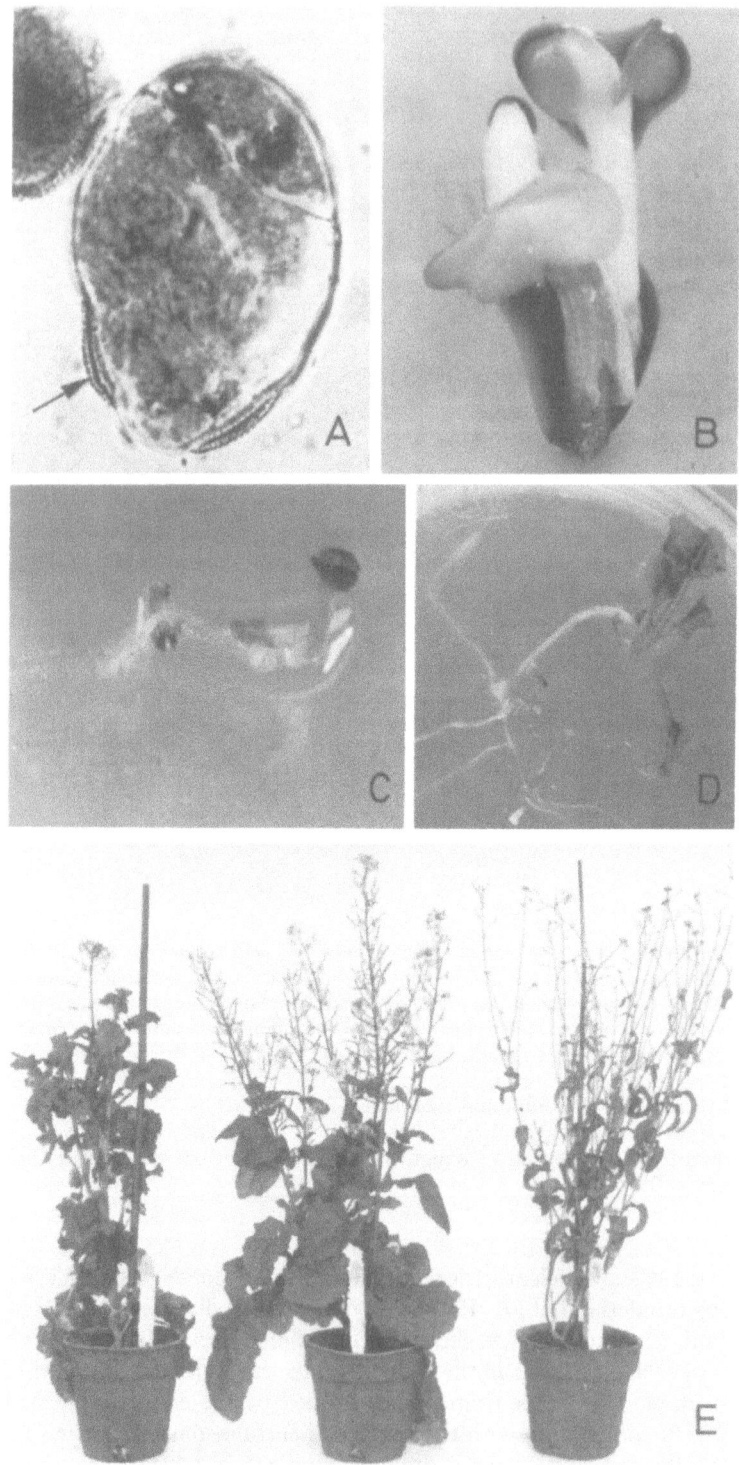

of $2n$, $4n$, and even $6n$ proembryoids could be explained on the same basis through occurrence of nuclear fusion following DNA replication in one or other of the two nuclei in the pollen.

One of the three alternative methods could be involved in the production of homozygous diploids, viz. spontaneous doubling, artificial diploidization by chemicals, and regeneration from callus. Spontaneous doubling of chromosomes under culture conditions has been observed as an inherent tendency in certain species like *Datura* (Narayanaswamy and Chandy 1971; Engvild et al. 1972; Sunderland et al. 1974), *Petunia* (Engvild 1973; Wagner and Hess 1974), *Atropa* (Zenkteler 1971; Narayanaswamy and George 1972), *Solanum* (Harn 1971; Raina and Iyer 1973), *Brassica* (Keller et al. 1975; Fig. 6 A–E; Keller and Armstrong 1977), and several other species. Homozygous diploids can be produced in large numbers by treating the haploid plants by colchicine. Since colchicine acts on dividing cells, it is ideal to use this chemical during the early stages of development when the cells are actively dividing. In *Nicotiana tabacum* more than half the treated plants were diploid (Sunderland 1970). Doubling at an early stage has the advantage that it requires minimum effort, and the duration of the whole procedure remains unaffected. In plants like *Nicotiana*, less than six months are required to obtain homozygous seeds through anther culture.

Regeneration directly from injured parts or from callus derived from stem or leaf or other parts of the pollen plant could also lead to the production of homozygous strains (Kochbar et al. 1971; Kasperbauer and Collins 1972; Pandey et al. 1975; Keller and Armstrong 1977).

An alternative to haploid induction by anther culture is the Bulbosum method which is based on interspecific hybridization and somatic chromosome elimination (see Kasha 1974). By this method it is possible to obtain a large number of haploids in cereals like barley (Kasha and Kao 1970) and wheat (Barclay 1975). The technique and its implications in cereal breeding have been reviewed recently (Jensen 1977).

Pollen Plantlets from Haploid, Triploid, and Tetraploid Species

Androgenic Plantlets from Haploid Datura

Studies on anther culture of diploid *Datura metel* had shown that plantlets of pollen-origin consisted of haploid and non-haploid types. All the plantlets were morphologically similar, but under field conditions they showed wide variations in vegetative as well as floral characteristics. The haploids were smaller in size compared to the diploids. Triploids and tetraploids were the most vigorous with larger leaves and flowers. Although flowering was regular, normal fruit-set occurred only in diploids.

Fig. 6 A–E. Development of embryoids and plantlets in a crucifer. **A** Remnants of pollen wall (*arrow*) on a developing embryoid of *Brassica campestris*, 10 days after culture. **B** Embryoids emerging from an anther after 20 days of culture. **C** Germinated embryoid on a low sucrose medium. **D** Young pollen plantlet of *B. campestris* prior to potting. **E** Hexaploid *left*, tetraploid *centre*, and diploid *right*, plants of *B. campestris* cv. Torch. (After Keller et al. 1975)

The haploids showed a single set of chromosomes in the somatic cells ($2n = 12$, Fig. 4D). The anthers contained a large percentage of abortive pollen as a result of irregular meiosis in pollen mother cells. Random segregation of univalents resulted in the formation of dyads, triads, irregular tetrads, and even monads. As a result the pollen grains were of assorted size and chromosome complement.

The yield of embryoids and plantlets from such anthers in culture was very low. Many of the embryoids degenerated within the anther sac before attaining maturity. Of the 80 plantlets examined, three were haploid ($2n = 12$) and the rest diploid ($2n = 24$). Plants originating from dyads and microspores having 12 chromosomes were naturally haploid. Diploid plantlets could arise from pollen with $2n$ chromosomes resulting from endoduplication (Chandy and Narayanaswamy 1971). Divisions were usually restricted to the large pollen grains, and it is possible that some of the plantlets with abnormal shoots may be aneuploid.

In *Nicotiana* no plantlets were obtained from the cultured anthers of haploid plants (Niizeki 1977).

Pollen Plantlets from Triploid and Tetraploid Datura

In anther culture of *Datura*, apart from plantlets with $1n$ and $2n$ number of chromosomes, triploids were also produced. Division stages in pollen mother cells of triploids ($2n = 36$) showed unequal segregation of chromosomes resulting in the development of pollen grains with assorted chromosome number. At the sporad stage, pollen grouped as triads, tetrads, and polyads was formed, and showed considerable variation in the size of pollen grain.

On culture of anthers of triploid plants, embryoids and plantlets were produced in 8 weeks of incubation. Although 10% anthers responded in culture, normal embryoids and plantlets were obtained from only 5%. The rest of the anthers gave rise to pollen calluses. Usually two or three plantlets developed from an anther.

Embryoid-bearing anthers and pollen calluses on transfer to fresh media of the same composition differentiated into plantlets. Both aberrant and normal plantlets were observed among the regenerants. Figure 7A–F shows the anatomy of pollen embryoids. Budding of proembryoids (pro. e) from periphery of embryonal callus and hypocotyl of embryoids was frequent.

Wide variations were observed in the chromosome make-up of the plantlets as ascertained from root tip squashes. Among the plantlets were hyperploids and hypoploids irrespective of the phenotypes. A similar observation has been made in anther cultures of triploid *Nicotiana* (Niizeki 1977).

It is reasonable to expect that when pollen of a triploid plant shows variation in chromosome number, its viability is as a consequence reduced. But many of the $2n + 1$ microspores and other abnormal ones were stimulated to divide under culture conditions leading to the development of haploid, trisomic, or nullisomic embryoids (Narayanaswamy and George 1974). Several different genotypes corresponding to the different chromosome number of pollen grains were produced both at the diploid and tetraploid level. These failed, however, to grow into mature plants probably because of the physiological imbalance caused by the deletion or addition of one or more chromosomes to the somatic complement.

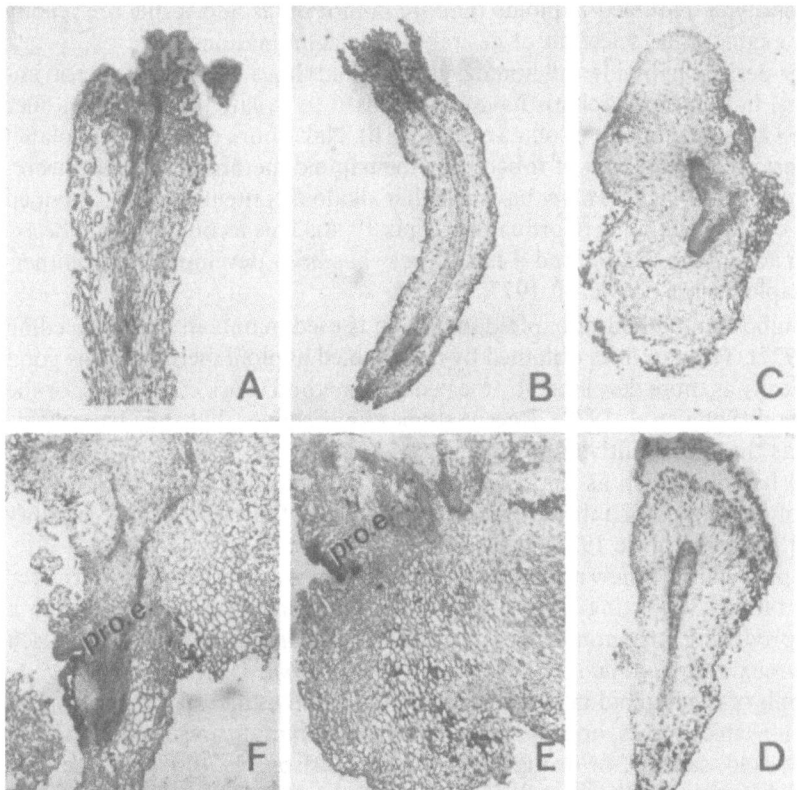

Fig. 7 A–F. Anatomy of abnormal embryoids of triploid *Datura metel* from pollen in vitro. **A** and **B** Embryoids showing aborted shoot apexes. Note well-developed primary root. **C** Median section of an embryoid showing cotyledonary lobes but devoid of hypocotyl or apical meristem. **D** L.s. embryoid showing vasculature but no apical meristem. **E** Budding of proembryoids (pro.e) from periphery of embryonal callus. **F** Same as above, from hypocotyl region. (After Narayanaswamy and George née Chandy 1974)

In anther culture of tetraploid *Datura*, 25% anthers showed positive response by producing embryoids and plantlets via pollen embryogeny. Among the progeny, diploids and tetraploids were present.

Concluding Remarks

Anther culture has become a topic of great contemporary interest mainly because of its implications in the speeding up of homozygosity, and in the analysis of genetic combinations. The totipotent nature of pollen grains has been exploited to advantage in the production of haploids in many plants of economic and ornamental value with varying success. Since haploids contain only one set of chromosomes, all the genes present in them, even the recessive ones, are able to express in the phenotype. Doubled haploids that are homozygous and fertile are readily obtained, enabling the selection of desirable gene combinations.

Anther-derived haploids and homozygous diploids have been used for crop improvement. In tobacco, haploids have been utilised to isolate four breeding lines differing in alkaloid content (Collins et al. 1974 b). Nakamura et al. (1974) isolated three superior breeding lines of tobacco by the haploid method. Homozygous recombinants of *Hyoscyamus niger* having higher alkaloid content could be obtained by anther culture technique (Corduan and Spix 1978). This technique could be exploited to advantage in rice, and a new variety has been developed from anther-derived haploids (see Scowcroft 1977).

The Bulbosum method of haploid induction is used mainly in barley breeding (Jensen 1977). The new lines obtained by the doubled haploid method are as good agronomically as those developed by the pedigree method (Park et al. 1976), or the bulk method (Song et al. 1978). Recent studies have shown that they have yields as stable as the check cultivars (Reinbergs et al. 1978).

Cereal haploids, such as that of maize, find yet another use in the one-step transfer of genotypes of inbred lines into cytoplasm that causes male sterility (Goodsell 1961; Kermicle 1973).

In the production of new mutant forms, haploids provide excellent material for experimentation. Subjecting the plants to X or gamma irradiation would be a means of producing large numbers of mutants as in *Nicotiana tabacum* (J.P. Nitsch 1969; Devreux and Nettancourt 1974) and *Brassica napus* (Hoffmann 1978).

Free cells of the haploid in suspension could be used as an experimental material for mutagenic studies, and to evolve biochemical mutants resistant to pathogens, heat, cold, salinity, or drought conditions. Carlson (1970) isolated several biochemical mutants of *Nicotiana* by treating the haploid cells with ethyl methanesulphonate. Haploid cell cultures have been used to isolate cell lines resistant to streptomycin (Binding et al. 1970), 5-bromodeoxyuridine (Maliga et al. 1973), aminopterin (Mastrangelo and Smith 1977), and sodium chloride (Dix and Street 1975). Tobacco plants resistant to methionine sulfoximine (Carlson 1973), and valine (Bourgin 1978) could be isolated from haploid protoplasts. Haploid tissues of *Arabidopsis* and *Lycopersicon* have been used for the transfer and subsequent expression of three systems of genes from *Escherichia coli* (Doy et al. 1972).

Anther-derived haploids have been extensively used in protoplast culture and somatic hybridization (Melchers and Labib 1974; Schieder 1978).

Culture of isolated microspores provides a novel experimental system for the study of factors controlling pollen embryogenesis of higher plants (Vasil and Nitsch 1975). This technique ensures the production of more isogenic progeny than anther culture, and therefore would prove more efficient in mutagenic studies.

An element of uncertainty regarding homozygosity of androgenic plants stems from their possible development from unreduced microspores or somatic cells. Extensive homozygosity tests must be performed before utilizing such spontaneous pollen diploids for breeding experiments.

The main obstacle facing anther culturists at present is the lack of response from pollen grains of several crop plants and other species of economic importance. In several instances, the pollen embryoids fail to attain maturity or yield only a few plantlets. Since anther response is influenced by various factors like light, temperature, and physiological status of the anther donor plants, it should be possible to increase the yield by controlling these factors. In *Nicotiana*, by subjecting the donor plants and excised flower buds to various stress conditions, an increase in anther response and embryo yields could be obtained (Sunderland 1978). Recent studies in *Paeonia, Tradescantia, Anemone*, and certain other species (Sunderland 1977) have shown that pollen dimorphism is of wide occurrence in plants and the anomalous units observed in vivo are the source of embryoids in cultured anthers. This suggests that pollen embryogenesis is predetermined and, by controlling the formation of potential embryogenic grains, improvement in embryo-yields and induction in recalcitrant species may be achieved (Sunderland 1978). Recently, Wernicke et al. (1978) developed a technique for the isolation of potential embryogenic grains from the microspore population, and this might help in increasing the yield of haploids in low-yielding species.

In retrospect, it is remarkable that microspores and young pollen grains of many angiosperms (whose haploid life cycle is short) are highly responsive for experimentation of different kinds, especially in the manipulation of their regenerative ability to form male gametophytes, female gametophytes (as for instance the pollen embryo sacs of *Hyacinthus:* Stow 1930), or sporophytes of different ploidy.

There have been additional examples of regeneration of plants through anther/pollen culture. The reader is referred to the extensive reviews by Wenzel (1979), Maheshwari et al. (1980), and Vasil (1980).

References

Abo-el-Nil MM, Hildebrandt AC (1971) Differentiation of virus-symptomless geranium plants from anther callus. Plant Dis Rep 55:1017–1020

Anagnostakis SL (1974) Haploid plants from anthers of tobacco: enhancement with charcoal. Planta 115:281–283

Babbar SB, Mittal A, Gupta SC (1980) In vitro induction of androgenesis, callus formation and organogenesis in *Iberis amara* Linn. anthers. Z. Pflanzenphysiol 100:409–414

Bajaj YPS (1976) In vitro induction of haploid plants. In: Evans PK (ed) Towards plant improvement by in vitro methods. Academic Press, London New York, in press

Bajaj YPS (1978) Regeneration of haploid tobacco plants from isolated pollen grown in drop culture. Indian J Exp Biol 16:407–409

Bajaj YPS (1981) Regeneration of plants from ultra-low frozen anthers of *Primula obconica*. Scientia Hort 14:93–95

Bajaj YPS, Pierik RLM (1974) Vegetative propagation of *Freesia* through callus cultures. Neth J Agric Sci 22:153–159

Bajaj YPS, Reinert J, Heberle E (1977) Factors enhancing in vitro production of haploid plants. In: Gautheret RJ (ed) La culture de tissus et des cellules de végétaux. Masson, Paris, pp 47–58

Ban Y, Kokubu T, Miyaji Y (1971) Production of haploid plants by anther culture of *Setaria italica*. Bull Fac Agric Kagoshima Univ 21:77–81

Barclay IR (1975) High frequencies of haploid production in wheat *(Triticum aestivum)* by chromosome elimination. Nature (London) 256:410–411

Bernhard S (1971) Développement d'embryons haploïdes a partir d'anthères cultivées in vitro. Etude cytologique comparée chez le tabac et le *Petunia*. Rev Cytol Biol Vég 34:165–188

Binding H, Binding K, Straub J (1970) Selektion in Gewebekulturen mit haploiden Zellen. Naturwissenschaften 57:138–139

Bourgin JP (1978) Valine-resistant plants from in vitro selected tobacco cells. Mol Gen Genet 161:225–230

Bourgin JP, Nitsch JP (1967) Obtention de Nicotiana haploïdes à partir d'étamines cultivées in vitro. Ann Physiol Vég 9:377–382

Carlson PS (1970) Induction and isolation of auxotrophic mutants in somatic cell cultures of Nicotiana tabacum. Science 168:487–489

Carlson PS (1973) Methionine sulfoximine-resistant mutants of tobacco. Science 180:1366

Chandy LP, Narayanaswamy S (1971) Diploid and haploid androgenic plantlets from haploid Datura in vitro. Indian J Exp Biol 9:472–475

Chen CH, Chen FT, Chien CF, Wang CH, Chang SJ, Hsu HE, Ou HH, Ho YT, Lu TM (1979) A process of obtaining pollen plants of Hevea brasiliensis Muell-Arg. Sci Sinica 22:81–90

Cheng-hua Chen, Fa-tsu Chen, Chang-fa Chien, Chuan-hua Wang (1978) Obtaining pollen plants of Hevea brasiliensis Muell. Arg. In: Proc Symp Plant Tissue Culture, Peking Sci Press, Peking, pp 11–21

Clapham D (1971) In vitro development of callus from the pollen of Lolium and Hordeum. Z Pflanzenzucht 65:285–292

Clapham D (1973) Haploid Hordeum plants from anthers in vitro. Z Pflanzenzücht 69:142–155

Clapham D (1977) Haploid induction in cereals. In: Reinert J, Bajaj YPS (eds) Applied and fundamental aspects of plant cell, tissue, and organ culture. Springer, Berlin Heidelberg New York, pp 279–340

Collins GB, Dunwell JM, Sunderland N (1974a) Irregular microspore formation in Datura innoxia and its relevance to anther culture. Protoplasma 82:365–378

Collins GB, Legg PD, Kasperbauer MJ (1974b) Use of anther-derived haploids in Nicotiana. I. Isolation of breeding lines differing in total alkaloid content. Crop Sci 14:77–80

Corduan G (1975) Regeneration of anther-derived plants of Hyoscyamus niger L. Planta 127:27–36

Corduan G, Spix C (1975) Haploid callus and regeneration of plants from anthers of Digitalis purpurea L. Planta 124:1–12

Corduan G, Spix C (1978) Anther culture of a hybrid of the genus Hyoscyamus, a rapid method to obtain homozygous recombinants. In: Alfermann AW, Reinhard E (eds) Production of natural compounds by cell culture methods. Int Symp Plant Cell Culture Munich, pp 295–302

Craig IL (1974) Haploid plants ($2n = 21$) from in vitro anther culture of Triticum aestivum. Can J Genet Cytol 16:697–700

Debergh P, Nitsch C (1973) Premiers résultats sur la culture in vitro de grains de pollen isolés chez la tomate. CR Acad Sci Paris 276:1281–1284

De Buyser J, Henry Y (1980) Induction of haploid and diploid plants through in vitro anther culture of haploid wheat ($n = 3 \times = 21$). Theor Appl Genet 57:57–58

Devreux M (1970) New possibilities for the in vitro cultivation of plant cells. Euro-Spectra 9:105–110

Devreux M, de Nettancourt D (1974) Screening mutations in haploid plants. In: Kasha KJ (ed) Haploids in higher plants: Advances and potential. Guelph Univ, Guelph, pp 309–322

Devreux M, Saccardo F (1971) Mutazioni sperimentali osservate su piante aploidi di tabacco ottenute per culture in vitro di antere irradiate. Am Ass Genet 16:69–71

Dix PJ, Street HE (1975) Sodium chloride resistant cultured cell lines from Nicotiana sylvestris and Capsicum annuum. Plant Sci Lett 5:231–237

Doreswamy R, Chacko EK (1973) Induction of plantlets and callus from anthers of Petunia axillaris (Lam.) BSP cultured in vitro. Hort Res 13:41–44

Doy CH, Gresshoff PM, Rolfe BG (1972) Transfer and expression of bacterial genes in plant cells. Search 3:447–448

Dunwell JM, Sunderland N (1973) Anther culture of Solanum tuberosum L. Euphytica 22:317–323

Engvild KC (1973) Triploid petunias from anther cultures. Hereditas 74:144–147

Engvild KC, Linde-Laursen I, Lundqvist A (1972) Anther cultures of *Datura innoxia:* flower bud stage and embryoid level of ploidy. Hereditas 72:331–332

Foroughi-Wehr B, Wilson HM, Mix G, Gaul H (1977) Monohaploid plants from anthers of a dihaploid genotype of *Solanum tuberosum* L. Euphytica 26:361–367

Gamborg OL, Constabel F, Miller RA (1970) Embryogenesis and production of albino plants from cell cultures of *Bromus inermis.* Planta 95:355–358

Geier T, Kohlenbach HW (1973) Entwicklung von Embryonen und embryogenem Kallus aus Pollenkörnern von *Datura meteloides* und *Datura innoxia.* Protoplasma 78:381–396

George L, Narayanaswamy S (1973) Haploid *Capsicum* through experimental androgenesis. Protoplasma 78:467–470

George L, Rao PS (1979) Experimental induction of triploid plants of *Physalis* through anther culture. Protoplasma 100:13–19

Goodsell SF (1961) Male sterility in corn by androgenesis. Crop Sci 1:227–228

Gresshoff PM, Doy CH (1972a) Haploid *Arabidopsis thaliana* callus and plants from anther culture. Aust J Biol Sci 25:259–264

Gresshoff PM, Doy CH (1972b) Development and differentiation of haploid *Lycopersicon esculentum* (tomato). Planta 107:161–170

Grunewaldt LJ, Malepszy S (1975) Observations on anther callus from *Hordeum vulgare.* Z Pflanzenphysiol 75:55–62

Guha S, Maheshwari SC (1964) In vitro production of embryos from anthers of *Datura.* Nature (London) 204:497

Guha S, Maheshwari SC (1966) Cell division and differentiation of embryos in the pollen grains of *Datura* in vitro. Nature (London) 212:97–98

Guha S, Maheshwari SC (1967) Development of embryoids from pollen grains of *Datura* in vitro. Phytomorphology 17:454–461

Guha S, Iyer RD, Gupta S, Swaminathan MS (1970) Totipotency of gametic cells and the production of haploids in rice. Curr Sci 39:174–176

Guha-Mukherjee S (1973) Genotypic differences in the in vitro formation of embryoids from rice pollen. J Exp Bot 24:139–144

Harn C (1971) Studies on anther culture in *Solanum nigrum.* SABRAO Newslett 3:39–42

Harn C, Kim MZ (1972) Induction of callus from anthers of *Prunus armeniaca.* Korean J Breed 4:49–53

Hoffmann F (1978) Mutation and selection of haploid cell culture systems of rape and rye. In: Alfermann AW, Reinhard E (eds) Production of natural compounds by cell culture methods. Proc Int Symp Plant Cell Culture, Munich, pp 319–329

Horner M, McComb JA, McComb AJ, Street HE (1977) Ethylene production and plantlet formation by *Nicotiana* anthers cultured in the presence and absence of charcoal. J Exp Bot 28:1365–1372

Hughes KW, Bell SL, Caponetti JD (1975) Anther-derived haploids of the African violet. Can J Bot 53:1442–1444

Iyer RD, Raina SK (1972) The early ontogeny of embryoids and callus from pollen and subsequent organogenesis in anther cultures of *Datura metel* and rice. Planta 104:146–156

Jensen CJ (1977) Monoploid production by chromosome elimination. In: Reinert J, Bajaj YPS (eds) Applied and fundamental aspects of plant cell, tissue, and organ culture. Springer, Berlin Heidelberg New York, pp 299–340

Kameya T, Hinata K (1970) Induction of haploid plants from pollen grains of *Brassica.* Jpn J Breed 20:82–87

Kasha KJ (ed) (1974) Haploids in higher plants: Advances and potential. Guelph Univ, Guelph

Kasha KJ, Kao KN (1970) High frequency haploid production in barley (*Hordeum vulgare* L.). Nature (London) 225:874–875

Kasperbauer MJ, Collins GB (1972) Reconstitution of diploids from leaf tissue of anther-derived haploids in tobacco. Crop Sci 12:98–101

Kasperbauer MJ, Buckner RC, Springer WD (1980) Haploid plants by anther-panicle culture of tall fescue. Crop Sci 20:103–106

Keller WA, Armstrong KC (1977) Embryogenesis and plant regeneration in *Brassica napus* anther cultures. Can J Bot 55:1383–1388

Keller WA, Armstrong KC (1978) High frequency production of microspore-derived plants from *Brassica napus* anther cultures. Z. Pflanzenzücht 80:100–108

Keller WA, Rajhathy T, Lacapra J (1975) In vitro production of plants from pollen in *Brassica campestris*. Can J Genet Cytol 17:655–666

Kermicle JL (1973) Androgenesis and the *indeterminate gametophyte* mutation: source of the cytoplasm. Maize Genet Coop Newslett 47:208–209

Kimato M, Sakamoto S (1972) Production of haploid albino plants of *Aegilops* by anther culture. Jpn J Genet 47:61–63

Kochbar T, Sabharwal P, Engelberg J (1971) Production of homozygous diploid plants by tissue culture technique. J Hered 62:59–61

Kohlenbach HW, Geier T (1972) Embryonen aus in vitro kultivierten Antheren von *Datura meteloides* Dun., *D. wrightii* Regel und *Solanum tuberosum* L. Z Pflanzenphysiol 67:161–165

Kuo JS, Wang YY, Chien NF, Ku SJ, Kung ML, Hsu UC (1973) Investigations on the anther culture in vitro of *Nicotiana tabacum* L. and *Capsicum annuum* L. Acta Bot Sinica 15:37–50

LaRue CD (1954) Studies on growth and regeneration in gametophytes and sporophytes of gymnosperms. Brookhaven Symp Biol 6:187–208

Li-Kung C (1978) Induction and culture of regenerated plantlets from anthers of genus *Populus*. Abstr. In: Proc Symp Plant Tissue Culture Peking. Science Press, Peking, p 241

Maheshwari SC, Tyagi AK, Malhotra K, Sopory SK (1980) Induction of haploidy from pollen grains in angiosperms: The current status. Theor Appl Genet 58:193–206

Malepszy ST, Grunewaldt J (1974) A contribution to the production of haploids in *Hordeum vulgare* L. Z Pflanzenzücht 72:206–211

Maliga P, Marton L, Sz-Breznovits A (1973) 5-bromodeoxyuridine-resistent cell lines from haploid tobacco. Plant Sci Lett 1:119–121

Mastrangelo IA, Smith HH (1977) Selection and differentiation of aminopterin resistant cells of *Datura innoxia*. Plant Sci Lett 10:171–179

Melchers G (1972) Haploid higher plants for plant breeding. Z Pflanzenzücht 67:19–32

Melchers G, Labib G (1970) Die Bedeutung haploider höherer Pflanzen für Pflanzenphysiologie und Pflanzenzüchtung. Ber Dtsch Bot Ges 83:129–150

Melchers G, Labib G (1974) Somatic hybridization of plants by fusion of protoplasts. I. Selection of light-resistant hybrids of "haploid" light-sensitive varieties of tobacco. Mol Gen Genet 135:277–294

Murakami M, Jakahashi N, Harada H (1973) Induction of haploid plant by anther culture in maize. I. Callus formation and root differentiation. Sci Rep Kyoto Prefect Univ Agric 24:1–8

Murashige T, Skoog F (1962) A revised medium for rapid growth and bioassay with tobacco tissue cultures. Physiol Plant 15:473–497

Nakamura A, Yamada T, Kadotani N, Itagaki R, Oka M (1974) Studies on the haploid method of breeding in tobacco. SABRAO J 6:107–131

Nakata K, Tanaka M (1968) Differentiation of embryoids from developing germ cells in anther culture of tobacco. Jpn J Genet 43:65–71

Narayanaswamy S, Chandy LP (1971) In vitro induction of haploid, diploid, and triploid androgenic embryoids and plantlets in *Datura metel* L. Ann Bot 35:535–542

Narayanaswamy S, George L (1972) Morphogenesis of belladonna (*Atropa belladonna* L.) plantlets from pollen in culture. Indian J Exp Biol 10:382–384

Narayanaswamy S, George L (1974) Experimental androgenesis in triploid and tetraploid *Datura metel*. Indian J Genet 34 A:270–276

Niizeki H, Oono K (1968) Induction of haploid rice plant from anther culture. Proc Jpn Acad 44:554–557

Niizeki H, Oono K (1971) Rice plants obtained by anther culture. In: Colloq Int, CNRS 193. Les cultures de tissus de plantes, pp 251–257

Niizeki M (1977) Haploid, polyploid and aneuploid plants from cultured anthers and calluses in species of *Nicotiana* and forage crops. J Fac Agric Hokkaido Univ 58:343–466

Nishi T, Mitsuoka S (1969) Occurrence of various ploidy plants from anther and ovary culture of rice plant. Jpn J Genet 44:341–346

Nitsch C (1974) La culture de pollen isolé sur milieu synthétique. CR Acad Sci Paris 278:1031–1034

Nitsch C, Norreel B (1973) Effect d' un choc thermique sur le pouvoir embryogène due pollen de *Datura innoxia* cultivé dans l' anthèreou isolé de l' anthère. CR Acad Sci Paris 276:303–306

Nitsch JP (1951) Growth and development in vitro of excised ovaries. Am J Bot 38:556–576

Nitsch JP (1969) Experimental androgenesis in *Nicotiana*. Phytomorphology 19:389–404

Nitsch JP, Nitsch C (1969) Haploid plants from pollen grains. Science 163:85–87

Nitsch JP, Nitsch C, Hamon S (1968) Réalisation expérimentale de l' androgènese chez divers *Nicotiana*. CR Soc Biol 162:369–372

Nitzsche W (1970) Herstellung haploider Pflanzen aus *Festuca-Lolium* Bastarden. Naturwissenschaften 57:199–200

Nitzsche W, Wenzel G (1977) Haploids in plant breeding. Parey, Berlin Hamburg

Ouyang T, Hu H, Chuang C, Tseng C (1973) Induction of pollen plants from anthers of *Triticum aestivum* L. cultured in vitro. Sci Sinica 16:79–90

Pagniez M, Demarly Y (1979) Obtention d'individus androgenetiques par culture in vitro d'antheres de Ray-grass d'Italie (*Lolium multiflorum* Lam.). Ann Amelior Plantes 29:631–637

Pandey KK (1973) Theory and practice of induced androgenesis. New Phytol 72:1129–1140

Pandey KK, Couchman P, De Lautour G, Williams E (1975) A tissue culture technique for rapid clonal multiplication of androgenic haploids of tobacco. N Z J Bot 13:317–320

Park SJ, Walsh EJ, Reinbergs E, Song LSP, Kasha KJ (1976) Field performance of doubled haploid barley lines in comparison with lines developed by the pedigree and single seed descent methods. Can J Plant Sci 56:467–474

Pelletier G, Raquin C, Simon G (1972) La culture in vitro d' anthéres d' asperge (*Asparagus officinalis*) CR Acad Sci Paris 274:848–851

Picard E, De Buyser J (1975) New results on anther culture of wheat (*Triticum aestivum* L.) in vitro: conditions of regeneration of haploid plantlets and production of homozygous lines. CR Acad Sci Paris 281:989

Preil W, Huhnke W, Engelhardt M, Hoffmann (1977) Haploide bei *Gerbera jamesonii* aus in vitro: Kulturen von Blutenkopfchen. Z. Pflanzenzücht 79:167–171

Radojevic L (1978) In vitro induction of androgenic plantlets in *Aesculus hippocastanum*. Protoplasma 96:369–374

Raghavan V (1975) Induction of haploid plants from anther cultures of henbane. Z Pflanzenphysiol 76:89–92

Raina SK, Iyer RD (1973) Differentiation of diploid plants from pollen callus in anther cultures of *Solanum melongena*. Z Pflanzenzücht 70:275–280

Raquin C, Pilet V (1972) Production de plantules à partir d' anthères de Pétunias cultivées in vitro. CR Acad Sci Paris 274:1019–1022

Rashid A, Street HE (1973) The development of haploid embryoids from anther cultures of *Atropa belladonna* L. Planta 113:263–270

Rashid A, Street HE (1974) Segmentation in microspores of *Nicotiana sylvestris* and *Nicotiana tabacum* which lead to embryoid formation in anther cultures. Protoplasma 80:323–334

Reinbergs E, Song LSP, Choo TM, Kasha KJ (1978) Yield stability of double haploid lines of barley. Can J Plant Sci 58:929–933

Reinert J, Bajaj YPS (1977) Anther culture: Haploid production and its significance. In: Reinert J, Bajaj YPS (eds) Applied and fundamental aspects of plant cell, tissue and organ culture. Springer, Berlin Heidelberg New York, pp 251–340

Reinert J, Bajaj YPS, Heberle E (1975) Induction of haploid tobacco plants from isolated pollen. Protoplasma 84:191–196

Rosati P, Devreux M, Laneri U (1975) Anther culture in strawberry. Hort Sci 10:119–120

Sangwan RS, Norreel B (1975) Induction of plants from pollen grains of *Petunia* cultured in vitro. Nature (London) 257:222–224

Schieder O (1978) Haploids from *Datura innoxia* as a tool for the production of homozygous lines with high content of scopolamine and for induction of mutants. In: Alfermann AW, Reinhard E (eds) Production of natural compounds by cell culture methods. Proc Int Symp Plant Cell Culture, Munich, pp 330–336

Scowcroft WR (1977) Somatic cell genetics and plant improvement. Adv Agron 29:39–81

Sharp WR, Raskin RS, Sommer HE (1972) The use of nurse culture in the development of haploid clones in tomato. Planta 104:357–361

Sharp WR, Caldas LS, Crocomo OJ (1973) Studies on the induction of *Coffea arabica* callus from both somatic and microsporogenous tissue, and subsequent embryoid and plantlet formation. Am J Bot 60:13

Sheldrake AR (1973) The production of hormones in higher plants. Biol Rev 48:509–559

Sink KC, Padmanabhan V (1977) Anther and pollen culture to produce haploids: progress and application for the plant breeder. Hort Sci 12:143–148

Song LSP, Park SJ, Reinbergs E, Choo TM, Kasha KJ (1978) Doubled haploid vs the bulk plot method for production of homozygous lines in barley. Z Pflanzenzücht 81:271–280

Stow I (1930) Experimental studies on the formation of the embryo sac-like giant pollen grain in the anther of *Hyacinthus orientalis*. Cytologia 1:417–439

Sun CS, Wang CC, Chu CC (1973) Cytological studies on the androgenesis of *Triticale*. Acta Bot Sinica 15:163–173

Sunderland N (1970) Pollen plants and their significance. New Sci 47:142–144

Sunderland N (1971) Anther culture: a progress report. Sci Prog Oxford 59:527–549

Sunderland N (1973) Pollen and anther culture. In: Street HE (ed) Plant tissue and cell culture. Blackwell, Oxford, pp 205–239

Sunderland N (1974) Anther culture as a means of haploid induction. In: Kasha KJ (ed) Haploids in higher plants: Advances and potential. Guelph Univ, Guelph, pp 91–122

Sunderland N (1977) Observations on anther culture of ornamental plants. In: Gautheret RJ (ed) La culture des tissues et des cellules de végétaux. Travaux dédiés à la mémoire de Georges Morel. Masson, Paris, pp 34–46

Sunderland N (1978) Strategies in the improvement of yields in anther culture. In: Proc Plant Tissue Culture, Peking, pp 65–86

Sunderland N, Roberts M (1977) A new approach to pollen culture. Nature (London) 270:236–238

Sunderland N, Wicks FM (1969) Cultivation of haploid plants from tobacco pollen. Nature (London) 224:1227–1229

Sunderland N, Wicks FM (1971) Embryoid formation in pollen grains of *Nicotiana tabacum*. J Exp Bot 22:213–226

Sunderland N, Collins GB, Dunwell JM (1974) The role of nuclear fusion in pollen embryogenesis of *Datura innoxia* Mill. Planta 117:227–241

Thomas E, Wenzel W (1975) Embryogenesis from microspores of *Brassica napus*. Z Pflanzenzücht 74:77–81

Thomas E, Hoffmann V, Wenzel G (1975) Haploid plantlets from microspores of rye. Z Pflanzenzücht 75:215–224

Vasil IK (1980) Androgenetic haploids. In: Vasil IK (ed) Perspectives in plant cell and tissue culture. Int Rev Cytol, Suppl 11 A. Academic Press, New York London, pp 195–223

Vasil IK, Nitsch C (1975) Experimental production of haploids and their uses. Z Pflanzenphysiol 76:191–212

Wagner G, Hess D (1974) Haploide, diploide, und triploide Pflanzen von *Petunia hybrida* aus Pollenkörnern. Z Pflanzenphysiol 73:273–276

Wang CC, Chu CC, Sun CS, Wu SH, Yin KC, Hsu C (1973) The androgenesis in wheat *(Triticum aestivum)* anthers cultured in vitro. Sci Sinica 16:218–222

Wang YY, Sun CS, Wang CC, Chien NF (1973) The induction of pollen plantlets of *Triticale* and *Capsicum annuum* from anther culture. Sci Sinica 16:147–151

Wenzel G (1978) Production of haploids of rape and rye. In: Alfermann AW, Reinhard E (ed) Production of natural compounds by cell culture methods. Proc Int Symp Plant Cell Culture, Munich, pp 312–318

Wenzel G (1979) Recent progress in microspore culture of crop plants. In: Proc 4th John Innes Symp, pp 185–196

Wenzel G, Hoffmann F, Thomas E (1977) Increased induction and chromosome doubling of androgenetic haploid rye. Theor Appl Genet 51:81–86

Wenzel G, Thomas E (1974) Observations on the growth in culture of anthers of *Secale cereale*. Z Pflanzenzücht 72:89–94

Wernicke W, Kohlenbach HW (1976) Investigations on liquid culture medium as a means of anther culture in *Nicotiana*. Z Pflanzenphysiol 79:189–198

Wernicke W, Kohlenbach HW (1977) Versuch zur Kultur isolierter Microsporen von *Nicotiana* und *Hyoscyamus*. Z Pflanzenphysiol 81:330–340

Wernicke W, Harms CT, Lorz H, Thomas E (1978) Selective enrichment of embryogenic microspore populations. Naturwissenschaften 65:540–541

White PR (1963) The cultivation of animal and plant cells. 2nd edition. Ronald Press, New York

Woo SC, Huang CY (1980) Anther culture of *Oryza glaberrima* Steud. and its hybrids with *Oryza sativa* L. Bot Bull Academia Sinica 21:75–79

Woo SC, Tung IJ (1972) Induction of rice plants from hybrid anthers of *indica* and *japonica* cross. Bot Bull Acad Sinica 13:67–69

Zenkteler M (1971) In vitro production of haploid plants from pollen grains of *Atropa belladonna* L. Experientia 27:1087

Zenkteler M (1972) Development of embryos and seedlings from pollen grains in *Lycium halimifolium* Mill. in the in vitro culture. Biol Plant Prague 14:420–422

Zenkteler M (1973) In vitro development of embryos and seedlings from pollen grains of *Solanum dulcamara*. Z Pflanzenphysiol 69:189–192

Zenkteler M, Misiura E (1974) Induction of androgenic embryos from cultured anthers of *Hordeum, Secale,* and *Festuca*. Biochem Physiol Pflanz 165:337–340

Zenkteler M, Misiura E, Ponitka A (1975) Induction of androgenetic embryoids in the in vitro cultured anthers of several species. Experientia 31:289–291

6. Ovary, Ovule, and Nucellus Culture

T.S. RANGAN

Following Laibach's (1925) finding that it is possible to excise the embryo at an early stage of development and grow it to maturity on a nutrient medium, the technique of embryo culture has been employed with the twin objectives of obtaining viable hybrids from normally unsuccessful crosses, and to overcome seed dormancy (Tukey 1938). Despite many years of experimentation, however, embryos at very early stages of development have not proved amenable for isolation and culture in vitro. The young embryos are difficult to excise and are liable to be mutilated or injured, and their nutritive requirements are more complex than those of mature embryos. They often tend to grow precociously and produce malformed seedlings. Also, in orchids, phanerogamic parasites such as *Cistanche, Orobanche*, and in some saprophytes, even the mature embryo is so minute that its excision is extremely difficult. Under such circumstances it is only logical to culture entire seeds. Thus, an understanding of factors that regulate the development of a zygote through organized stages to a mature embryo is much facilitated through ovule and embryo culture. This chapter deals with some aspects of ovary, ovule, and nucellus culture.

Ovaries

In angiosperms, the effect of pollination and fertilization extends beyond the formation of embryo and endosperm, eventually resulting in the enlargement and maturation of ovary into a fruit. When working with a whole plant, it is very difficult to study the precise requirement for fruit development. As an approach to study this and other related problems in fruit physiology, a method was devised by which young flowers or fruits could be excised and grown under aseptic conditions on a synthetic medium.

Effect of Growth Substances

LaRue (1942) was the first to attempt the culture of ovaries of a few angiosperms, and obtained root formation from the pedicels. Jansen and Bonner (1949) grew the ovaries of *Lycopersicon pimpinellifolium* on a medium supplemented with casein hydrolysate (CH), IAA, and a mixture of B vitamins. Although the ovaries enlarged, viable seeds were not produced. Nitsch (1949, 1951) successfully cultured the ovaries of *Lycopersicon esculentum, Cucumis anguria, Phaseolus vulgaris, Fragaria* sp., and *Nicotiana tabacum*. On a medium supplemented with auxins such as 2,4-D, 2,4,5-T, or naphtoxyacetic acid (NoA), ovaries excised 2 or more days after pollination and even unpollinated ovaries of *N. tabacum* developed into small but seedless fruits. Nitsch concluded that certain growth substances can replace the stimuli of pollination and seed development on the growth of ovaries into fruits, and also suggested that the technique of ovary culture could be a very useful tool in understanding the physiology of fruit growth.

At the University of Delhi, considerable success has been achieved in culturing the pollinated ovaries of several species such as *Tropaeolum majus* (Sachar and Kanta 1958), *Linaria maroccana* (Sachar and Baldev 1958), *Althaea rosea* (Chopra 1958), apomictic species of *Aerva tomentosa* (Puri 1963), *Zephyranthes* (Sachar and Kapoor 1959), *Iberis amara* (N. Maheshwari and Lal 1961 a), *Hyoscyamus niger* (Bajaj 1966), *Anethum graveolens*, *Foeniculum vulgare*, and *Trachyspermum ammi* (Johri and Sehgal 1966), and *Allium cepa* (Guha and Johri 1966). In *Zephyranthes* Kapoor (1959) observed that ovaries excised two days after pollination developed into normal fruits on the basal medium itself.

With a view to study the effect of chemicals on the pattern of development of embryo and endosperm, Rau (1956) cultured the ovaries of *Phlox drummondii* on Nitsch's medium (NM). Both embryo and endosperm showed normal development. However, when cultured on a medium containing colchicine, the endosperm showed aberrant divisions and nuclear fusion, and eventually degenerated. Similarly, the embryo also showed certain cytological abnormalities, and if the ovaries were retained on the colchicine medium for more than two weeks, the embryo aborted.

Sachar and Kanta (1958) studied the influence of 2,4-D, indolebutyric acid (IBA), IAA, KN, GA, CH, and yeast extract (YE) on the growth of ovaries of *Tropaeolum majus*. Ovaries excised two days after pollination were cultured on NM containing vitamins, glycine, and one of the above growth substances. Although to begin with the ovaries grew normally, the test-tube fruits were smaller than those in vivo even when the medium was fortified with one or more of the above growth substances.

The influence of growth substances on fruit development was also studied in *Linaria maroccana* (Sachar and Baldev 1958). Pollinated ovaries when grown on NM developed into fruits which were smaller than the in vivo control fruits. The addition of growth substances such as KN, IAA, IBA, 2,4-D, or adenine only slightly improved the growth. On addition of YE, however, the fruits not only reached the natural size but also matured earlier (15–17 days) than those in vivo (21–23 days).

Chopra (1958, 1962) studied the effects of IAA, IBA, KN, and GA on both unpollinated and pollinated ovaries of *Althaea rosea*. On NM supplemented with IBA (20 mg/l) the unpollinated ovaries developed into parthenocarpic fruits comparable in size with those which developed in vivo. Although the addition of KN or IAA alone had no pronounced effect, KN and IAA used in conjunction acted synergistically in inducing parthenocarpy.

Guha and Johri (1966) studied the minimum requirement of developing fruits of *Allium cepa*. The maximum growth of fruit occurred on NM supplemented with GA, IAA, and KN. While only a few viable seeds were produced on the basal medium, the addition of IAA or tryptophan increased seed-set to 20% and 30%, respectively.

Often, in culture, the ovaries fail to grow into full-size fruits in the restricted space of culture vial. To overcome this Ito (1961, 1966) devised a partial sterile culture method in which, instead of culturing the entire pistil, only the flower stalk is inserted into the aseptic nutrient medium through an opening in the stopper, thus leaving the ovary free to grow outside the culture vial. Using this technique, Ito studied the growth requirements of the ovaries of *Dendrobium nobile*. The ovaries grew well on NM containing only inorganic salts and sugar. Of the sugars tested, maltose and lactose proved superior to sucrose, and while vitamin B 1 and B 6 stimulated ovary growth, vitamin E (tocopherol acetate) increased seed fertility.

Role of Floral Organs

N. Maheshwari and Lal (1961 a) cultured the flowers of *Iberis amara* on NM supplemented with B vitamins. Flowers cultured 1 day after pollination produced normal fruits only if the calyx was not removed before culture. If the calyx was removed, addition of sucrose (5%) was necessary to obtain satisfactory growth. However, if the flowers were excised and cultured 8 days after pollination, removal of the calyx did not affect the growth of ovaries in vitro.

Similar observations that the floral organs play a significant role in fruit development are also reported by other investigators. In ovary cultures of *Althaea rosea*, Chopra (1958, 1962) observed that the growth of ovaries was considerably influenced by the presence of the calyx. If the calyx was removed prior to culture, the fruits grew to a diameter of only 12 mm, the endosperm remained free-nuclear, and the embryo reached only the heart-shaped stage. If the calyx was left intact, however, the fruits enlarged to a diameter of 19 mm, the endosperm became cellular, and the embryo showed normal development. Likewise, in *Hyoscyamus niger* (Bajaj 1966) ovaries excised 3 days after pollination (with the calyx intact) matured into fruits within 4 weeks, on White's medium (WM). If the calyx was removed, however, the fruits remained much smaller, malformed, and seedless.

La Croix et al. (1962) pointed out that in barley the development of proembryo proceeded normally in florets whose lemma and palea were kept intact, but became impaired if these were removed. In cultures of excised spikes, the proembryos grew normally even when all the florets were devoid of lemma and palea, provided a single leaf was retained on the spike. La Croix et al. concluded that a "hull factor" was necessary for normal development of the embryo, and that it is supplied by the tissues external to the ovary.

Guha and Johri (1966) also observed that in *Allium cepa* the growth of ovaries devoid of perianth was markedly inhibited, and the addition of auxin and other growth substances only partially restored growth. Since young embryos are less amenable to excision, and their nutritional requirement is more complex, Rédei and Rédei (1955) cultured the ovaries, flowers, and spikelets of *Triticum aestivum* and *T. spelta* (excised 4–6 days after pollination) on a nutrient medium containing YE. When ovaries deprived of all the other floral organs were cultured, the embryos seldom continued growth, but in ovaries cultured with palea intact the embryos showed normal development. The growth of the embryo was better when both rachillae and paleae were retained. After growing the ovaries for 8–12 days,

Rédei and Rédei excised the embryo and cultured them on a medium supplemented with CH (0.5%). The embryos continued their development and germinated to form normal seedlings. The above results suggest that the floral envelopes do play an important role in fruit development.

Incompatibility Barriers in Hybridisation

In hybridisation the plant breeder is often confronted with various impediments, such as the failure of pollen germination on the stigma, or the slow and insufficient growth of pollen tube, as well as the precocious abscission of flowers. Several methods have been devised from time to time to circumvent these obstacles. Dulieu (1963, 1966) grew unpollinated pistils of *Nicotiana tabacum* on a nutrient medium and pollinated them artificially 24 h later. Fertilization took place in about 25% ovules, and 10–15 seeds were produced in each ovary. Some of the seeds germinated while still in the ovary and produced normal seedlings. Likewise, in *N. rustica* Rao (1965) observed that the process of fertilization, and embryo and endosperm formation progressed normally in test-tube pollinated ovaries, and mature seeds with viable embryos were obtained. Shivanna (1965) also reported the successful in vitro fertilization and seed formation in in vitro-pollinated ovaries of *Petunia violacea*, a self-incompatible species. While the self-pollinated pistil turned brown and ceased growth, the cross-pollinated pistil continued its growth and produced viable seeds.

Ovary culture has also been successfully employed to obtain hybrids of diploid *Brassica chinensis* and autotetraploid *B. pekinensis* (Inomata 1968), which are normally cross-incompatible. The ovaries excised 4 days after pollination were cultured on NM supplemented with various growth substances such as CH, GA, IAA, KN, YE, tomato juice, and coconut milk. Hybrid seeds were produced on a medium enriched with YE and CH. The triploid hybrid plants thus obtained showed intermediate characteristics of their parents.

Induction of Polyembryony

Ovary culture has also been successful in inducing polyembryony in *Ranunculus sceleratus* (Sachar and Guha 1962), *Anethum graveolens, Foeniculum vulgare*, and *Trachyspermum ammi* (Johri and Sehgal 1966), which normally bear monoembryonate seeds, *Haworthia turgida* (Majumdar 1970), *Citrus* (Mitra and Chaturvedi 1970), *Ammi majus* (Sehgal 1972a), and *Coriandrum sativum* (Sehgal 1972b).

In *Ranunculus sceleratus* (Sachar and Guha 1962) the ovaries were cultured with a view to study the development of embryos. The embryo usually developed normally on the basal medium, but in a medium containing CH it exhibited a tendency to form additional embryo-like structures from the hypocotyl.

The ovaries of *Anethum* (Johri and Sehgal 1963) were excised and cultured 3 days after pollination, when the ovules contained only the zygote and free-nuclear endosperm. On a medium supplemented with CH, IAA, and YE the zygote

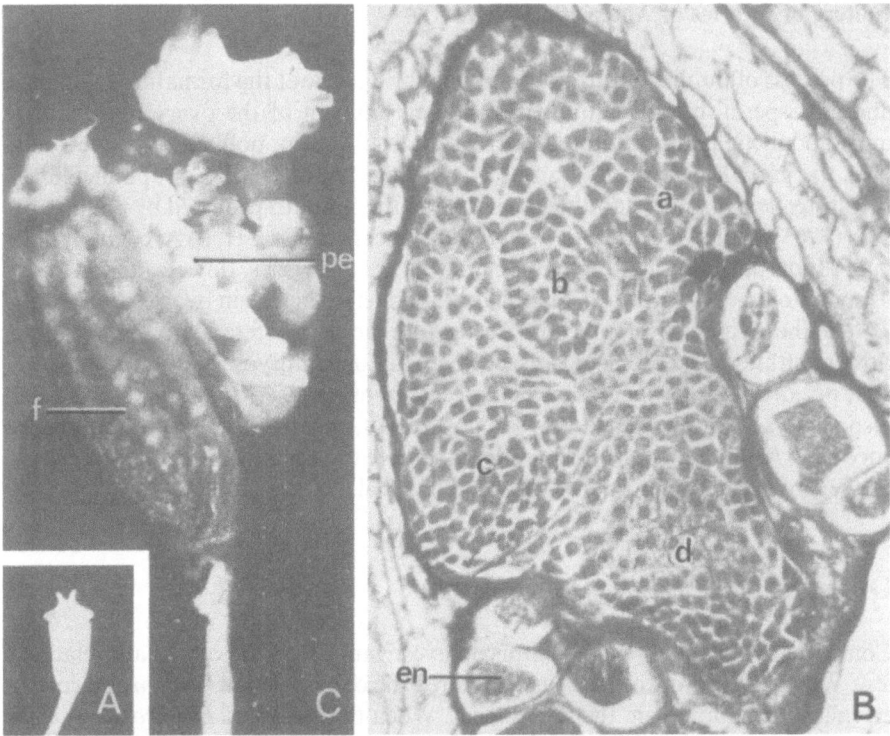

Fig. 1 A–C. Ovary culture of *Anethum graveolens* (*en* endosperm; *f* fruit; *pe* polyembryonal mass). **A** Ovary at culture. **B** L.s. ovary (4 weeks after culture) on WM + CH (1,000 mg/l) showing four *a–d* embryonal masses. **C** 20-week-old culture showing ruptured fruit, and a polyembryonal mass. (After Johri and Sehgal 1963)

cleavaged and budded to form 15–52 embryo-like structures which exhibited cotyledonary abnormalities. The polyembryonal mass projected outside rupturing the developing ovary, and produced multiple shoots (Fig. 1 A–C). When the regenerants were isolated and cultured, they developed into normal plants. Sehgal also observed similar responses in cultures of ovaries of *Foeniculum vulgare, Trachyspermum ammi, Ammi majus,* and *Coriandrum sativum* (Sehgal 1972a, b; see also Johri and Sehgal 1966).

Nishii and Mitsuoka (1969) and Majumdar (1970) reported the formation of plantlets from unpollinated ovary cultures of *Oryza sativa* and *Haworthia turgida,* respectively. After 30–35 days in culture, many of the ovaries of *H. turgida* produced leafy shoots as well as the formation of callus tissue from the wall. Subsequently, the callus differentiated into many plantlets, and also a few roots (Majumdar 1970). The in vitro plantlets were successfully transferred to soil in pots where they developed normally. Mitra and Chaturvedi (1972) also reported that unpollinated ovaries of *Citrus aurantifolia* and *C. sinensis* produced embryoids and plantlets from the ovary wall.

Culture of Ovaries of Apomictic Species

The process of double fertilization not only brings about the formation of embryo and endosperm, but also stimulates the development of the ovary into fruit. In most apomictic plants, although there is no fertilization, pollination alone stimulates the growth of the ovary and seed. The culture of ovaries of apomicts may, therefore, help in understanding the nature of the stimulus provided by pollination. Puri (1963) studied the effect of growth substances such as CH, IAA, coconut milk, and YE on the growth of embryos in the culture of ovaries, flowers, and portions of spikes of *Aerva tomentosa*, an obligate apomict. Whereas in ovary cultures only 7% of the ovules developed into seeds with poorly differentiated embryos, in flower cultures nearly 15% of the seeds bore mature embryos. The addition of YE, CH, or coconut milk increased the percentage of seed-set only slightly. The best response was, however, elicited when portions of spike were cultured. In the presence of IAA and YE the seed-set was comparable to that in Nature, and the growth of the embryo and endosperm was quite normal.

Ovules

Compared with studies on the embryo, investigations on the ovules are relatively few. White (1932), for the first time, cultured the ovules of *Antirrhinum* and reported callus formation from the integument. When the ovules of *Erythronium americanum* were grown on WM fortified with IAA, LaRue (1942) observed that there was an increase in size but the seeds failed to mature. By culturing the ovules of *Epidendrum cochleatum* × *E. tampense* and of a hybrid of *Cattleya octave* × *C. mossiae*, Withner (1942, 1943) successfully shortened the time lag between pollination and the maturation of seeds, thereby hastening the production of seedlings. Subsequently, several investigators studied the nutritive factors which affect the growth of orchid ovules in culture. Spoerl (1948) noted that arginine and aspartic acid support satisfactory growth of unripe and mature orchid ovules, respectively. Vacin and Went (1949) observed that a better differentiation and growth of embryos was obtained in a medium supplemented with tomato juice or "prominogen" (a protein hydrolysate). In his extensive work on orchids, Ito (1961) reported that in the absence of peptone in the medium the germination of seeds and growth of the protocorms was unsatisfactory. Professional orchid growers are now very familiar with several media and growth supplements for obtaining seedlings through ovule culture (see Withner 1959; Valmayor and Sagawa 1967).

Effect of Growth Substances

Significant contributions on ovule culture have come from the Department of Botany, University of Delhi. N. Maheshwari (1958) obtained normal, mature seeds of *Papaver somniferum* by culturing the fertilized ovules (containing zygote or two-celled proembryo and a few endosperm nuclei) on NM fortified with KN. Subsequently, N. Maheshwari and Lal (1961 b) observed that KN accelerated growth and differentiation of proembryo in the ovule. In ovules grown on KN-medium for 10 days, the embryos grew to a length of 0.45 mm, surpassing that attained by embryos in vivo, and possessed well-developed cotyledons and stem tip. However, this initial rate of growth was not maintained, and the final length of the embryo (0.54 mm) in cultured ovules was less than that of embryos in Nature. When the ovules were maintained in culture for 50 days, they germinated.

Sachar and Kapoor (1959) reared ovules of *Zephyranthes* (with the zygote and primary endosperm nucleus) on NM and reported that the embryo failed to develop beyond the globular stage, and the formation of endosperm was also inhibited. Subsequently, Kapoor (1959) successfully obtained complete development of embryo and endosperm. On a medium supplemented with Seitz-filtered coconut milk (25%), the ovules showed a globular proembryo and cellular endosperm within 4 days after culture, and a mature embryo in 12 days. Within 20 days the seeds also germinated in situ and produced normal seedlings. When the medium contained casamino acids instead of coconut milk, the growth of ovules was slower and the seeds germinated after 24 days, producing normal seedlings. Further studies to determine the specific active constituent of the casamino acids revealed that amino acids histidine, arginine, and leucine were the most favourable for growth of embryo. In cultures of ovules of *Citrus microcarpa*, Rangaswamy (1961) observed that the growth and development of nucellar embryos occurred even on WM containing 5% sucrose, while the embryos dissected from the ovule showed normal growth only on addition of CH (400 mg/l) to the medium.

Bajaj (1964) cultured the ovules of *Abelmoschus esculentus* 2 and 4 days after pollination (containing four-celled proembryo with free nuclear endosperm and globular proembryo with cellular endosperm, respectively) and observed that ovules of both ages developed satisfactorily on medium containing CH (500 mg/l) + coconut milk (10%–25%). The ovules matured in 6 weeks and also germinated.

To study morphogenetic effects of the placenta, Chopra and Sabharwal (1963) cultured the pollinated ovules of *Gynandropsis gynandra*. They observed that the placental tissue had a beneficial effect on the growth and maturation of seeds, and suggested that this effect could either be due merely to an increased surface absorption, or the influence of some growth substances present in it. In *Opuntia dillenii*, however, Sachar and Iyer (1959) demonstrated that the ovules failed to develop even when cultured intact with the placenta. But on a medium containing IAA and KN, the placenta proliferated to form a callus of limited growth. Pontovich and Sveshnikova (1966) investigated the effect of various factors which stimulate the formation and development of embryo and endosperm in ovule cultures of *Papaver somniferum*. Normal development of ovules occurred when cultured with placenta in the presence of growth supplements such as adenine and birch sap. In this context, the findings of Melnick et al. (1964) are significant. They demonstrated that young fertilized ovules of different species, genera, and even families could be transplanted on to the placenta of fruits of *Capsicum* and grown in vivo to mature viable seeds. This indicates that the physiological requirement for the growth of these ovules are not species-specific.

Beasley and his co-workers employed the ovule culture method in their extensive studies on embryogenesis, and the effect of phytohormones on fibre and seed development in cotton (Beasley 1971; Beasley et al. 1974). Fertilized ovules excised

Fig. 2. Ovule culture of cotton. Two-week-old culture of cotton ovules, showing (left to right) normal development, callus formation, and reduced number of ovules with fibres due to limited fertilization. (After Beasley and Ting 1974a)

on the second day post-anthesis grew and developed fibres only when floated on a liquid medium. Submerged ovules or ovules placed on identical but agar medium callussed or failed to grow (Fig. 2). If cultured on the first day of post-anthesis, only a few ovules produced fibres, and none when cultured on the day of anthesis. The capacity of ovules to develop fibres and grow normally in vitro also seemed to be dependent upon having ideal growing conditions for parent plants. For example, ovules from greenhouse-grown plants collected in midsummer grow more and produce more fibre than those collected in winter. Subsequently, Beasley and Ting (1973, 1974a) studied the effect of plant growth substances on in vitro fibre development of cotton ovules. Gibberellic acid induced a marked stimulation of fibre development from fertilized ovules. Both KN and ABA inhibited the development of fibres. GA could also overcome the KN- and ABA-induced inhibition. Beasley and Ting (1974b) also observed that unfertilized ovules enlarged in the presence of KN, and enlarged as well as produced fibres in the presence of IAA or GA. Indoleacetic acid induced greater stimulation of fibre production than GA, and a combination of GA and IAA yielded additional amounts of fibre (Fig. 3).

Poddubnaya-Arnoldi (1959, 1960) simplified the ovule culture technique for studying the developmental phases following pollination. Using a 10% solution of sucrose, she successfully grew the ovules from pollinated ovaries of *Calanthe veitchii, Cypripedium insigne, Dendrobium nobile,* and *Phalaenopsis schilleriana,* traced the events beginning with the entry of the pollen tube up to the development of embryo, and described the histochemical changes accompanying these events.

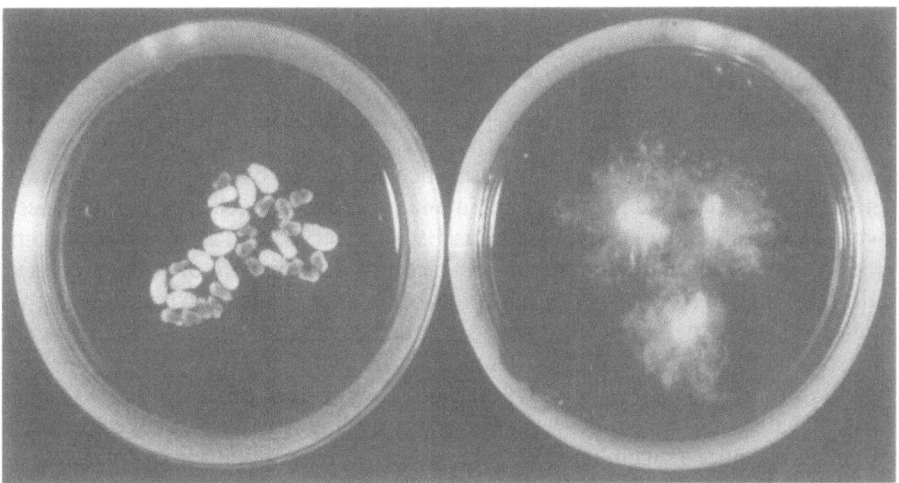

Fig. 3. Ovule culture of cotton. White-fibreless and brown shrunken cotton ovules (all from a single ovary) cultured on basal medium *left*, and ovules showing fibre development on basal medium supplemented with 5.0 μ*M* of IAA and 0.5 μ*M* GA *right*. (After Beasley 1977)

Effect of Physical Factors

The extent of embryo growth is related to the age of the ovule at culture. In *Nicotiana tabacum* ovules excised up to 9 days after pollination failed to grow in culture (Siddiqui 1964). Only ovules cultured 12 days after pollination (with globular proembryo) showed a good response, but the growth of the embryos was significantly slower than in vivo. In Nature, embryos 12 + 5 days after pollination possessed well-differentiated cotyledons, while in culture, ovules of similar age showed only heart-shaped embryos. However, Dulieu (1966) reported that the ovules of *N. tabacum* excised 4 or 5 days after fertilization (with 10-celled proembryo) could be successfully reared in vitro on a medium containing mineral salts, sucrose, and B vitamins. He observed that with 4% sucrose the embryos attained appreciable length and a maximum number of seeds germinated in situ. In *Allium cepa* the ovules cultured 13 days after pollination developed normally, and germinated in situ after 20 days (Guha and Johri 1966).

Eid et al. (1973) report that in cotton, when 5-day-old ovules (with 2- to 10-celled proembryo) were cultured, the embryo grew only up to the early dicotyledonous stage. But the addition of GA and CM improved embryo growth. Although the embryos germinated readily, the seedlings were abnormal. Ten-day-old ovules (with globular proembryo and free-nuclear endosperm), when grown for 60 days, produced normal embryos but smaller than those in Nature. In 15- and 20-day-old ovule cultures, the embryos not only developed to the same extent as those in vivo, but also germinated in ovulo. Eid et al. concluded that the difference in the rate of growth is associated with the developmental stage of endosperm (free-nuclear in 5- and 10-day-old, and cellular in 15- and 20-day-old ovules) present in the ovule at culture.

It has been demonstrated that the osmotic concentration significantly influences the growth of excised embryo (Mauney 1961; P. Maheshwari and Rangaswamy 1965). Ryczkowski (1962 a, b) observed that during the early stages of embryo development the central vacuolar sap of the ovule showed an increase in the osmotic value, but with further growth of the embryo, the value decreased considerably. He suggested that a study of osmotic value of the ovules during its growth and development would prove helpful in understanding the requirement of excised ovules in vitro. Wakizuka and Nakajima (1974) observed that the osmotic concentration of the medium considerably influences the development of young ovules in culture. Ovules of *Petunia hybrida* cultured 4 days after pollination (with the zygote and several endosperm nuclei) developed satisfactorily on a medium with 6% sucrose, while those excised 3 days after pollination (with only the zygote and primary endosperm nucleus) required 8% sucrose. They further observed that when the osmotic value of the medium was adjusted by adding appropriate concentrations of cucumber juice (cucumber fruit juice prepared from young fruits of *Cucumis sativus* L.), the ovules of both ages developed equally well. Even ovules excised and cultured immediately after fertilization could also be grown successfully on an osmotically balanced medium. Subsequently, Wakizuka and Nakajima (1975) reported that although ovules excised 3 and 4 days after pollination developed up to early globular proembryo on a simple medium (containing 8% and 6% sucrose, respectively), further development of the embryo and eventual formation of seedlings required the presence of cucumber juice (see also Nakajima et al. 1969).

Ovule Culture in Hybridization

In many interspecific and intergeneric crosses, the F_1 hybrid embryo frequently aborts in the developing seed, or the seeds produced are incapable of supporting the development of embryo. In several such instances, embryo culture has been successfully employed to obtain hybrid seedlings (see P. Maheshwari and Rangaswamy 1965). In their attempts on interspecific hybridization in *Abelmoschus*, Gadwal et al. (1968) observed that the hybrid embryo failed to develop beyond the heart- or torpedo-shaped stage. By resorting to both ovule and embryo culture, they successfully obtained viable hybrids in three out of five crosses attempted, namely, *A. esculentus* × *A. ficulneus*, *A. esculentus* × *A. moschatus*, and *A. tuberculatus* × *A. moschatus*. Similarly, in *Gossypium*, in repeated attempts to hybridize the species of the New and the Old World, it has been reported that the embryo constantly aborts (Weaver 1958). Efforts to excise and grow the young embryos have met with only limited success (Mauney 1961; Mauney et al. 1967); the difficulties have been attributed to their sensitivity to physical conditions as well as to complex nutrient requirements. As an alternative, ovule culture has been suggested and some success achieved. Joshi (1962), and Joshi and Johri (1972) studied the effect of IAA, KN, GA, CH, and YE on the in vitro growth of ovules of *Gossypium hirsutum*. The ovules containing a 12-celled proembryo and about 500 free endosperm nuclei, were excised and cultured 6 days after pollination. On WM supplemented with CH (1,000 mg/l) and GA (5 mg/l) the ovules surpassed the size of in vivo-developed ovules, 98 days after pollination. The addition of CH did not

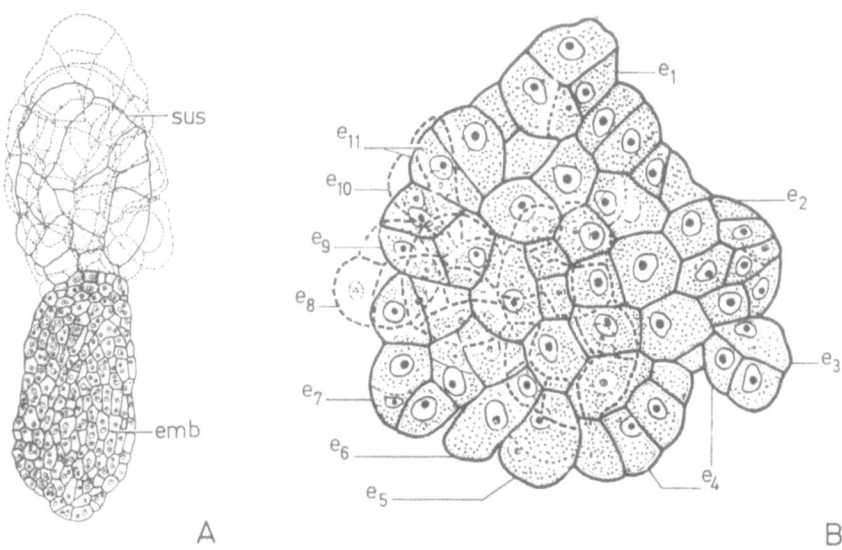

Fig. 4A–B. Ovule culture of cotton (*emb* embryo; *sus* suspensor). **A** Embryo isolated from a nearly 10-week-old culture of ovule on WM + IAA (1.5 mg/l) showing profuse proliferation of suspensor. **B** Embryo isolated from 3-week-old ovule culture on WM + ovule extract (20%) showing 11 distinct embryonal masses e_1–e_{11}. (After Joshi and Johri 1972)

markedly favour the growth of embryo. Whereas a lower concentration of CH (250 mg/l) did not promote the growth of embryo beyond the globular stage, higher concentrations (1,000 and 2,000 mg/l) supported its growth up to the early dicotyledonous stage.

The addition of IAA did not favour embryo development, but the cells of the suspensor proliferated to form a mass of 100–110 cells (Fig. 4A). The integuments also exhibited proliferation and, compared with the cells of the outer integument, those of the inner integument showed greater capacity for callusing.

Polyembryony due to cleavage or budding of embryo was also observed in certain treatments. The addition of ovule extract (20%) to the medium resulted in the formation of as many as 11 embryonal masses (Fig. 4B).

Joshi and Pundir (1966) cultured ovules from the cross *G. arboreum* × *G. hirsutum* but without much success [1]. The ovules were excised 3, 20, and 22 days after pollination and cultured on WM supplemented with IAA (1 mg/l) + GA (4 mg/l) + KN (1 mg/l) + CH (1,000 mg/l). The ovules 3 days after pollination (containing the zygote or 2-celled proembryo, and a few endosperm nuclei) showed a poor response, turned brown, and finally degenerated. Ovules cultured 20–22 days after pollination (containing globular proembryo, heart-shaped or well-differentiated embryo with degenerated endosperm) did not show any growth. Some of the

---

[1] They subsequently attempted the culture of hybrid embryos (late-globular to heart-shaped) on a medium containing CH + GA + KN. Differentiation of small cotyledons, hypocotyl, and radicle was observed in such cultures. After about 30 days triploid seedlings developed. However, these seedlings could not be maintained in a healthy condition, and could not be established in soil

ovules increased in size while others burst and secreted a mucilaginous substance. After about 4 weeks in culture, a few ovules proliferated and produced a callus.

Nitzsche and Hennig (1976) successfully obtained a hybrid between *Lolium perenne* and *Festuca rubra* by means of ovule culture. The ovules, excised 5 days after pollination, were cultured on modified Murashige and Skoog's (MS) medium containing 6-benzylaminopurine and later transferred to Gamborg's medium. The developing plant showed certain morphological features of *F. rubra*, thereby giving credence to its hybrid origin.

Culture of Unfertilized Ovules

Uchimiya et al. (1971) attempted culturing unfertilized ovules of *Solanum melongena*, and obtained vigorous callus formation on a medium supplemented with IAA and KN. Although the origin of the callus was not determined, a cytological examination revealed it to be haploid in nature.

A significant achievement of research on ovule culture has been the development of the technique of test-tube pollination and fertilization, and of much interest in this connection are the papers by Kanta et al. (1962), Kanta and P. Maheshwari (1963), and P. Maheshwari and Kanta (1964). They excised unfertilized ovules of *Argemone mexicana*, *Eschscholtzia californica*, *Papaver somniferum*, *Nicotiana tabacum*, and *N. rustica*, and cultured them along with ripe pollen grains. All stages from germination of pollen to double fertilization were observed, and mature seeds containing viable embryos were obtained (see also Zenkteler 1965; Kameya et al. 1966). Using the method, Zenkteler (1967) succeeded in fertilizing the ovules of *Melandrium album* with pollen from other species of Caryophyllaceae and subsequently even with pollen of *Datura stramonium* (Zenkteler 1970). While Kameya and Hinata (1970) obtained true hybrids between *Brassica chinensis* and *B. pekinensis*, Rangaswamy and Shivanna (1967) employed this technique to overcome incompatibility barriers in *Petunia axillaris*.

Similarly, work on the culture of ovules and seeds of angiospermic parasites is comparatively recent. It is generally believed that in obligate root parasites such as *Orobanche* and *Striga*, the formation of seedlings is dependent on some stimulus from the host root. However, studies on *Orobanche aegyptiaca* (Rangaswamy 1963) and *Cistanche tubulosa* (Rangan and Rangaswamy 1968) have demonstrated that the differentiation of shoots in vitro can be induced in the absence of any stimulus from the host.

Nucellus

Normally, the embryo originates from the zygote, but sometimes in plants such as *Citrus* and mango, the embryos arise adventitiously from cells of the nucellus or integument. The adventive embryos are of considerable importance to the horticulturist, in that they are genetically uniform and reproduce the characters of the maternal parent without inheriting the variations brought about by gametic fusion. In *Citrus*, on the one hand there are species, varieties, and hybrids which are extremely polyembryonic, and on the other there are those which are essentially monoembryonic. In monoembryonic *Citrus* taxa a serious problem is that of obtaining disease-free clones. It is generally believed that most of the viruses of citrus are not transmitted through seedlings, whether of zygotic or nucellar origin. Accordingly, in contrast to clones established from cuttings or by conventional methods of propagation, plants derived from nucellar embryos are free from most viruses. Moreover, many of the desirable plant vigour and fruiting characteristics associated with juvenility are restored in trees newly established from nucellar seedlings. This is of great advantage in citriculture, and has been used on a commercial scale for propagating desirable varieties of citrus.

Culture of Nucellus from Fertilized Ovules

In recent years the technique of tissue culture has been employed to study factors responsible for the formation of adventive embryos. Using a novel method for the first time, Rangaswamy (1958) studied nucellar embryony in vitro. He excised the nucellus from pollinated ovules of *Citrus microcarpa*, a natural polyembryonic species, and cultured it on a modified WM supplemented with CH. The nucellus proliferated into a callus mass, and differentiated into embryo-like structures (termed pseudobulbils) which eventually developed into plants. Rangaswamy (1961) concluded that "freed from the restraining influence of the integuments and grown on a suitable medium, the nucellar tissue of *Citrus microcarpa* was activated to unlimited growth", and could be made to yield a continuous supply of nucellar embryos. Sabharwal (1963) obtained similar results in *Citrus reticulata*, but believed that the pseudobulbils arise not from nucellus per se but from nucellar embryos.

Rangan et al. (1968, 1969) demonstrated that it is possible to induce nucellar embryogenesis in monoembryonic citrus species as well. Adventive embryos were successfully initiated in nucellus cultures of *Citrus grandis* (shaddock), *C. limon* (ponderosa lemon), and *C. reticulata* × *C. sinensis* (temple orange) on MS-medium supplemented with either malt extract (500 mg/l) or a combination of adenine sulphate (25 mg/l) + NAA (0.5 mg/l) + orange juice (5%). Unlike in *C. microcarpa* or *C. reticulata*, the nucellar explants did not produce callus or pseudobulbils but gave rise directly to highly organized multiple embryos and seedlings (Fig. 5 A–E). Bitters et al. (1972) extended these studies to several other mono- and polyembryonic as well as seedless varieties of *Citrus* such as *C. temple* (temple tangor), *C. reticulata* (clementine mandarin), *C. limon* (meyer lemon), *C. maxima* (chandler shaddock), *C. sinensis* (Robertson navel), *C. latipes* (Khasi papeda), and *C. latifolia* (Bearss lime, a triploid seedless variety), to determine whether adventive embryos could be initiated in isolated nucellus and if such embryos could give rise to virus-free plants. It was observed that embryos and plants can be obtained, although the percentage of successful cultures varied from species to species. The developmental state of the ovule at which the nucellus was excised and cultured also influenced the initiation of adventive embryos, and the appropriate developmental stage varied with the cultivar. For example, with temple tangor (a monoembryonic variety) no success was obtained until the 10th week after pollination, but successful cultures were obtained from the 10th through the 18th week. In contrast, in Robertson navel (a

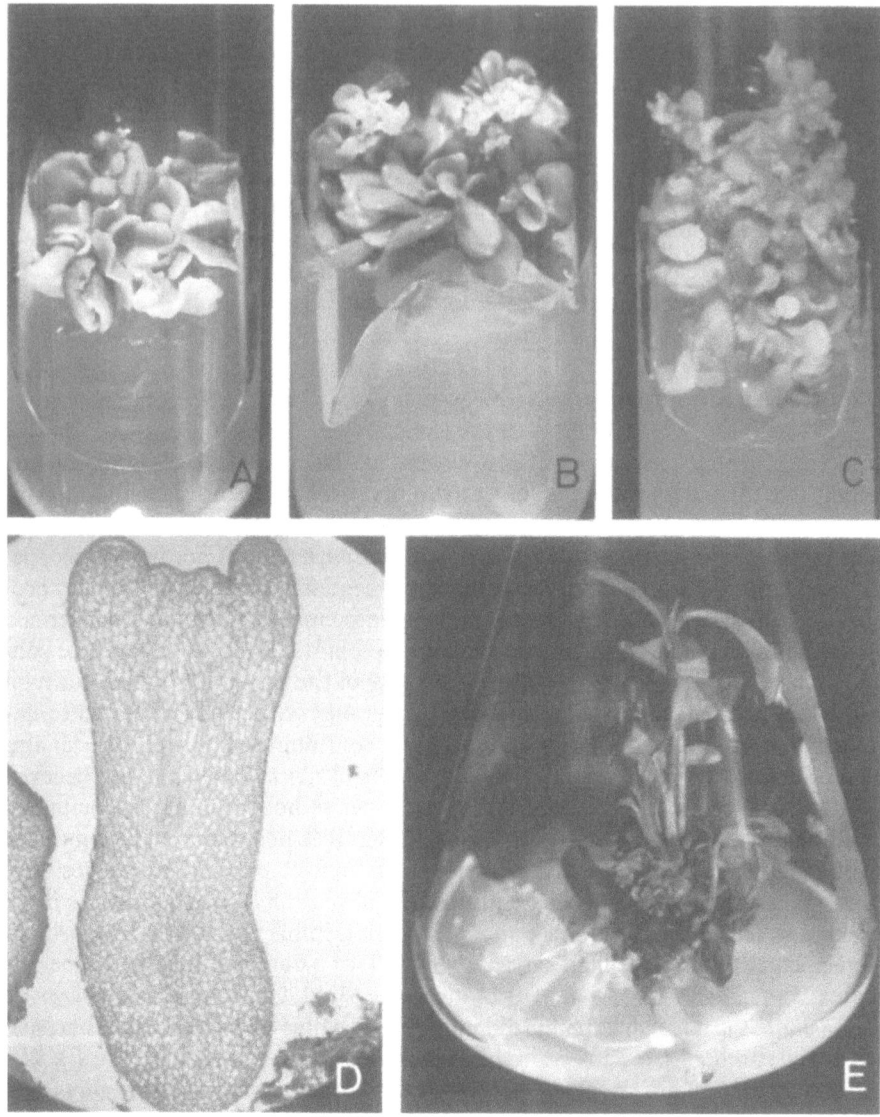

Fig. 5 A–E. Nucellar embryogenesis in *Citrus*. **A–C** Six-week-old polyembryonal masses of *Citrus aurantium* (polyembryonic), *C. reticulata* × *C. sinensis* (monoembryonic), and *C. sinensis* (seedless variety with abortive ovules). **D** L.s. nucellar embryo formed in vitro. **E** Nucellar seedlings of *C. grandis* (monoembryonic) approximately 10 weeks after isolation and culture of nucellus. (**A, C,** and **D** after Rangan unpublished; **B** and **E** after Rangan et al. 1968)

seedless variety with abortive ovules), embryos were produced from explants taken at the time of pollination to 6 weeks after pollination. Maximum success was achieved with nucellus isolated 6 weeks after pollination, but beyond 6 weeks the ovules had completely degenerated in the fruit.

The mother tree of Robertson navel orange had carried several viruses, including those of exocortis and psorosis, while that of temple tangor had carried tristeza, vein enation, and exocortis viruses. A check of seedlings by means of "Etrog" citron test [2] and other indexing procedures indicated that the viruses present in the mother plants were eliminated in the nucellar seedlings obtained in vitro (Bitters et al. 1972).

It is significant that adventive embryos will not develop in nucellus taken from fertilized ovules of all monoembryonic cultivars (Bitters et al. 1972). An active antiembryogenic substance from nucellus of monoembryonic *Citrus* cultivars has recently been discovered by Esan (see Murashige 1974), and the presence of this substance could be a major factor in limiting the initiation of nucellar embryogenesis in vitro.

Culture of Nucellus from Unfertilized Ovules

Pollination and fertilization are generally considered as essential prerequisites for the induction of nucellar embryogenesis (Frost and Soost 1968). In previous investigations (Rangaswamy 1961; Sabharwal 1963; Rangan et al. 1968, 1969) explants were taken only from fertilized ovules. However, Button and Bornman (1971), Kochba et al. (1972), and Mitra and Chaturvedi (1972) succeeded in inducing embryogenesis in vitro from nucelli of unfertilized ovules of Washington navel, Shamouti orange *(Citrus sinensis)*, and *C. aurantifolia*, respectively. While Button and Bornman found that in Washington navel embryos arose directly from the nucellar tissue, Kochba et al. and Mitra and Chaturvedi reported that embryos arose directly or from callus which proliferated from the nucellus. These investigators also reported the production of further embryos from those formed initially. Such studies prompt the conclusion that the stimulus of pollination and/or fertilization can no longer be considered as a limiting factor for nucellar embryogenesis.

Factors Affecting Nucellar Embryogenesis in Vitro

Malt extract invariably promoted adventive embryogenesis in nucelli excised from fertilized (Rangan et al. 1968, 1969; Bitters et al. 1972) and unfertilized ovules (Button and Bornman 1971; Kochba et al. 1972; Mitra and Chaturvedi 1972). The active fraction(s) of malt extract are as yet unknown, but are likely to include one or more amino acids and possibly a cytokinin (van Staden 1974). There are other substances which have promoted embryogenesis in nucellar cultures, including adenine (Button and Bornman 1971), a combination of adenine sulphate and KN (Mitra and Chaturvedi 1972), or a combination of orange juice, NAA, and adenine sulphate (Rangan et al. 1968, 1969).

---

2 The "Etrog" citron test is carried out by grafting a citron bud into a healthy, vigorous rough lemon or Mexican lime seedling about ½ cm in diameter, and forcing the citron bud to grow as fast as possible. Simultaneously, two buds from any tree that is to be tested for the presence of virus are grafted, one below and one above the citron bud. If the virus is present, symptoms appear in the sensitive citron from 1–6 months after budding

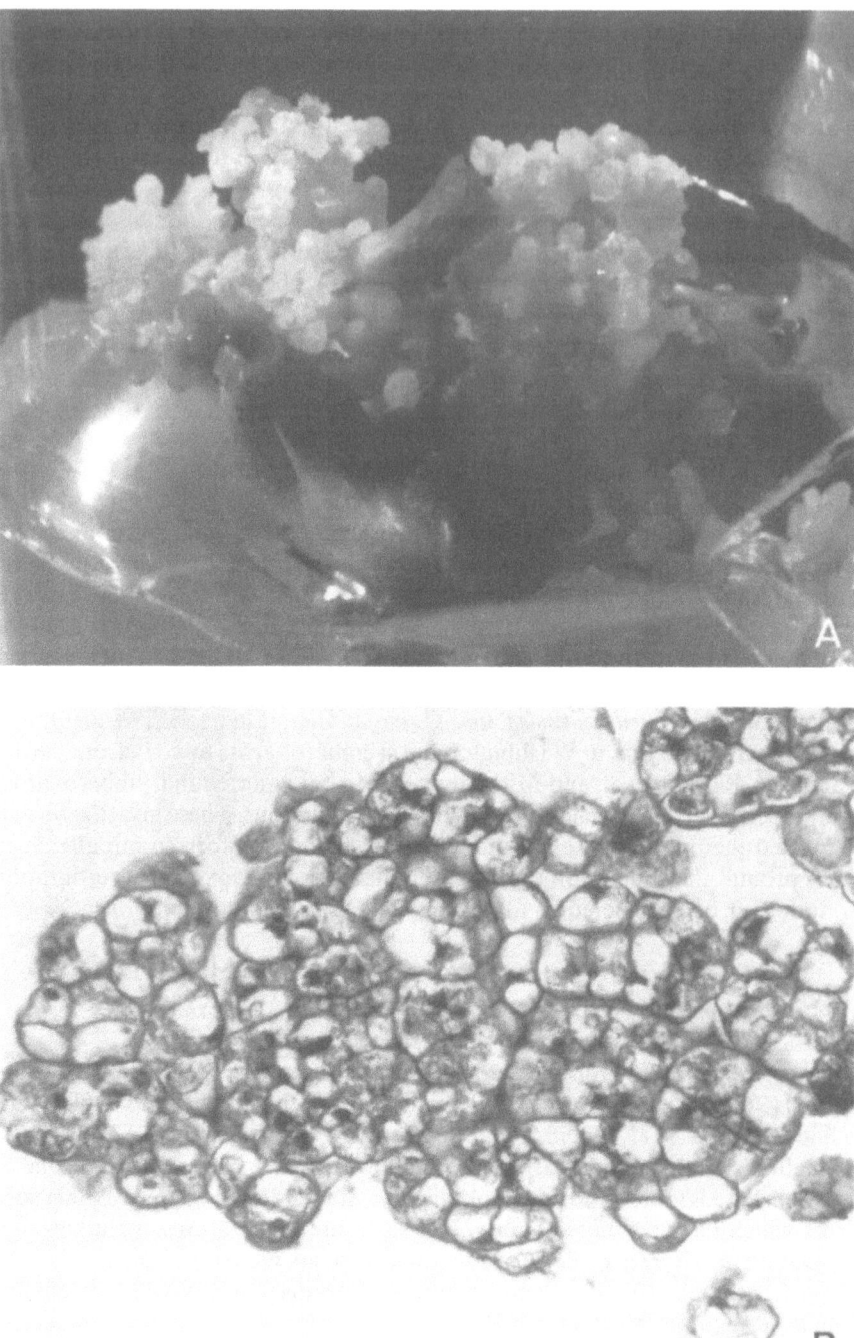

Fig. 6 A–D. Nucellar embryogenesis in *Citrus* ("Shamouti" orange). **A** Embryogenic calli of "Shamouti" orange. **B** A section through the embryogenic calli showing clusters of embryoids. **C** A three to four-celled proembryo derived from a single cell and developing from

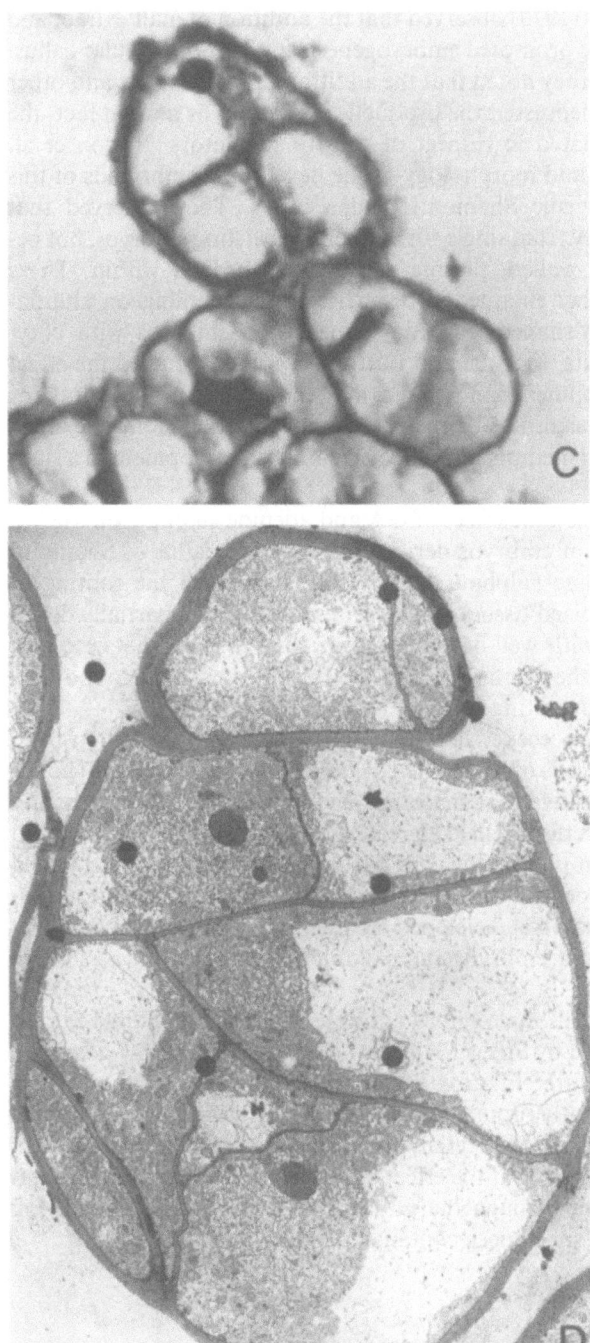

the periphery of another embryo. **D** Electron micrograph showing a globular proembryo which has broken away from the parent embryogenic mass. The remnants of cell wall of presumably adjoining cells are visible. (After Button et al. 1974)

Kochba and Spiegel-Roy (1973) observed that the addition of malt extract and low concentration of adenine promoted embryogenesis in Shamouti ovular callus. After a number of passages, they noted that the addition of malt extract and other growth substances actually depressed the production of embryos and, in fact, the cultures had become habituated in respect of growth regulators. Button et al. (1974) studied the ontogeny and morphology of the developing embryoids of this unique, habituated embryogenic Shamouti ovular callus. They observed that proembryos developed mainly from single surface cells of existing embryos, but occassionally also from thick-walled, plasma-rich cells embedded within. These proembryos (Fig. 6 A–D) either enlarged into spherical pseudobulbils or, alternatively, developed into heart-shaped, torpedo-shaped, and embryos with cotyledons, and thence to plantlets. Kochba and Button (1974) also studied the effect of ageing on embryo development, and found that the ageing of the callus for 14 weeks prior to subculture significantly increased embryo production. The omission of sucrose from the basal medium also affected embryo development in a similar way.

Kochba et al. (1974) studied the effect of GA and adenine sulphate on the initiation and growth of roots of embryos derived from ovular callus of Shamouti orange. Both GA and adenine sulphate significantly stimulated the rooting of small embryos (with unorganized tissue) and larger embryos (with partially developed root zone). Embryos with well-developed root zone formed roots even on basal medium. Interestingly, the formation of shoots on embryos (prior to root formation) completely inhibited rooting.

Button and Botha (1975) succeeded in obtaining regeneration of plantlets from enzymatically isolated single cells of habituated callus of Shamouti orange. The enzyme macerase was more effective than macerozyme in macerating the callus, and higher concentrations of macerase yielded a greater number of single cells than did the lower concentrations. On plating the cells on an agar medium supplemented with either malt extract or coconut milk, proliferation and subsequent organization occurred. While pseudobulbils developed frequently on malt extract-supplemented medium, embryos and embryogenic callus predominated on a medium containing coconut milk.

The effect of gamma irradiation on embryo formation in Shamouti orange ovular callus has been investigated by Spiegel-Roy and Kochba (1973). Low doses (up to 2kR) suppressed embryo differentiation and stimulated callus growth, whereas higher doses (16 kR) significantly promoted embryo formation but inhibited callus development. A dose of 32 kR proved lethal for both callus and embryo growth. Spiegel-Roy and Kochba suggest that the effect of radiation on the differentiation of embryos may be mediated through the nutrient medium, possibly involving some changes in the growth substances contained in it.

Nucellar Embryogenesis in Vitro in Species Other than Citrus

All attempts to induce and study nucellar polyembryony in vitro appear so far to have been confined only to *Citrus* species. An obvious question would be, can ad-

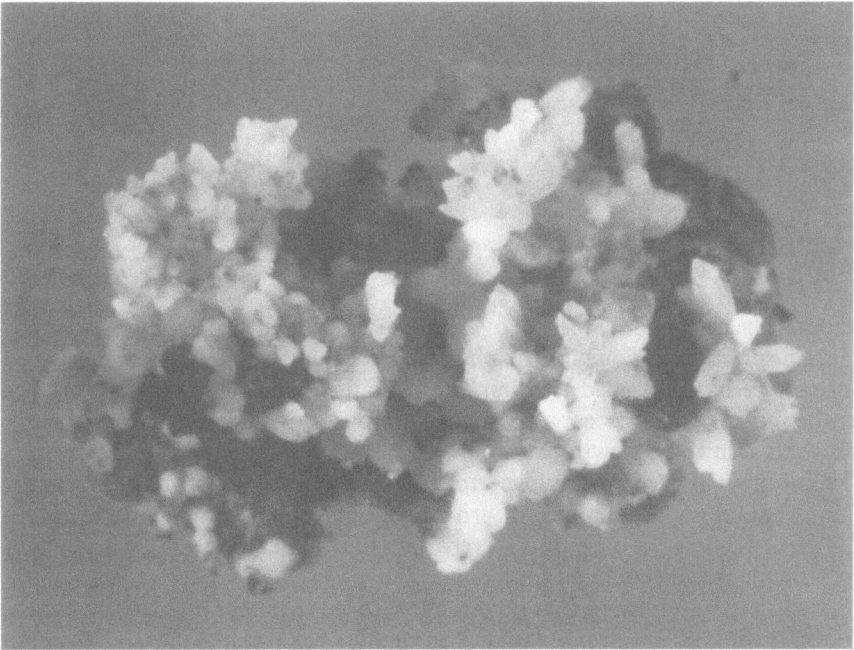

Fig. 7. Development of numerous embryoids in ovule cultures of red currant (*Ribes rubrum* L.). (After Zatyko et al. 1975)

ventive embryos be induced in plants which are not naturally polyembryonic? Rangaswamy and Shivanna (1975) established nucellus cultures of *Luffa cylindrica* and *Trichosanthes anguina*. Although no embryogenesis was observed, the callus cultures showed the differentiation of tracheary elements. The presence or absence of 2,4-D significantly affected xylogenesis. In the absence of 2,4-D xylogenesis was not only earlier but also profuse, whereas its presence resulted in delayed and less profuse formation of tracheary elements.

Mullins and Srinivasan (1976) succeeded in inducing nucellar embryony in ovule cultures of *Vitis vinifera* L. cv. Cabernet-Sauvignon, a normally monoembryonic species. Unfertilized ovules grown on NM supplemented with benzyladenine and B-NOA formed a nucellar callus which subsequently produced embryoids. These embryoids which were slightly larger than the zygotic embryos, when transferred to a semi-solid medium containing GA and isopentenyladenine produced plantlets which exhibited the characteristic morphology of grape vine seedlings such as spiral phyllotaxy and absence of tendrils.

In unpollinated ovary cultures of *Cynanchum vincetoxicum*. Haccius and Hausner (1976) observed the development of green plantlets through the ovary wall. Histological studies proved them to be adventive embryos originating from the nucellar cells of unpollinated ovules. Some of these in vitro-obtained nucellar plantlets were planted in soil where they developed into flowering plants.

Sladky (1974) reported the formation of adventive embryos and plantlets in cultures of perisperm of *Agrostemma githago*, on NM supplemented with coconut milk. The induction of polyembryony in ovule cultures of normally monoembryonic *Ribes rubrum* is also reported (Zatyko et al. 1975). Ovules cultured on Miller's medium containing KN developed a number of embryo-like structures. The individual embryos were capable of proliferating into an embryonal mass containing numerous embryos even on a medium devoid of plant growth hormones (Fig. 7).

Concluding Remarks

Studies on the culture of embryos, ovules, ovaries, and anthers have considerably advanced our understanding of the physiology of these organs. Investigations on ovary culture have revealed two important points: First, that ovaries detached from the mother plant are capable of autonomous growth and bear fertile seeds; and second, that their growth pattern in vitro is essentially similar to that in vivo. Ovary culture may also be employed to study the physiology of fruit maturation and fruit pathology. The failure of growth of ovules and embryos on synthetic media can sometimes be overcome by resorting to ovary culture, and hybrid plants from difficult crosses could be obtained.

The difficulty of growing very young embryos led to attempts to culture ovules. Work on these lines by several workers on orchids has resulted in a wealth of information regarding the requirement of immature orchid embryos. This has helped in the application of this technique by orchid growers, and could prove a valuable tool in the propagation of self-sterile orchid species such as *Oncidium* and *Renanthera*. From the viewpoint of physiology and morphogenesis, ovule culture of orchids, phanerogamic parasites, and saprophytes (where it is difficult to excise the embryo) will no doubt prove very rewarding.

That entire ovules can be cultured as early as the zygote, or two to four-celled proembryo stage, is of considerable importance. For plant breeders, ovule culture can be a boon in obtaining seedlings from crosses which are normally unsuccessful because of abortive embryos. Although to date, successful crosses between different species of cotton have not been achieved through ovule culture, such a possibility is now greatly enhanced. It has been demonstrated in cotton that the growth of the ovule and fibre is regulated by exogenous hormone, and in this respect ovule culture also offers a unique method for studies on mechanism of cell wall biosynthesis and cellular growth. Similarly, success with test-tube fertilization opens up new vistas in hybridization programs, especially in overcoming incompatibility barriers.

In horticultural practices the artificial induction of polyembryony holds a great potential. Pseudobulbils of *Citrus*, therefore, represent tissue "banks" capable of initiating clones of adventive embryos. That the nucellus of monoembryonic *Citrus* can be induced to form adventive embryos is significant. For these varieties of *Citrus*, which have been impossible to free of virus by other means, the nucellus culture offers a practical approach. Although several embryos and plants can be obtained from single nucellus culture, the percentage of success varies from species to species. There is a need, therefore, to examine further the physiology underlying the process of embryogenesis in order to make the technique more viable and widely applicable. However, this method is still useful, as all one needs is a few virus-free plants for clonal propagation. It would also be desirable to have as many plants as possible, in as much as genetic variants, including polyploids, have been known to occur among many plants propagated through tissue culture. The formation of nucellar embryos from unpollinated ovules is also of considerable interest in that it can be employed to import seedless *Citrus* varieties without the risk of introducing new viral diseases, thereby avoiding the necessity of subjecting vegetative material to long periods of quarantine. While the artificial induction of embryos has proved decisively advantageous, at times it becomes desirable to eliminate them. In polyembryonic *Citrus*, for instance, the large number of nucellar embryos poses a major problem in breeding and variety improvement programs. The need to induce adventive embryony in those plants in which it does not occur in Nature, and its control in those in which it exists as a normal feature offers much scope for experimental work.

A parthenogenetic development of the egg cell has been achieved in many animals and a few cryptogams, but attempts to induce parthenogenetic development of egg cell in angiosperms have failed. Although haploids do occur in Nature, little is known about the causal factors responsible for their occurrence. A direct handling of the egg cell in higher plants is by no means easy, because of the problems involved in its excision without injury. The next logical step is to induce the egg cell to divide within its ovular environment, as the excised ovules cultured in vitro may be more amenable to physical and chemical stimulation than those borne in Nature. The know-how of its achievments, apart from its practical significance, would provide meaningful information about the physiological processes that govern plant reproduction.

Srivastava et al. (1980) cultured fertilized ovules of *Pisum sativum*, with and without pods, three, five and seven days after anthesis, on Liquid White's medium (as modified by Rangaswamy, 1961). The growth of pods was always inhibited in vitro but, from pods fifth day after anthesis, the ovules developed into seeds of normal size. Isolated single ovules, five days after anthesis, did not show any growth. Probably, they required some unknown factor/s from the pod. Seven-day-old ovules were grown successfully on an enriched medium under green light (to simulate the condition in pods), and were comparable to in vivo-grown ovules. With seven-day-old pods, the ovules developed into seeds in 15 days, and germinated in situ.

Eichholtz et al. (1979) reported the formation of adventive embryos in nucellus cultures of apple (*Malus domestica* Borkh var Golden Delicious). Micropylar halves of nucellus formed embryo-like structures 50 days after culture in darkness. Subsequent reculture of these embryos on a fresh medium resulted in the formation of secondary embryo-like structures from the cotyledons.

References

Bajaj YPS (1964) Development of ovules of *Abelmoschus esculentus* var. Pusa Sawani in vitro. Proc Natl Inst Sci 30 B:175–185

Bajaj YPS (1966) Growth of *Hyoscyamus niger* ovaries in culture. Φyton (Argentina) 23:57–62

Beasley CA (1971) In vitro culture of fertilized cotton ovule. Bioscience 21:906–907

Beasley CA (1973) Hormonal regulation of growth in unfertilized cotton ovules. Science 179:1003–1005

Beasley CA (1977) Ovule culture: Fundamental and pragmatic research for the cotton industry. In: Reinert J, Bajaj YPS (eds) Applied and fundamental aspects of plant cell, tissue, and organ culture. Springer, Berlin Heidelberg New York, pp 160–178

Beasley CA, Ting IP (1973) The effects of plant growth substances on in vitro fibre development from fertilized cotton ovules. Am J Bot 60:130–139

Beasley CA, Ting IP (1974a) Effects of plant growth substances on in vitro fibre development from unfertilized cotton ovules. Am J Bot 61:188–194

Beasley CA, Ting IP (1974b) Phytohormone effects on in vitro cotton seed development. In: Plant growth substances. Hirokawa, Tokyo, pp 917–924

Beasley CA, Ting IP, Linkins AE, Birnbaum EH (1974) Cotton ovule culture: A review of progress and a preview of potential. In: Street HE (ed) Tissue culture and plant science. Academic Press, London New York, pp 169–192

Bitters WP, Murashige T, Rangan TS, Nauer E (1972) Investigation on establishing virus-free citrus plants through tissue culture. In: Pierce WC (ed) Proc 5th Conf Int Organ Citrus Virologists, Univ Florida Press, Gainesville, pp 267–271

Button J, Bornman CH (1971) Development of nucellar plants from unpollinated and unfertilized ovules of the Washington navel orange in vitro. J S Afr Bot 37:127–134

Button J, Botha CEJ (1975) Enzymic maceration of *Citrus* callus and the regeneration of plants from single cells. J Exp Bot 26:723–729

Button J, Kochba J, Bornman CH (1974) Fine structure of and embryoid development from embryogenic callus of "Shamouti" orange (*Citrus sinensis* Osb.). J Exp Bot 25:446–458

Chopra RN (1958) In vitro culture of ovaries of *Althaea rosea* Cav. In: Maheshwari P (ed) Proc Seminar Mod Dev Plant Physiol, Univ Delhi, Delhi, pp 87–89

Chopra RN (1962) Effect of some growth substances and calyx on fruit and seed development of *Althaea rosea* Cav. In: Plant embryology – A symposium. Council of Scientific and Industrial Research, New Delhi, pp 170–181

Chopra RN, Sabharwal PS (1963) In vitro culture of ovules of *Gynandropsis gynandra* (L.) Briq. and *Impatiens balsamina* L. In: Maheshwari P, Rangaswamy NS (eds) Plant tissue and organ culture – A symposium. Int Soc Plant Morphol, Univ Delhi, Delhi, pp 257–264

Dulieu HL (1963) Sur la fécondation in vitro chez le *Nicotiana tabacum* L. CR Acad Sci Paris 256:3344–3346

Dulieu HL (1966) Pollination of excised ovaries and culture of ovules of *Nicotiana tabacum* L. Phytomorphology 16:69–75

Eichholtz D, Robitaille HA, Hasegawa PM (1979) Adventive embryology in apple. Hort Sci 14:699–700

Eid AAH, DeLanghe E, Waterkeyn L (1973) In vitro culture of fertilized cotton ovules. The growth of cotton embryos. Cellule 69:361–371

Frost HB, Soost RK (1968) Seed reproduction: Development of gametes and embryos. In: Reuther W, Batchelor LD, Webber HJ (eds) The citrus industry, vol II. Univ California, Berkeley, pp 290–324

Gadwal VR, Joshi AB, Iyer RD (1968) Inter-specific hybrids in *Abelmoschus* through ovule and embryo culture. Indian J Genet Plant Breed 28:269–274

Guha S, Johri BM (1966) In vitro development of ovary and ovule of *Allium cepa* L. Phytomorphology 16:353–364

Haccius B, Hausner G (1976) Ein transplantierbarer embryogener Callus aus Nucellusgewebe von *Cynanchum vincetoxicum* und die Rolle globulärer Vorstadien in der Entwicklungsgeschichte nicht-zygotischer Embryonen. Protoplasma 90:265–282

Inomata N (1968) In vitro culture of ovaries of *Brassica* hybrids between 2X and 4X. I. Culture medium. Jpn J Breed 18:139–148

Ito I (1961) In vitro culture of ovary and seed in orchids. PhD Thesis, Kyoto Prefect Univ, Kyoto

Ito I (1966) In vitro culture of ovary in orchids (1). Effects of sugar, peptone, and coconut milk upon the growth of ovary of *Dendrobium nobile*. Sci Rep Kyoto Prefect Univ Agric 18:38–50

Jansen LL, Bonner J (1949) Development of fruits from excised flowers in sterile culture. Abstr Am J Bot 36:826

Johri BM, Sehgal CB (1963) Chemical induction of polyembyony in *Anethum graveolens* L. Naturwissenschaften 50:47–48

Johri BM, Sehgal CB (1966) Growth response of ovaries of *Anethum, Foeniculum*, and *Trachyspermum*. Phytomorphology 16:364–378

Joshi PC (1962) In vitro growth of cotton ovules. In: Plant embryology – A symposium. Council of Scientific and Industrial Research, New Delhi, pp 199–204

Joshi PC, Johri BM (1972) In vitro growth of ovules of *Gossypium hirsutum*. Phytomorphology 22:195–209

Joshi PC, Pundir NS (1966) Growth of ovules in the cross *Gossypium arboreum* × *G. hirsutum* in vivo and in vitro. Indian Cotton J 20:23–29

Kameya T, Hinata K (1970) Test-tube fertilization of excised ovules in *Brassica*. Jpn J Breed 20:253–260

Kameya T, Hinata K, Mizushima U (1966) Fertilization in vitro of excised ovules treated with $CaCl_2$ in *Brassica oleracea* L. Proc Jap Acad 42:165–167

Kanta K, Maheshwari P (1963) Test-tube fertilization in some angiosperms. Phytomorphology 13:230–237

Kanta K, Rangaswamy NS, Maheshwari P (1962) Test-tube fertilization in a flowering plant. Nature (London) 194:1214–1217

Kapoor M (1959) Influence of growth substances on the ovules of *Zephyranthes*. Phytomorphology 9:313–315

Kochba J, Button J (1974) The stimulation of embryogenesis and embryoid development in habituated ovular callus from the "Shamouti" orange *(Citrus sinensis)* as affected by tissue age and sucrose concentration. Z Pflanzenphysiol 73:415–421

Kochba J, Spiegel-Roy P (1973) Effect of culture media on embryoid formation from ovular callus of "Shamouti" orange *(Citrus sinensis)*. Z Pflanzenphysiol 69:156–162

Kochba J, Spiegel-Roy P, Safran H (1972) Adventive plants from ovules and nucelli in *Citrus*. Planta 106:237–245

Kochba J, Button J, Spiegel-Roy P, Bornman CH, Kochba M (1974) Stimulation of rooting of *Citrus* embryoids by gibberellic acid and adenine sulphate. Ann Bot 38:795–802

La Croix LJ, Naylor JM, Larter EN (1962) Factors controlling embryo growth and development in barley (*Hordeum vulgare* L.). Can J Bot 40:1515–1523

Laibach F (1925) Das Taubwerden von Bastardsamen und die künstliche Aufzucht früh absterbender Bastardembryonen. Z Bot 17:417–459

LaRue CD (1942) The rooting of flowers in sterile cultures. Bull Torrey Bot Club 69:332–341

Maheshwari N (1958) In vitro culture of excised ovules of *Papaver somniferum*. Science 127:342

Maheshwari N, Lal M (1961 a) In vitro culture of ovaries of *Iberis amara* L. Phytomorphology 11:17–23

Maheshwari N, Lal M (1961 b) In vitro culture of excised ovules of *Papaver somniferum* L. Phytomorphology 11:307–314

Maheshwari P, Kanta K (1964) Control of fertilization. In: Linskens HF (ed) Pollen physiology and fertilization. North-Holland Publ, Amsterdam, pp 187–193

Maheshwari P, Rangaswamy NS (1965) Embryology in relation to physiology and genetics. Adv Bot Res 2:219–321

Majumdar SK (1970) Production of plantlets from the ovary wall of *Haworthia turgida* var. *pallidifolia*. Planta 90:212–214

Mauney JR (1961) The culture in vitro of immature cotton embryos. Bot Gaz 122:205–209

Mauney JR, Chappel J, Ward BJ (1967) Effects of malic acid salts on growth of young cotton embryos in vitro. Bot Gaz 128:198–200

Melnick VM, Holm L, Struckmeyer BE (1964) Physiological studies on fruit development by means of ovule transplantation in vivo. Science 145:609–611

Mitra GC, Chaturvedi HC (1972) Embryoids and complete plants from unpollinated ovaries and from ovules of in vivo-grown emasculated flower buds of *Citrus* spp. Bull Torrey Bot Club 99:184–189

Mullins MG, Srinivasan C (1976) Somatic embryos and plantlets from an ancient clone of the grapevine (cv. Cabernet-Sauvignon) by apomixis in vitro. J Exp Bot 27:1022–1030

Murashige T (1974) Plant propagation through tissue culture. A Rev Plant Physiol 25:135–165

Nakajima T, Doyama Y, Matsumoto H (1969) In vitro culture of excised ovules of white clover, *Trifolium repens* L. Jpn J Breed 19:373–379

Nishii T, Mitsuoka S (1969) Occurrence of various ploidy plants from anthers and ovary culture of rice plant. Jpn J Genet 44:341–346

Nitsch JP (1949) Culture of fruits in vitro. Science 110:499

Nitsch JP (1951) Growth and development in vitro of excised ovaries. Am J Bot 38:566–577

Nitzsche W, Hennig L (1976) Fruchtknotenkultur bei Gräsern. Z Pflanzenzücht 77:80–82

Poddubnaya-Arnoldi VA (1959) Study of fertilization and embryogenesis in certain angiosperms using living material. Am Nat 93:161–169

Poddubnaya-Arnoldi VA (1960) Study of fertilization in the living material of some angiosperms. Phytomorphology 10:185–198

Pontovich VE, Sveshnikova IN (1966) Formation of *Papaver somniferum* L. embryos during cultivation of the ovules in vitro. Fiziol Rast Akad Nauk SSSR 13:105–113

Puri P (1963) Growth in vitro of parthogenetic embryos of *Aerva tomentosa* Forsk. In: Maheshwari P, Rangaswamy NS (eds) Plant tissue and organ culture – A symposium. Int Soc Plant Morphol Univ Delhi, Delhi, pp 281–291

Rangan TS, Rangaswamy NS (1968) Morphogenic investigations on parasitic angiosperms. I. *Cistanche tubulosa* Wight (family Orobanchaceae). Can J Bot 46:263–266

Rangan TS, Murashige T, Bitters WP (1968) In vitro initiation of nucellar embryos in monoembryonic *Citrus*. Hort Sci 3:226–227

Rangan TS, Murashige T, Bitters WP (1969) In vitro studies on zygotic and nucellar embryogenesis in *Citrus*. In: Chapman HD (ed) Proc 1st Int Citrus Symp, vol I. Univ California, Riverside USA, pp 225–229

Rangaswamy NS (1958) Culture of nucellar tissue of *Citrus* in vitro. Experientia 14:111–112

Rangaswamy NS (1961) Experimental studies on female reproductive structures of *Citrus microcarpa* Bunge. Phytomorphology 11:109–127

Rangaswamy NS (1963) Studies on culturing seeds of *Orobanche aegyptiaca* Pers. In: Maheshwari P, Rangaswamy NS (eds) Plant tissue and organ culture – A symposium. Int Soc Plant Morphol Univ Delhi, Delhi, pp 345–354

Rangaswamy NS, Shivanna KR (1967) Induction of gamete compatibility and seed formation in axenic cultures of diploid self-incompatible species of *Petunia*. Nature (London) 216:937–939

Rangaswamy NS, Shivanna KR (1975) Nucellus culture of two cucurbits – *Luffa cylindrica* and *Trichosanthes anguina*. Ann Bot 39:193–196

Rao PS (1965) The in vitro fertilization and seed formation in *Nicotiana rustica* L. Фyton (Argentina) 22:165–167

Rau MA (1956) Studies in the growth in vitro of excised ovaries – I. Influence of colchicine on the embryo and endosperm in *Phlox drummondii* Hock. Phytomorphology 6:90–96

Rédei G, Rédei G (1955) Rearing wheats from ovaries cultured in vitro. Acta Bot Acad Sci Hung 2:183–186

Ryczkowski M (1962a) Changes in the osmotic value of the central vacuolar sap in developing ovules (Dicotyledonous perennial plants). Bull Acad Pol Sci 10:371–374

Ryczkowski M (1962b) Changes in the osmotic value of the sap from embryos, the central vacuole and the cellular endosperm during development of the ovules. Bull Acad Pol Sci 10:375–380

Sabharwal PS (1963) In vitro culture of ovules, nucelli, and embryos of *Citrus reticulata* Blanco var. Nagpuri. In: Maheshwari P, Rangaswamy NS (eds) Plant tissue and organ culture – A symposium. Int Soc Plant Morphol Univ Delhi, Delhi, pp 265–274

Sachar RC, Baldev B (1958) In vitro growth of ovaries of *Linaria maroccana* Hook. Curr Sci 27:104–105

Sachar RC, Guha S (1962) In vitro growth of achenes of *Ranunculus sceleratus* L. In: Plant embryology – A symposium. Council of Scientific and Industrial Research, New Delhi, pp 244–253

Sachar RC, Iyer RD (1959) Effect of auxin, kinetin, and gibberellin on the placental tissue of *Opuntia dillenii* Haw. cultured in vitro. Phytomorphology 9:1–3

Sachar RC, Kanta K (1958) Influence of growth substances on artificially cultured ovaries of *Tropaeolum majus* L. Phytomorphology 8:202–218

Sachar RC, Kapoor M (1959) In vitro culture of ovules of *Zephyranthes*. Phytomorphology 9:147–156

Sehgal CB (1972a) In vitro induction of polyembryony in *Ammi majus* L. Curr Sci 41:263–264

Sehgal CB (1972b) Experimental induction of zygotic multiple embryos in *Coriandrum sativum* L. Indian J Exp Biol 10:457–459

Shivanna KR (1965) In vitro fertilization and seed formation in *Petunia violacea* Lindl. Phytomorphology 15:183–185

Siddiqui SA (1964) In vitro culture of ovules of *Nicotiana tabacum* L. var. N.P. 31. Naturwissenschaften 51:517

Sladky Z (1974) Organogenese v kulture perispermu *Agrostemma githago* L. In: Univerzita Komfnskeho v Bratislave Kolokvium o rastlinnych. Explantatovych kulturach stupy pri pezinku, pp 11–17

Spiegel-Roy P, Kochba J (1973) Stimulation of differentiation in orange *(Citrus sinensis)* ovular callus in relation to irradiation of the media. Rad Bot 13:97–104

Spoerl E (1948) Amino acids as sources of nitrogen for orchid embryos. Am J Bot 35:88–95

Srivastava PS, Varga A, Bruinsma J (1980) Growth in vitro of fertilized ovules of pea, *Pisum sativum* L., with and without pods. Z. Pflanzenphysiol 98:347–354

Staden J van (1974) Evidence for the presence of cytokinins in malt and yeast extracts. Physiol Plant 30:182–184

Tukey HB (1938) Growth patterns of plants developed from immature embryos in artificial culture. Bot Gaz 99:630–665

Uchimiya H, Kameya T, Takahashi N (1971) In vitro culture of unfertilized ovules in *Solanum melongena* and ovaries in *Zea mays*. Jpn J Breed 21:247–250

Vacin EF, Went FW (1949) Use of tomato juice in the asymbiotic germination of orchid seeds. Bot Gaz 111:175–183

Valmayor HL, Sagawa Y (1967) Ovule culture in some orchids. Bull Am Orchid Soc 36:766–769

Wakizuka T, Nakajima T (1974) Effect of cultural condition on the in vitro development of ovules in *Petunia hybrida* Vilm. Jpn J Breed 24:182–187

Wakizuka T, Nakajima T (1975) Development of proembryo in cultured ovules of *Petunia hybrida* Vilm. Jpn J Breed 25:161–167

Weaver JB (1958) Embryological studies following interspecific crosses in *Gossypium*. II. *G. arboreum* × *G. hirsutum*. Am J Bot 45:10–16

White PR (1932) Plant tissue culture: A preliminary report of results obtained in the culturing of certain plant meristems. Arch Exp Zellforsch 12:602–620

Withner CL (1942) Nutrition experiments with orchid seedlings. Bull Am Orchid Soc 1:112–114

Withner CL (1943) Ovule culture: A new method for starting orchid seedlings. Bull Am Orchid Soc 11:261–263

Withner CL (1959) The orchids. A scientific survey. Ronald Press, New York

Zatyko JM, Simon I, Szabo CS (1975) Induction of polyembryony in cultivated ovules of red currant. Plant Sci Lett 4:281–283

Zenkteler M (1965) Test-tube fertilization in *Dianthus caryophyllus* Linn. Naturwissenschaften 23:645–646

Zenkteler M (1967) Test-tube fertilization in ovules in *Melandrium album* Mill. with pollen grains of several species of the Caryophyllaceae family. Experientia 23:775

Zenkteler M (1970) Test-tube fertilization of ovules in *Melandrium album* Mill. with pollen grains of *Datura stramonium*. Experientia 26:661–662

7. Pollen-Pistil Interaction and Control of Fertilization

K.R. Shivanna

Fertilization in flowering plants is a unique phenomenon and involves two fusion processes – syngamy between one of the male gametes and egg, and triple fusion between the other male gamete and fusion nucleus. Successful fertilization leads to fruit and seed development. The processes leading to fertilization are initiated on the stigma. Soon after pollination the crucial decision of recognition followed by subsequent acceptance or rejection of the male partner is made by the pistil. A chain of sequential, integrated processes in the pistil following pollen recognition regulate the post-pollination behaviour of the pollen. If the pollination is compatible, the pollen grain germinates on the stigma, and the pollen tube grows through the style and eventually reaches the embryo sac where it discharges the male gametes. If the pollination is incompatible, the pistil will effectively prevent fertilization, by inhibiting either pollen germination, entry of the pollen tube into the stigma, or growth of pollen tube in the style. Pollen-stigma interaction, therefore, is of paramount importance in the biology of sexual reproduction, because the vital function of selection of the male gametes in flowering plants is performed not by the egg, but by the pistillate tissue (see Shivanna 1979). For a rational understanding of the pollen-pistil interaction and fertilization it is important to understand the structure of the pollen grain and the pistil in relation to their function of recognition and rejection.

Although the basic events of pollen-pistil interaction leading to fertilization became known by the end of the nineteenth century, studies remained basically descriptive until 1950. Following the realization of the significance of plant hybridization programme in crop improvement, limitations of these traditional approaches to studies on fertilization became apparent. This realization, coupled with the general trend of plant sciences to become experimental, resulted in the interaction of histochemical, ultrastructural, and biochemical techniques to unravel the mysteries of fertilization. Significant progress has been made on these lines during the last two decades. This chapter deals largely with these latter developments, on the processes from pollination to fertilization. Sexual incompatibility (both intra- and interspecific), and methods of overcoming incompatibility are also discussed in greater detail. For earlier literature the reader is referred to Maheshwari (1950), Rangaswamy (1964), Maheshwari and Rangaswamy (1965), Lewis (1954), Linskens and Kroh (1967), Arasu (1968a), and Nettancourt (1977).

The Pollen Grain

Pollen grains are surrounded by a two-layered wall, the exine composed of sporopollenin and the pectocellulosic intine. The exine is a more elaborate structure dif-

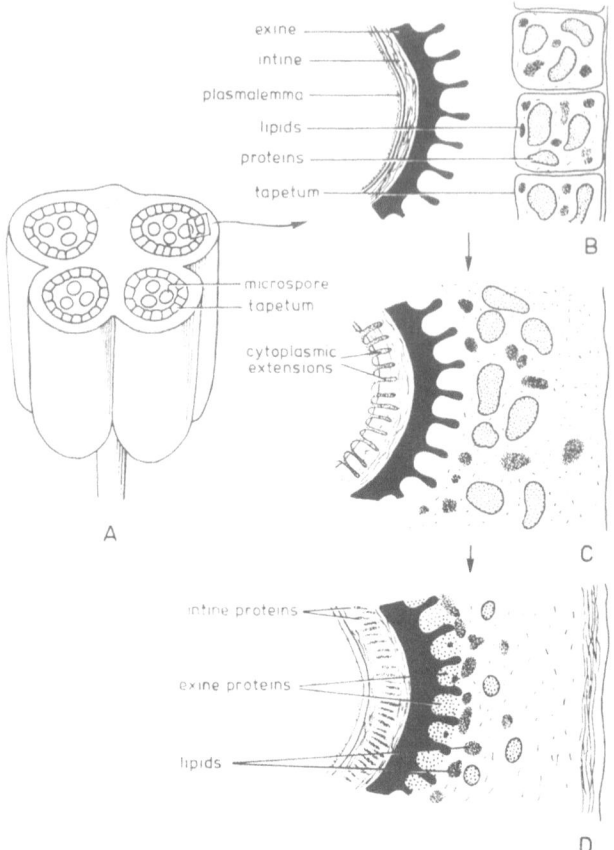

Fig. 1 A–D. Origin of the intine and exine proteins. **A** Anther cut transversely to show young microspores surrounded by a layer of tapetum. Only a part of the microspore wall and tapetum are shown in **B–D.** Note the origin of intine proteins in the pollen cytoplasm, and of exine proteins in the tapetum. (After Shivanna 1977)

ferentiated into an outer sculptured layer (the sexine), and an inner non-sculptured layer (the nexine). The sculptured layer is made up of radially directed rods, the baculae, which may be enclosed above to form a tectum (tectate grain), or stand free or joint together to form various patterns non-tectate grain). In tectate grain the tectum is invariably perforated by micropores.

Most of the classic studies were made with acetolyzed pollen and concerned with the morphology of exine as a means of identification and phylogeny. Recent progress in our understanding on the structure of pollen grain comes from the studies of the unacetolyzed pollen wall. Both wall layers of the pollen grain, the exine and intine, have now been demonstrated to contain a large amount of mobile proteins (Knox and Heslop-Harrison 1969, 1970, 1971 a; Heslop-Harrison et al. 1973, 1975 a). These include many hydrolytic enzymes and proteins responsible for pollen allergy (Knox et al. 1970). The intine proteins are present in the form of tubules or leaflets, and are generally concentrated near the germpores. In the exine the proteins are located in the sculptured region of the exine in the chambers between the baculae (tectate grains) and surface depressions (non-tectate grains). These wall proteins are readily released into the medium on moistening.

The origin of these pollen wall proteins is diagrammatically represented in Fig. 1. The intine proteins are produced in the pollen cytoplasm and incorporated into the intine during its development (see Heslop-Harrison 1975a). As the deposition of intine progresses soon after the release of microspores from the tetrad, the plasmalemma of pollen cytoplasm sends out radially oriented tubules into the developing intine. Eventually these tubules, with their protein inclusions, are cut off from the plasmalemma and sealed off from the cell surface by the deposition of a layer of intine free from tubules. In aperturate pollen the intine-proteins are principally concentrated in the region of germpore (Heslop-Harrison et al. 1973); in non-aperture monocotyledonous pollen they are distributed throughout the intine (Knox 1971; Heslop-Harrison 1975a). Similar tubules in the intine, apparently derived from the plasmalemma, have been described in ultrastructural studies of many other species (see Heslop-Harrison 1975a). In *Cosmos bipinnatus* the incorporation of intine proteins is rather unique. Instead of tubular evaginations, leaflets of plasmalemma with their protein inclusions get separated from the cytoplasm and become incorporated in the intine in the form of a series of tangentially oriented lamellae (Knox and Heslop-Harrison 1970; Heslop-Harrison 1975a).

The exine proteins, on the other hand, originate in the cells of the surrounding tapetum, a sporophytic tissue (Heslop-Harrison et al. 1974). During meiosis of microspore mother cells, proteins, and lipids accumulate in the tapetal cells. When the tapetal wall breaks down towards the end of pollen development, proteins and lipids are released into the thecal cavity and, eventually, become deposited in the surface depressions of exine. In tectate grains the protein fraction passes through the micropores of the tectum and accumulates in the spaces between the baculae; the lipids remain on the surface of tectum (Heslop-Harrison et al. 1973).

Thus, the intine proteins are the products of pollen cytoplasm, the male gametophyte; and the exine proteins the products of tapetum, the sporophytic tissue. The significance of this differential origin of the exine and intine proteins in controlling the breeding behaviour of the species is discussed later.

The Pistil

Traditionally, two basic types of stigma have been recognised: the wet type in which the receptive surface becomes covered (to various degrees) with the exudate; and the dry type in which the receptive surface is free from any apparent exudate.

Recently, the variations in these two major categories of stigma have been divided into further groups (J. Heslop-Harrison 1976; J. Heslop-Harrison et al. 1975b). Y. Heslop-Harrison and Shivanna (1977) studied and classified the stigmas of about 1,000 species covering 900 genera belonging to over 250 families (Table 1).

The details of the origin and composition of the exudate have been studied only in a few systems. In *Petunia* the exudate is secreted by the cells of stigmatic tissue,

Table 1. General classification of angiosperm stigma types based on the morphology of the receptive surface, and the amount of secretion present during receptive period. (After J. Heslop-Harrison 1976; Y. Heslop-Harrison and Shivanna 1977). Some examples for each group are given in parenthesis

---

Dry stigmas (without apparent fluid secretions)

Group I   – Plumose, with receptive cells dispersed on multiseriate branches (Gramineae)
Group II – Receptive cells concentrated in distinct ridges, zones or heads
     A – Surface non-papillate (Acanthaceae)
     B – Surface distinctly papillate
          1. Papillae unicellular (Cruciferae, Compositae)
          11. Papillae multicellular
              (a) Papillae uniseriate (Amaranthaceae)
              (b) Papillae multiseriate (Bromeliaceae, Oxalidaceae)

Wet stigmas (surface secretions present during receptive period)

Group III – Receptive surface with low to medium papillae; secretion fluid flooding interstices (some Rosaceae, some Liliaceae)
Group IV – Receptive surface non-papillate; cells often necrotic at maturity; usually with more surface fluid than Group III (Umbelliferae)

---

and accumulates on the stigma surface by rupture of the cuticle (Konar and Linskens 1966a). In *Lilium*, on the other hand, the cells of the stigma are non-secretory; the exudate is presumably secreted by the canal cells of the hollow style and extruded on to the stigmatic surface through the stylar canal (Dashek et al. 1971; Labarca and Loewus 1973). The composition of the exudate is highly variable, and contains varying proportions of lipids, carbohydrates, phenolic compounds, and proteins (Martin 1969; Konar and Linskens 1966b; Labarca et al. 1970, Dumas 1974; Rosen 1971; Heslop-Harrison 1975b; Shivanna and Sastri 1976). In *Petunia* the exudate is largely lipoidal, and has no nutritive role for the germinating pollen. In *Lilium* the exudate is aqueous consisting largely of carbohydrates and a small amount of proteins (Kroh et al. 1970; Rosenfield and Loewus 1975). Arabinose is one of the major carbohydrate of the exudate and forms about 28% of the dry weight. When the exudate is added to the pollen culture medium, it stimulates pollen germination and tube growth, and is taken up by the pollen tubes.

The lipoidal substances of the exudate help in trapping the pollen grains and to protect the stigma from desiccation and wetting. The phenolic compounds are presumed to be helpful in protecting the stigma from microbes and pests. Recently, there have been evidences to implicate phenolic compounds in pollen nutrition, and in selective promotion, or inhibition, of pollen grains on the stigma (Martin 1970, 1972; Martin and Ruberte 1972; Tara and Namboodiri 1976; Sedgley 1975).

The dry stigmas are free from any apparent secretory products, and are generally covered by an unruptured cuticle of varying thickness. Recently, the presence of a hydrated proteinaceous covering (together with lipoidal fraction), termed pellicle, has been shown to be invariably present on the surface of cuticle (Mattsson et al. 1974; Y. Heslop-Harrison and Shivanna 1977). The pellicle can be easily localized histochemically by its intense non-specific esterase activity (Fig. 2 A–D),

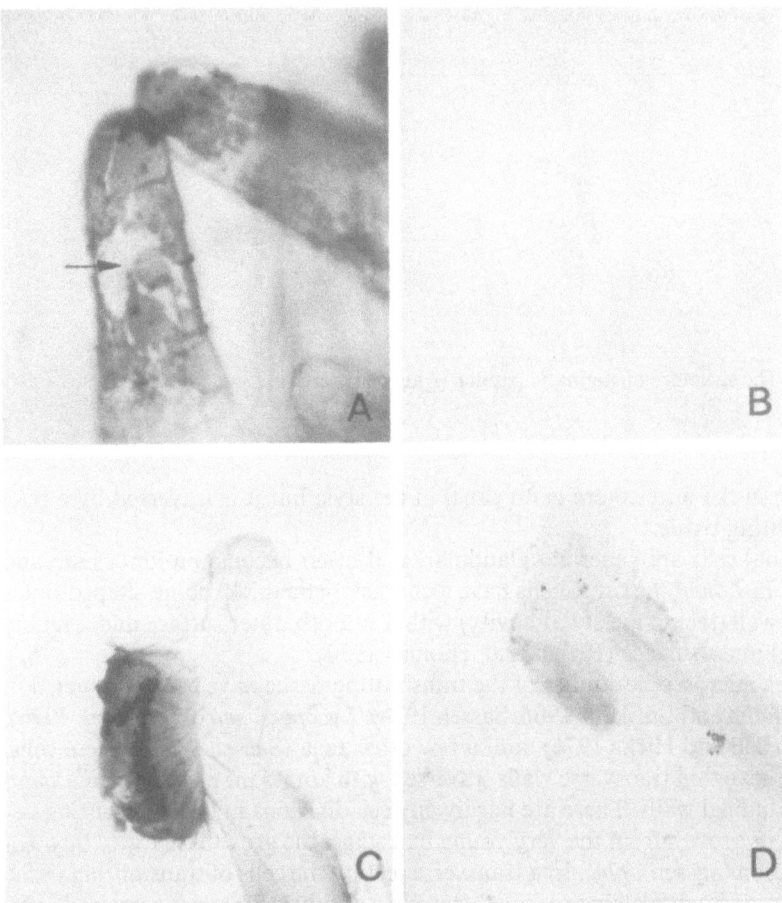

Fig. 2 A–D. Histochemical localization of stigma-surface esterases using α-napthylacetate as a substrate in a coupling reaction with fast blue B salt. **A, B** Stigmatic papillae of *Acidanthera bicolor* (dry stigma, group II B). **A** with substrate. The pellicle is seen as a dark sheath investing the cuticle, but is torn at places *arrow* and reveals the underlaying cuticle. **B** Control for **A** (without the substrate). **C, D** Stigma of *Vigna unguiculata* (wet stigma, group III). Esterase activity is clearly seen on the exudate in **C**. **D** Control for **C** (without the substrate). (After Shivanna 1979)

and sensitivity to pronase digestion. Although the details of the origin of pellicle are yet to be investigated, the cuticle has been shown to have discontinuities through which the pellicle is presumed to be extruded on to the surface of papillae (Fig. 3). The "dry" stigma is, therefore, not really dry as it was thought earlier, but covered with the pellicle which is physiologically comparable to the exudate of the wet stigma. The surface proteins, either as a component of the exudate or the pellicle, have an important role during pollen-stigma interaction (discussed later).

The styles are of two main types – hollow (open) and solid (closed). In the former the style is traversed by a stylar canal and lined with a layer of glandular cells,

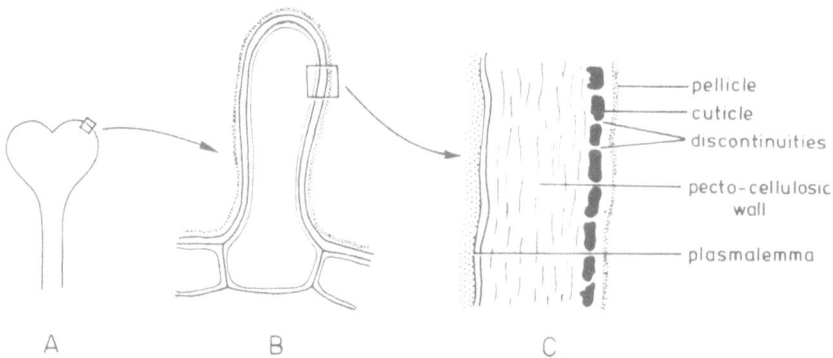

Fig. 3 A–C. The structure of stigmatic papilla. (After Shivanna 1977)

canal cells; in the latter there is no canal in the style but it is traversed by a core of transmitting tissue.

The canal cells are generally glandular, and often become multinucleate and polyploid. In *Lilium* the canal cells have a characteristic thick, dome-shaped outer tangential wall (facing the stylar cavity) with a smooth outer surface and a highly convoluted inner surface (Rosen and Thomas 1970).

Electron microscopic studies of the transmitting tissue have been conducted in *Petunia* (Pluijm and Linskens 1966; Sassen 1974), *Lycopersicum* (Cresti et al. 1976), *Nicotiana* (Bell and Hicks 1976), and a few other taxa (Sassen 1974). These cells, in general, have thin transverse walls traversed with abundant plasmodesmata and thick longitudinal walls. There are hardly any cell divisions in the transmitting tissue during its growth from the very young bud stage; the growth is largely through cell elongation (Sassen 1974). In a transverse section the cells of transmitting tissue are circular, and completely separated from one another. They are surrounded by an intercellular substance of different electron density than the cell wall. The intercellular substance, a secretion product of the transmitting tissue, is not comparable to the middle lamella. It is more complex than the middle lamella, and comparable to the secretion fluid found in the stylar canal (Sassen 1974). Recently, the intercellular substance in *Lycopersicum peruvianum* has been shown, histochemically, to contain proteins (Cresti et al. 1976). According to Sassen (1974), there are no basic differences between the structure of solid and hollow styles. The cells of the transmitting tissue of cotton, however, are not surrounded by any intercellular substance. The lateral walls are 7–10 µm thick, and consist of four distinct layers, including the outermost pectinaceous middle lamella (Jensen and Fischer 1969).

Pollen-Pistil Interaction

The first visible change in the pollen, soon after it lands on the stigma, is hydration. Simultaneously with hydration, pollen wall proteins, first the exine proteins and then the intine proteins, are released on the stigmatic surface (Knox and J. Heslop-

Harrison 1971 a; Knox 1973; J. Heslop-Harrison et al. 1975 b). The outflow of the pollen wall proteins can be readily detected by immunofluorescence, as well as by histochemical techniques. In dry stigma the pellicle is the receptor site for the pollen-wall proteins. Initially, the wall proteins are not bound to the stigma and hence may be easily removed by saline leaching. But soon the wall proteins bind to the pellicle establishing a close interaction, and the pellicle looses its identity as a discrete layer at this point. Recognition of the pollen appears to take place during this interaction, and results in the activation of the male gametophyte and papilla (Heslop-Harrison 1975 a).

If the pollen grain is compatible, the pollen tube soon emerges and comes in contact with the cuticle of the stigmatic papillae. The cuticle is eroded at the point of contact; the pollen tube enters the inner pecto-cellulosic wall and grows down the papillae. The events from pollination to pollen tube entry into the papillae are rapid and completed in less than 30 min in many taxa (J. Heslop-Harrison et al. 1975 b).

In cotton, which is also characterised by dry stigma with a covering of cuticle, the pollen tubes have been reported to grow down the papilla on the surface of the cuticle without entering the cuticle (Jensen and Fischer 1969). However, recent studies have clearly shown that in cotton as well as in many other related taxa, pollen tubes invariably enter the cuticle, and grow through the wall of papillae (Shivanna and Y. Heslon-Harrison, unpublished).

Pollen Tube Growth

In taxa characterised by wet stigma the cuticle is generally ruptured during the deposition of exudate, and the pollen tubes enter into stigmatic tissue through pecto-cellulosic layer. In taxa characterized by a hollow style the stigmatic surface is in direct contact with stylar canal. In *Lilium* the cuticle of stylar canal is ruptured during the secretion of exudate and, hence, the pollen tubes grow on the surface of canal cells bathed in exudate. In *Crocus*, in which cuticle remains intact on the papillae, as well as on the cells of stylar canal, pollen tubes bore through the cuticle of stigma and grow down the stylar canal between cuticle and canal cells (Y. Heslop-Harrison 1975 a).

In *Petunia* (Pluijm and Linskens 1966), and other taxa characterized by solid styles, pollen tubes grow through intercellular substance. In cotton it grows through wall layer three and not through the middle lamella (Jensen and Fischer 1969).

The pollen tube utilizes nutrients from the pistil for its growth. In many hollow-styled members such as *Aegle marmelos*, *Fritillaria*, and *Lilium* stylar tissue of the unpollinated pistils contain abundant starch. Following pollination and pollen tube growth the starch is broken down, apparently taken up by the growing pollen tubes (see Vasil 1974). By studying the incorporation of labelled glucose and myoinositol in the detached pistils of *Lilium*, Loewus and his associates (Kroh et al. 1970; Labarca et al. 1970; Labarca and Loewus 1973) have demonstrated that both glucose and myoinositol are taken up by the growing pollen tubes from the pistil and utilized for the synthesis of pollen wall materials. Evidences for the utilization of stylar carbohydrates by the growing pollen tubes have also been obtained in *Nicotiana* (Tupý 1961), *Oenothera* (Kumar and Hecht 1970), and *Petunia* (Kroh and Helsper 1974) from labeled studies.

In cotton there is no evidence to indicate the utilization of stylar carbohydrates (Jensen and Fischer 1969), as no depletion of starch or lipid takes place in the tissues of style following the growth of tube. On the other hand, the growth of pollen tube through the transmitting tissue results in the deposition of callose in the pit fields sealing off the cells adjacent to the pollen tubes.

Pollination initiates many physiological changes in the pistil. It increases respiratory activity, changes patterns of RNA and protein synthesis (Linskens 1975), and initiates marked increase in the activity of several enzymes (Roggen 1967). Some of these physiological changes are initiated even in those regions of pistil in which the pollen tubes have not yet entered. For example, in *Petunia* a wave of enzyme activity precedes the growing pollen tubes (Roggen 1967).

Also, within a few minutes after pollination an electrical signal characteristic of compatible or incompatible pollination can be measured at the base of style (Linskens and Spanjers 1973). In cotton (Jensen and Fischer 1968), and many other taxa, one of the synergids begins to degenerate even before the pollen tube enters the ovule. Thus, some stimulus precedes the growing pollen tube in the pistil, and initiates responses suitable for the normal growth of pollen tube.

Generally, a larger number of pollen tubes enter the ovary than the number of ovules, although only one pollen tube enters each embryo sac. There are a few reports of more than one pollen tube entering the embryo sac. However, in *Triticum durum* (Rudramuniyappa and Panchaksharappa 1974) and *Persea americana* (Sedgley 1976) – both uniovulate systems – only one pollen tube enters the ovary. In *P. americana* though, on an average, over 66 pollen grains germinate on the stigma, most of the tubes cease growth after traversing up to various distances in the style, and only one tube reaches the ovary. In twin embryo sacs generally two pollen tubes reach the ovary. Based on these studies, Sedgley (1976) suggested that the embryo sac has a control over the growth of pollen tubes in the style.

Chemotropism

One of the controversial aspects of progamic phase is the nature of chemotropic factors which guide the pollen tubes from the stigma to the embryo sac. Many investigators have demonstrated, using in vitro experiments, positive chemotropic effect of various parts of the pistil to pollen tubes (Iwanami 1959; Welk et al. 1965; Rosen 1961, 1971). In cultured pollen grains of *Antirrhinum majus* Mascarenhas and Machlis (1962, 1964) demonstrated calcium to be the chemotropic factor. They also found a gradation in the distribution of total calcium from the stigma to ovules in the pistil of *A. majus*. Based on these evidences they suggested that calcium may be the universal chemotropic factor in pistils of angiosperms. However, calcium is chemotropically inactive for pollen tubes of *Lilium* (Rosen 1971) and *Oenothera* (Glenk et al. 1970). Subsequently, Mascarenhas (1966) studied, cytochemically, the distribution of soluble calcium in the pistils of *Antirrhinum majus* and found it to be almost constant throughout the length of style. The micropyle and the embryo sac did not show higher concentrations of calcium. Placenta and ovary wall, on the other hand, showed the highest concentration of calcium. This is contrary to expectations if calcium is to be a chemotropic factor guiding the pollen tube to the embryo sac. Following these studies, Mascarenhas (1966) suggested that, in addition to calcium, some other factors are involved in directing pollen tubes. Glenk et al. (1970) also reported lack of calcium gradient from the stigma to embryo sac in the pistils of *Oenothera*. Kwack (1969) observed that calcium did not play any role in chemotropism in *Clivia* and *Crinum*, although it promoted germination in both taxa. According to Jensen and Fischer (1969), cells of the transmitting tissue in cotton, with their file-like arrangement, provide a path of least mechanical resistance for the growth of pollen tubes and, thus, there is no need for a chemotropic gradient in the style.

Lack of chemotropic gradient in the style was also demonstrated by Iwanami (1959), by a series of elegant experiments in *Lilium*. He demonstrated by stylar grafts that pollen tubes grow down the grafted style irrespective of its orientation. Also, when pollen grains were put in the stylar cavity by making a window, the pollen tubes grew in both directions – towards the pistil as well as the ovary. Ascher (1977) has shown that in *Lilium*, even when the ovarian end of the excised style was pollinated, pollen tubes grew normally towards the stigma.

Recently, Mascarenhas (1975) proposed a hypothesis according to which any growth factor may act as a chemotropic factor by causing a shift in the angle made by the centre of growing tip of the pollen tube with respect to the rest of the tube. Once this change in the direction of the growth takes place, the tube would continue to grow straight without the presence of a concentration gradient until another shift in the direction occurs. According to this hypothesis, a chemotropic factor is necessary in the pistil only for changing the direction of the pollen tube growth, from the placenta towards the micropyle.

Many suggestions have been put forward to explain the nature of the chemotropic factors originating in the ovule itself. Some investigators consider synergids, particularly the filiform apparatus, to be the source of chemotropic factor (Coe 1954). According to Chao (1971) the chemotropic substance in *Paspalum arbiculare* is produced by the dissolution of integumentary cells at the region of micropyle (see also Chao 1977).

Fertilization

The male gametes are released in the synergid through a pore formed at the tip of the pollen tube, or by rupture of the tube tip. Many hypotheses have been proposed to explain the causative factor for the rupture of pollen tube in the embryo sac. The only explanation which has some experimental basis is the one proposed by Stanley and Linskens (1967). According to this hypothesis, the rupture of the pollen tube is caused by low oxygen tension in the embryo sac. Experimental evidence for this concept comes from the studies of Linskens and Schrauwen (1966), and Stanley and Linskens (1967). Linskens and Schrauwen (1966) measured oxygen tension in the stylar canal in *Hippeastrum hybridum*. Unpollinated pistils showed high oxygen tension in the stigma and style, and a sharp drop in the lowermost 5 mm of style

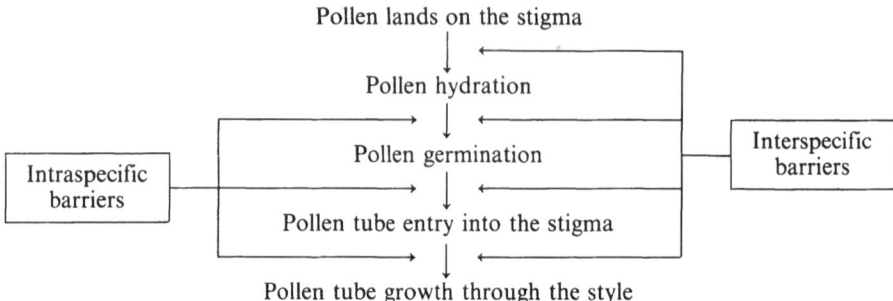

Fig. 4. Major post-pollination events that occur during pollen–pistil interaction. Intraspecific barriers operate at the last three stages, and interspecific barriers operate at all stages

and ovary. In the pollinated pistil there was a drop in oxygen tension which progressed along the growth of pollen tubes. These results have been interpreted as evidences to indicate that pollen tubes grow aerobically up to the base of style, and switch over to anaerobic state soon after entering the ovary. Stanley and Linskens (1967) could induce rupturing of the pollen tube tips of *Lilium* in vitro by decreasing oxygen tension in the medium. This led to the suggestion that, probably, lower oxygen tension in the embryo sac sets up cell wall stress in the pollen tube tip causing it to rupture. However, reducing oxygen tension in the medium did not induce the rupture of pollen tube in cultured pollen grains of *Petunia*. Thus, oxygen tension as the cause for the rupture of pollen tube is not tenable in solid-styled systems.

Apart from the classical studies on the process of double fertilization by Gerassimova (1933) and others, not much progress has been made in recent years on the process of fertilization per se. The only progress that has been made on these lines is the fine structural details of fertilization in a few taxa (see Kapil and Bhatnagar 1975 for details). One of the important features of fertilization that has been observed in many systems is the presence of gaps in the cell wall between the synergid and egg, and between the synergid and central cell. The male gamete enters into the cytoplasm of egg and central cell through these gaps. Eventually, the nuclear membranes of the male gamete and the egg fuse, resulting in nuclear fusion (for details see Jensen 1973).

Incompatibility

The details of the progamic phase discussed so far are those following compatible pollination in which pollen germination and pollen tube growth proceed normally, and result in double fertilization. However, there are many instances in which pollinations are incompatible, i.e. they do not lead to fertilization. This is due to the arrest of post-pollination events at different levels (Fig. 4). Incompatibility occurs between species (interspecific incompatibility), as well as within the species (intraspecific incompatibility or self-incompatibility). Whereas interspecific incompatibility prevents fertilization between gametes of distantly related species, intraspe-

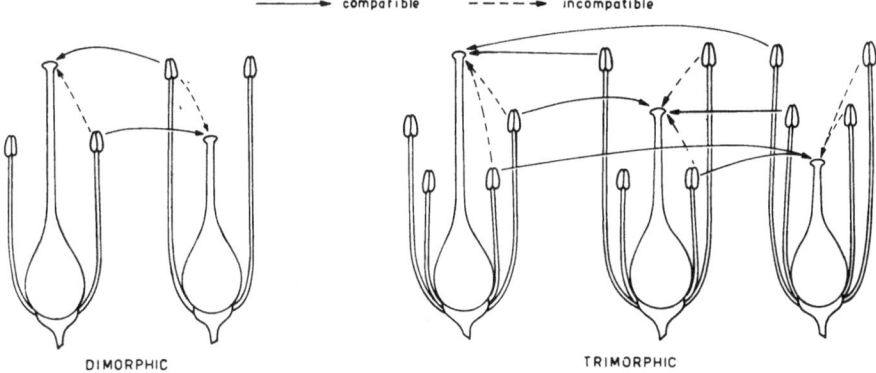

Fig. 5. Heteromorphic intraspecific incompatibility

cific incompatibility prevents fertilization between gametes of the same or other individuals of the same species.

There are two types of intraspecific incompatibility – heteromorphic and homomorphic. In heteromorphic incompatibility different individuals of the species produce either two or three types of flowers differing in the length of stamens and style. Each plant produces only one type of flower. Pollen grains either from the same plant, or any other plant bearing the same type of flower, will be nonfunctional; only the pollen grains from other plant bearing any other type of flower are functional. These are diagrammatically represented in Fig. 5.

In homomorphic type of incompatibility, on the other hand, all individuals of the species produce only one type of flower. Homomorphic incompatibility is governed by multiple alleles termed S alleles. Pollen tube having a particular S allele is inhibited in the style carrying the same S allele. In the majority of taxa incompatibility is determined by multiple alleles at one locus; in grasses incompatibility in controlled by multiple alleles at two independent loci, S and Z. In some members of Ranunculaceae and Chenopodiaceae (Lundquist 1975; Larsen 1977), and Cruciferae (Lewis 1977) incompatibility is controlled by multiple alleles at three or four loci.

Incompatibility reaction in pollen may be controlled by the S allele present in the pollen itself (gametophytic), or by both the S alleles of the parent sporophyte (sporophytic). Determination of the genetics of self-incompatibility is a laborious process involving extensive breeding programme for many generations; hence, the detailed genetical analysis has been carried out only on a few systems. Gametophytic incompatibility is found in members of Solanaceae, Leguminoseae, and Gramineae, and sporophytic incompatibility in members of Compositae, Cruciferae, and Convolvulaceae. In heteromorphic systems also, incompatibility in pollen is determined by the genotype of the parent, and hence is of sporophytic type.

The cytology of the pollen (two- or three-celled at the time of shedding) shows an interesting correlation with the zone of inhibition and the genetics of self-incompatibility. The taxa characterized by two-celled pollen show gametophytic type of incompatibility and the pollen tubes are inhibited in the style. The taxa characterized by the three-celled pollen, on the other hand, show sporophytic type of incompatibility and the pollen tubes are inhibited on the stigma (Brewbaker 1957; Lewis 1956). Although this correlation holds good in a large number of taxa, there are many exceptions. For example, members of Gramineae are characterized by three-celled pollen with gametophytic type of incompatibility. In species of *Oenothera* the incompatible pollen tubes are inhibited in the stigma, but pollen grains are two-celled and incompatibility is of the gametophytic type. In many heterostylic taxa also, this correlation does not hold good. J. Helson-Harrison et al. (1975 b) and Y. Heslop-Harrison and Shivanna (1977) analysed the morphology of pistil and observed that sporophytic type of incompatibility is invariably associated with taxa characterised by the dry type of stigma; whereas the gametophytic type may be associated either with wet or dry type. Intraspecific incompatibility is invariably a prefertilization barrier and is due to active inhibition of pollen tubes.

Studies on cytological details of post-pollination events following interspecific pollination are limited. The arrest of post-pollination events may occur at any level (Fig. 4) depending on the extent of reproductive isolation of the two parents. For example, the stigma of *Gladiolus* (Knox et al. 1976) allows hydration and germination of the pollen of *Crocosmia* (belonging to the same family, Iridaceae), but not the entry of pollen tubes into the papillae. When the stigma of *Gladiolus* was pollinated with pollen of *Gloriosa* (belonging to a different family, Liliaceae) pollen hydration itself is inhibited.

From the above discussions it is apparent that the pistil has mechanisms to recognize compatible pollen from incompatible pollen, and to reject effectively incompatible pollen. In recent years significant progress has been made in understanding the details of recognition and rejection.

Factors Involved in Recognition

Pollen-Wall Proteins

As described earlier, pollen-wall proteins have dual origin; the intine proteins are gametophytic and exine proteins sporophytic. This demonstration led to the suggestion that the exine proteins (sporophytic) are involved in sporophytic type of incompatibility and the intine proteins are involved in the gametophytic type of incompatibility (J. Heslop-Harrison et al. 1973). Subsequent experimental investigations have produced strong evidences to support this concept.

Upon pollination, whether compatible or incompatible, pollen-wall proteins are released on to the stigma and come in contact with the stigma-surface proteins. If compatible, pollen grains germinate and the tubes enter the stigma, and grow through the style. If incompatible, the pistil will initiate rejection reaction. The details of the rejection reaction vary from species to species. In *Raphanus* (Dickinson and Lewis 1973 a, b), *Iberis* (J. Heslop-Harrison et al. 1974), and *Cosmos* (Howlett et al. 1975) – all characterised by sporophytic type of incompatibility – the incompatible pollen either fails to germinate or produces a short tube which will, at the most, erode the cuticle of the papilla but fails to make any growth in the papilla. Generally, pollen tube tip gets plugged with callose. Thus, the cuticle is not the ef-

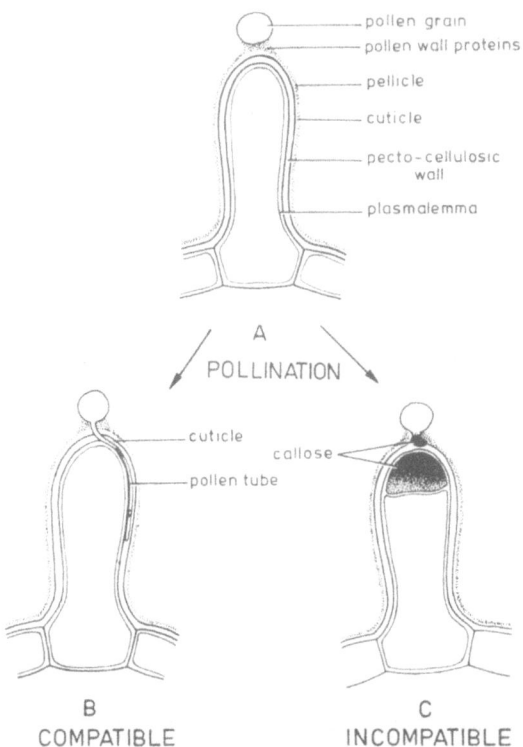

Fig. 6 A–C. Pollen-stigma interaction in Cruciferae, and Compositae. Pollen-wall proteins are released on to the pellicle **A** where recognition takes place. The compatible pollen tube penetrates the cuticle and grows down the papilla. The incompatible pollen tube, although penetrates the cuticle, is inhibited from further growth. Note the deposition of callose in the stigmatic papilla below the incompatible pollen. (After Shivanna 1977)

fective barrier in preventing pollen tube entry, as many of the incompatible tubes do penetrate the cuticle. Interestingly, the stigmatic papilla also shows characteristic rejection reaction. Just at the region of contact of pollen grain, the papilla shows deposition of callose between the plasmalemma and pectocellulosic layer (Fig. 6). This callose deposition is specific to incompatible pollen, and develops within 3–6 h after pollination (Fig. 7 A, B). In *Cosmos bipinnatus* the callose plug can be seen even 15 min after pollination (Howlett et al. 1975).

The characteristic rejection reaction of the stigma has been used as a bioassay for identifying the components of pollen involved in incompatibility. In species of *Iberis*, J. Heslop-Harrison et al. (1974) successfully induced rejection reaction on the stigma by deposition of an agarose film into which exine proteins were allowed to diffuse. More importantly, even the isolated fragments of the tapetum (before releasing their protein loads into the thecal cavity) have been shown to be effective in inducing characteristic rejection reaction on the stigma. The efficacy of exine-borne material in inducing rejection reaction has also been demonstrated in *Raphanus* (Dickinson and Lewis 1973 b). These elegant experiments provide direct evidence for involving exine proteins in recognition and rejection reactions. As the exine proteins are the products of sporophytic tissue, the tapetum, the genetic basis of sporophytic incompatibility is satisfactorily explained (see J. Heslop-Harrison 1975 b).

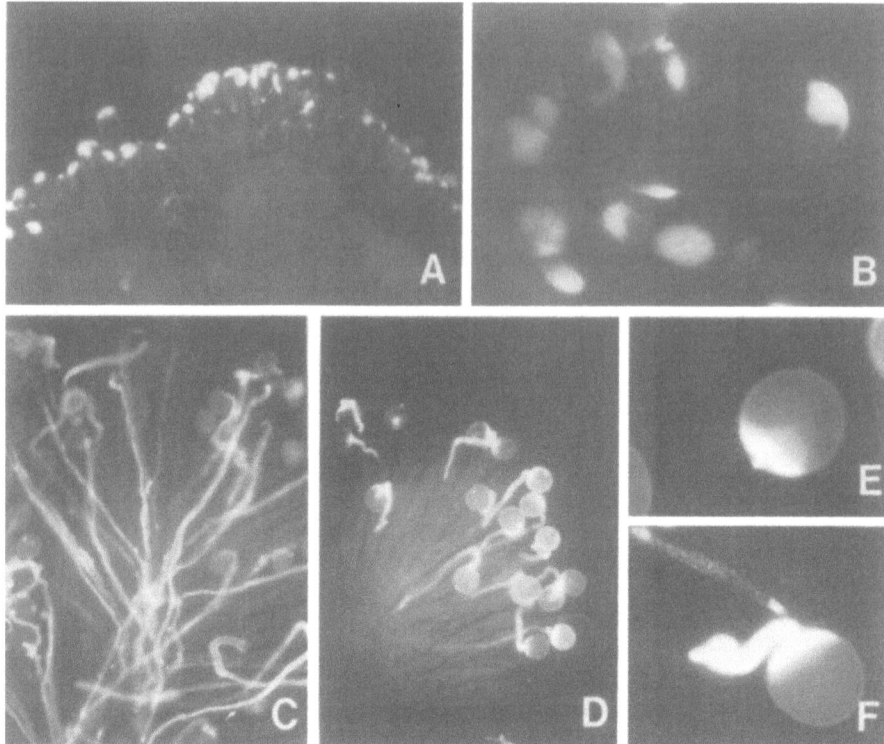

Fig. 7 A–F. Fluorescence micrographs of stigma 6 h after pollination following staining with decolourised aniline blue. **A, B** *Brassica campestris* portions of stigma 6 h after self-pollination. Observe characteristic rejection reaction at the tips of papillae. A few papillae are shown at higher magnification in **B**. **C–F** *Saccharum bengalensis.* **C** Part of the compatibly pollinated stigma; observe profuse growth of pollen tubes. **D** Part of the incompatibly pollinated stigma. Pollen tubes are inhibited after growing a short distance in the stigma. **E, F** Incompatible pollen grains. **E** Germ pore itself is blocked with callose inhibiting germination. **F** Pollen tube is arrested before entering the papilla. (After Shivanna 1979)

In a majority of taxa having gametophytic incompatibility, the pollen tubes are not inhibited on the stigma, but in the style. It is, therefore, difficult to carry out comparable experiments on gametophytic systems showing pollen tube inhibition on the stigma. However, grasses are the well-known exception with gametophytic system showing pollen tube inhibition in the stigma. This makes them suitable for conducting experiments to test the role of pollen-wall proteins in incompatibility. Investigations have been carried out on species of *Gaudinia, Saccharum,* and a few other graminaceous taxa (J. Heslop-Harrison 1975c; Shivanna et al. 1976; see also Shivanna 1977, 1979). In grasses there is no callose deposition on the stigmatic papillae subsequent to incompatible pollination. However, the callose deposition is very conspicuous in pollen tube (Fig. 7 C–F). In extreme cases the pollen germination itself is inhibited by deposition of callose in the germpore. In all the grasses investigated, release of only the exine components on to the stigma does not initiate rejection reaction. However, release of intine proteins triggers off rejection reaction immediately. The recognition reaction in grasses is completed within a few minutes after pollination, and subsequent acceptance or rejection become distinct in less than 10 min. Thus, in grasses there are strong evidences to implicate intine proteins in incompatibility. No data are available on any other taxon having gametophytic incompatibility.

Stigma-Surface Proteins

Most of the investigations implicating stigma-surface proteins are confined to the taxa with dry type of stigma. In members of Caryophyllaceae (J. Heslop-Harrison and Y. Heslop-Harrison 1975) and *Gladiolus* (Knox et al. 1976) digestion of stigma-surface proteins with pronase does not affect pollen germination but inhibits the entry of pollen tubes into the papillae. Some factors in the pellicle, therefore, are required for effective operation of the cutinase system in the pollen. In *Raphanus sativus* also, enzymatic digestion of the pellicle reduces pollen germination and inhibits the entry of even the compatible tubes into the papilla (Shivanna et al. 1978).

Studies of Knox et al. (1976) on *Gladiolus* provide more direct evidences for the involvement of stigma-surface proteins in pollen-pistil interaction. They showed that concanavalin A (con A) binds specifically to the pellicle. Stigmas of very young buds free from pellicle did not bind to con A. Following con A binding stigmatic surface did not inhibit pollen germination, but prevented the entry of pollen tubes. Therefore, components involved in con A binding were necessary for pollen tube entry, and not for pollen germination. Washing of the stigma with sodium deoxycholate removed the ability of the stigma to support pollen germination as well as its ability to bind to con A. Knox et al. (1976) suggested that stigma-surface receptors are composed of many components; some of them are involved in pollen germination and others in entry of tubes.

Studies on intraspecific incompatibility also have implicated stigma-surface proteins in pollen recognition and rejection. S allele-specific proteins have been shown to diffuse from intact stigma (Nasrallah and Wallace 1967). Also, water-soluble substances released from the undamaged stigmas selectively inhibit in vitro germination of self-pollen, but not of cross-pollen (Ferrari and Wallace 1976). Obviously, these proteins emanate from the pellicle.

Sequence of Recognition and Rejection

Intraspecific Incompatibility

As no data is available for heteromorphic systems, the discussion is confined to homomorphic systems only. A probable sequence of recognition and rejection that occurs during pollen-pistil interactions are presented in Fig. 8. This is based largely on the site of recognition and rejection, and the factors involved in recognition. In taxa characterised by sporophytic type of incompatibility, recognition of incompatible pollen as well as rejection are completed on the stigma. As pointed out earlier, exine proteins and pellicle proteins seem to be involved in these processes.

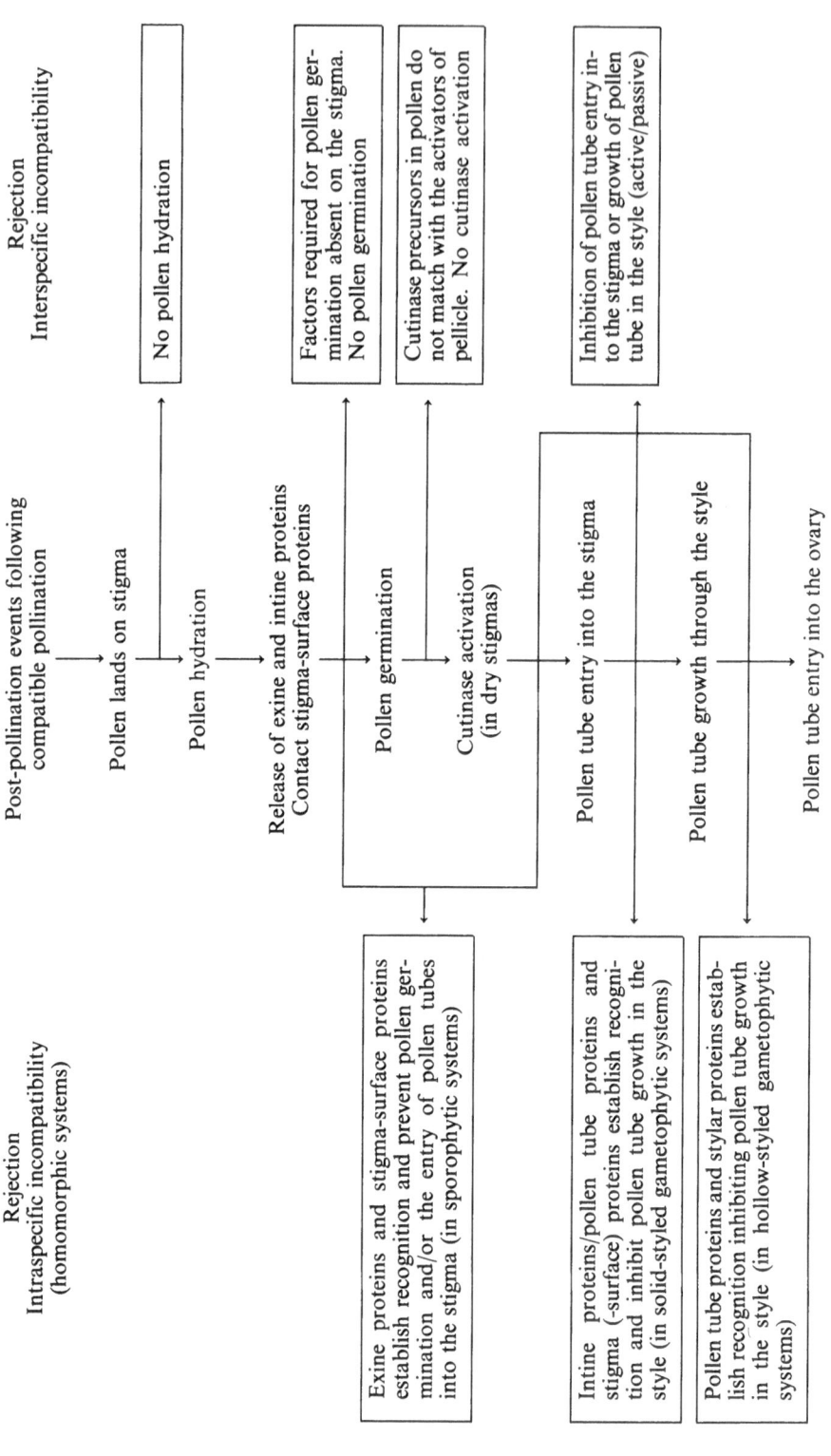

Fig. 8. Probable sequence of recognition and rejection during pollen-pistil interaction. (After Shivanna 1979)

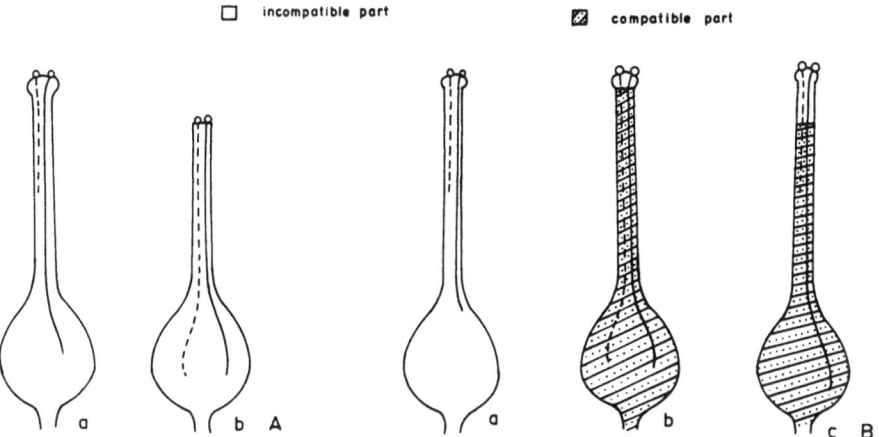

□ incompatible part   ▨ compatible part

Fig. 9 A, B. Responses of self- *broken line* and cross- *solid line* pollination to various treatments in *Prunus avium* **A**, and *Lilium henryi* **B**. In both taxa incompatible pollen tubes are inhibited in the style. In *P. avium* removal of stigma was effective in overcoming pollen tube inhibition indicating that incompatible message is received in the stigma. In *L. henryi* grafting of incompatible stigma on compatible style was not effective in inhibiting incompatible pollen. However, when incompatible stigma together with upper quarter of the style was grafted on to compatible style, incompatible pollen tubes were promptly inhibited. (A modified from Raff and Knox 1977; **B** from Lawson and Dickinson 1975)

    In taxa characterised by the inhibition of pollen tubes in the style, the sites of recognition and the factors involved in recognition are not apparent. In *Petunia hybrida* distinct differences have been observed between self- and cross-pollinated styles in the synthesis of RNA, DNA, and the size of free nucleotide pool within 3 h after pollination (van der Donk 1974, 1975). It was suggested that these differences were the result of recognition of pollen taking place much earlier, on the stigma itself, soon after pollination. In *Prunus avium* (Raff and Knox 1977) the presence of stigma is necessary for pollen tube inhibition (Fig. 9 A). When pollen grains were deposited on an artificial stigma made on the cut stump of the style (after removing the stigma), both compatible and incompatible tubes grew through the style and reached the ovary. Thus, in *Petunia*, *Prunus*, and probably other genera with solid style recognition and rejection are separated in time and space, recognition takes place soon after pollination in the stigma, and rejection is completed in the style about 24 h later. Intine proteins and stigma/stigma-surface proteins are likely to be involved in recognition.
    In *Lilium*, which has a hollow style, the stigma does not seem to play any role in pollen recognition. Details of grafting experiments conducted in *L. henryi* (Lawson and Dickinson 1975) have shown that the incompatible message is not received in the stigma, but is received only after the pollen tube has grown through a quarter of the style (Fig. 9 B). Similar results have been obtained in *L. longiflorum* also (Gladding and Paxton 1975). Further, hot water treatment to the pistil of *L. longiflorum* is known to inactivate incompatibility. Application of hot water

treatment only to stigma was not effective in overcoming incompatibility, but hot water treatment only to the lower part of the style was effective in overcoming incompatibility (Fett et al. 1976). These studies clearly demonstrate that in *Lilium* both recognition and rejection events are confined to the style, and stigma has no role in these processes. In *L. longiflorum* the removal of loosely bound wall materials (proteins and carbohydrates) does not affect pollen germination and pollen tube growth as well as self-incompatibility reaction (Fett et al. 1976). Thus, it appears that in *Lilium*, and probably other hollow-styled systems, pollen wall proteins and stigma proteins are not involved in recognition. Both these processes occur in the style, and proteins synthesized in the pollen tubes and those present/synthesized in the style are involved in recognition and rejection function.

Interspecific Incompatibility

The data on interspecific incompatibility are fragmentary. It is apparent that the operation of interspecific incompatibility is more diverse and much less understood as compared with intraspecific incompatibility (Fig. 8). Pollen-wall proteins and stigma-surface proteins are probably involved in inhibition of pollen hydration, pollen germination, and cutinase activation.

Besides stigma-surface proteins, non-proteinaceous factors such as phenolic compounds and carbohydrates may also play a role in pollen hydration and pollen germination. Phenolic compounds have now been shown to promote or inhibit selectively pollen germination (Martin 1970, 1972; Martin and Ruberte 1972). In *Impatiens balsamina* (Tara and Namboodiri 1976) failure of the stigma of a mutant to support pollen germination has been correlated to the absence of some of the phenolic compounds (present in other varieties, and which support pollen germination). Thus, lack of pollen germination following interspecific pollinations may be due to the absence of some substance(s) on the stigma (needed for germination). Such mechanisms represent passive rejection, and there is no need for pollen recognition.

Another method of passive rejection would be the inability of pollen tubes to penetrate the cuticle of the stigma, as has been shown in the cross involving *Gladiolus* and *Crocosmia* (Knox et al. 1976; see also J. Heslop-Harrison 1975b). These passive rejections are more like a "lock and key" mechanism: absence of a suitable key with one of the partners for the lock present with the other partner results in incompatibility. Closely related species would have the right key at the stigma level, but may not be able to grow through the style.

No information is available on the factors involved in pollen tube inhibition in the style. Inability of the pollen tubes to utilise stylar nutrients, which may be due to the lack of positive recognition, may often be the reason. In very closely related species it may be due to active inhibition.

The sequence of recognition events, presented in Fig. 8, is a tentative one and indicate prominently the lacunae in our understanding of the details of pollen-pistil interaction. As more data become available the scheme is bound to be appended and/or modified.

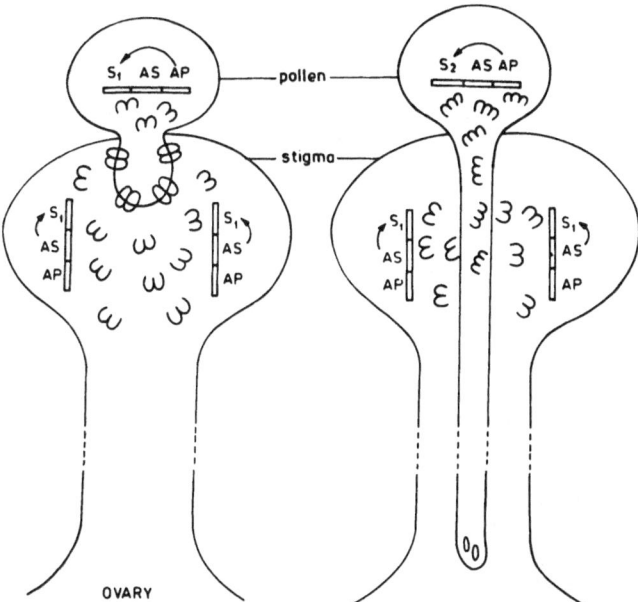

Fig. 10. A generalised model to explain the function of S allele. This is based on the tripartite nature of the S locus (Lewis 1960) having a specificity part *S* common to both pollen and style, and two activity parts controlling the reaction in pollen *AP* and style *AS*. (Modified from Nettancourt 1972)

Mechanism of Inhibition

Intraspecific Incompatibility

Many models have been proposed to explain the mechanism of inhibition (Linskens 1965; Lewis 1965; Ascher 1966; see Nettancourt 1977). Basically, these models envisage the production of S allele-specific polypeptide identical in pollen and style. When the polypeptide of the pollen comes in contact with the identical polypeptide in the style, it dimerises on the surface of the pollen tube to form a repressor which inhibits the growth of the pollen tube (Fig. 10).

Production of S allele-specific proteins has been demonstrated in *Oenothera* (Lewis 1952; Lewis et al. 1967), *Petunia* (Linskens 1960), and *Brassica* (Nasrallah and Wallace 1967). Similarity of S allele-specific proteins in pollen and pistil has been shown only in *Petunia;* in *Oenothera* the investigations are confined only to pollen, and in *Brassica* S allele-specific proteins have been found only in the pistil and not in pollen. Detailed studies on *Brassica oleracea* by Nasrallah and his associates (Nasrallah and Wallace 1967; Nasrallah et al. 1970, 1972; Nasrallah 1974) have shown that each S allele-specific protein had a different electrophoretic mobility and, hence, each of them could be localized to a specific band on the gel. These proteins were heritable as shown by the presence, in the heterozygous plants, of both S allele-specific bands of the parents (Nasrallah et al. 1970). Studies of F_1 and F_2 progenies, involving crosses of different homozygous self-incompatible

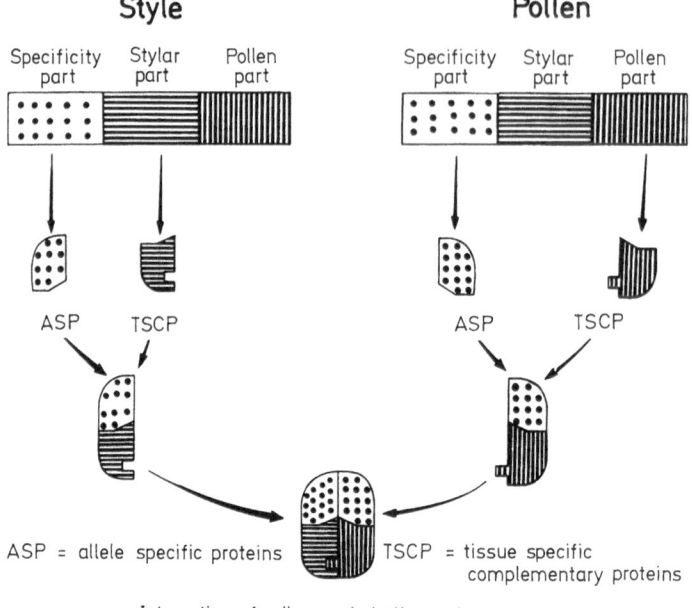

Fig. 11. The hypothesis concerning S-gene products. Identical specificity proteins in the pollen and style, form a complex with tissue-specific adaptive proteins. This model explains the formation of mutually reactive S allele-specific proteins in the pollen and style. (Modified from Pandey 1975)

genotypes, showed that S allele-specific proteins present in the progenies were exactly correlated with the segregation of S alleles as determined by genetical analysis (Nasrallah et al. 1972). Recent studies of Nishio and Hinata (1977) on isoelectric focussing of stigma proteins of *B. oleracea* have indicated that S allele produces several fractions of proteins rather than a single protein, and S allele specificity is expressed by a combination of the protein fractions.

One of the limitations of these models, which envisage identical S allele-specific proteins in pollen and style, is the lack of explanation for the reactivity of identical proteins (of the pollen and style), and the mechanism that prevents the formation of repressors by polypeptides of the pollen or the style themselves. To overcome these difficulties, Pandey (1975) modified the concept of Lewis (1960) concerning S gene determination based on the tripartite nature of S gene. According to this hypothesis the S allele proteins in the pollen and style have an identical specificity protein and a tissue specific adaptive protein (Fig. 11). This model explains the failure of the stylar or the pollen proteins to react amongst themselves, but allow for the interaction between them to produce a repressor of the genes involved in pollen tube growth.

The mechanism by which the repressor inhibits the growth of the pollen tube is not clear (see Nettancourt 1977; Shivanna 1979). The concept that self-incompatibility inhibition is mediated through the inhibition of protein synthesis in pollen tubes is in agreement with present knowledge on pollen tube inhibition (Nettancourt 1977; Nettancourt et al. 1974, 1975; Cresti et al. 1977). Although earlier investigators considered the mechanism of inhibition in sporophytic systems different from that in gametophytic systems (Christ 1959; Kroh 1966), recent evidences have indicated that the basic mechanism in both systems may be similar (Ferrari and Wallace 1977; Shivanna 1979). Ferrrari and Wallace (1977) put forward a model, based on the inhibition of protein synthesis, to explain self-incompatibility inhibition in both sporophytic and gametophytic systems.

Interspecific Incompatibility

Our knowledge on the details of inhibition following interspecific pollination are meagre. As pointed out earlier, in a large number of instances it may be a passive inhibition due to the lack of some factor(s) in the pistil, required for pollen germination or pollen tube growth. In crosses involving closely related species, particularly in instances of unilateral incompatibility, the inhibition may be active.

According to some investigators the S allele has a dual function and is involved in both intra- and interspecific incompatibility (Lewis and Crowe 1958; Pandey 1968). However, many recent genetical studies suggest that interspecific incompatibility is controlled by genes different from S genes (Abdalla 1974; Takahashi 1974). Hogenboom (1975) includes all interpopulational incompatibility, not controlled by S alleles, under incongruity. Incongruity is due to the lack of genetic information in one partner about some relevant character of the other. It is a by-product of evolutionary divergence and, hence, varies from system to system depending on the extent of evolutionary divergence. It may also be concerned with the post-fertilization barrier. Thus, incongruity represents a passive rejection, whereas incompatibility involves active rejection as a result of S gene action.

The hypotheses to explain unilateral and interspecific incompatibility are based largely on genetical studies. Very little work has been done on the physiological and biochemical aspects. A few ultrastructural and physiological data available are not unequivocal. Nettancourt et al. (1973 a, b, 1974, 1975) studied ultrastructural details of pollen tubes inhibited in self-incompatible and unilateral incompatible reactions. They observed that, in both reactions, the pollen tubes showed accumulation of a large number of bipartite particles at the tip, and the formation of characteristic concentric endoplasmic reticulum. The only difference which is consistent between the tubes inhibited in self-incompatibility and those inhibited in unilateral incompatibility, was the nature of the outer wall of the pollen tube; it was very thick in the former but rather thin in the latter. Based on these and other genetical studies, Nettancourt et al. (1975) concluded that interspecific unilateral incompatibility results from the interaction of S elements in the pollen grains with one or several non-identified stylar genes. Roggen and Linskens (1967) studied pollen tube growth and respiration pattern following intergeneric crosses between *Petunia hybrida* and *Salpiglossis sinuata*. The morphological abnormalities of the incompatible pollen tubes (branching and swelling of tube tip, increased callose deposition, etc.), and the respiration pattern following intergeneric crosses were similar to those following selfing. They suggested that the mechanism of inhibition of pollen tubes is similar in both self- and intergeneric incompatibility. However, in *Lilium*, unlike self-incompatibility which can be overcome by giving hot water treatment to the pistil, interspecific incompatibility could not be overcome by hot water treatment (Ascher and Peloquin 1968), indicating that the mechanism of action in both types of incompatibility is different. Further physiological and biochemical studies on interspecific incompatibility are necessary for a better understanding of the mechanism of inhibition.

Methods of Overcoming Incompatibility

There are many effective methods to overcome intraspecific incompatibility. Success in overcoming interspecific incompatibility is limited only to a few systems. Also, interspecific incompatibility in many of the closely related crosses operates after fertilization, and the techniques of embryo culture have been effectively used to overcome post-fertilization barriers (see Raghavan 1976). Table 2 lists most of the well-established techniques used to overcome barriers to fertilization, with some examples. The details of many of these techniques have been adequately covered by Nettancourt (1977). A few of the recent techniques are dealt here in greater detail.

Recognition Pollen/Mentor Pollen

As discussed earlier, pollen wall proteins have been implied in intraspecific incompatibility. Knox and his associates indicated that they are involved in controlling interspecific incompatibility also. Knox et al. (1972 a, b) attempted to cross *Populus deltoides* and *P. alba* by mixing killed compatible pollen (either by irradiation, or by treatment with organic solvents) with live incompatible pollen. By this method they could obtain a significant number of hybrids. Even the proteinaceous diffusates obtained from the wall of compatible pollen were effective in overcoming interspecific incompatibility. They presumed that the proteinaceous recognition factors released from the wall of killed compatible pollen mask the rejection reaction on the recipient stigma, thus making it ineffective in inhibiting incompatible pollen tubes. Sastri and Shivanna (1976 b) attempted to overcome interspecific incompatibility between *Sesamum indicum* and *S. mulayanum* by this technique of recognition pollen. Although recognition pollen was effective in overcoming incompatibility on the stigma, it failed to overcome incompatibility barrier in the style. These experiments provide rational explanation for the success of distant hybridisation by the use of mixed pollen and mentor pollen by many investigators, particularly in Russia.

As pollen-wall proteins are involved in controlling intraspecific incompatibility, at least in sporophytic systems, it should be possible to overcome intraspecific incompatibility by using the technique of recognition pollen. This has been achieved in many taxa. In *Cosmos bipinnatus* (Howlett et al. 1975), a strictly self-incompatible species with a sporophytic incompatibility, the application of exine diffusate of compatible pollen on the stigma before self-pollination, resulted in sig-

Table 2. Effective techniques to overcome incompatibility[a]

Methods	Species	Authors
Intraspecific incompatibility:		
Induced mutations	*Oenothera*	Lewis (1949)
	Prunus	Lewis and Crowe (1953)
	Petunia	Brewbaker and Natarajan (1960)
	Trifolium	Pandey (1956), Denward (1963)
	Nicotiana	Pandey (1965), Gastel and Nettancourt (1975)
Induction of autotetraploidy	*Prunus*	Crane and Lawrence (1929)
	Pyrus	Crane and Lewis (1942)
	Petunia	Stout and Chandler (1941)
	Tradescantia	Annerstedt and Lundqvist (1967)
	Lolium	Ahloowalia (1973)
Irradiation of pistils (X-rays/UV rays)	*Petunia*	Linskens et al. (1960)
	Lilium	Hopper and Peloquin (1968)
	Rubus	Arasu (1968b)
	Ribes	Arasu (1968b)
	Lycopersicum	Nettancourt and Ecochard (1968)
	Nicotiana	Gastel and Nettancourt (1975)
Bud pollination	*Brassica*	Kakizaki (1930), Sears (1937), Attia (1950), Odland and Noll (1950)
	Raphanus	Kakizaki (1930)
	Petunia	Yasuda (1934), Eyster (1941), Shivanna and Rangaswamy (1969)
	Nicotiana	Sears (1937), Pandey (1963)
Delayed pollination	*Brassica*	Kakizaki (1930)
	Lilium	Ascher and Peloquin (1966a)
Hot water/high temperature treatment	*Malus*	Lewis (1942)
	Pyrus	Lewis (1942)
	Prunus	Lewis (1942)
	Oenothera	Lewis (1942), Hecht (1960, 1961), Bali (1963)
	Trifolium	Laffel (1963), Townsend (1965, 1966, 1971), Kendall and Taylor (1969)
	Brassica	El Muraba (1957), Visser (1977), Roggen and van Dijk (1976)
	Raphanus	El Muraba (1957)
	Lycopersicum	Hogenboom (1972)
	Lilium	Ascher and Peloquin (1966b, 1970)
	Secale	Wricke (1974)
	Nemesia	Campbell and Ascher (1972)
Treatment of stigma with organic solvents (hexane)	*Brassica*	Ockendon (1978)
Increased atmospheric humidity	*Brassica*	Carter and McNeilly (1973), Ockendon (1978)
Mutilation of the stigma	*Brassica*	Sears (1937), Roggen and van Dijk (1972)
Application of growth substances	*Lilium*	Emsweller and Stuart (1948), Emsweller et al. (1960), Matsubara (1973)

Table 2 (continued)

Methods	Species	Authors
	Petunia ⎫ *Tagetes* ⎪ *Trifolium* ⎬	Eyster (1941)
	Brassica ⎭ *Lycopersicum*	Nettancourt et al. (1971)
Electric-aided pollination	*Brassica*	Roggen et al. (1972)
End-season pollination	*Nicotiana*	East (1934)
	Petunia	Yasuda (1934)
	Abutilon	Pandey (1960)
Use of recognition pollen (mentor pollen)	*Theobroma*	Glendinning (1960), Opeke and Jacob (1969)
	Cosmos	Howlett et al. (1975)
	Brassica	Roggen (1975)
	Malus	Dayton (1974)
	Petunia	Sastri and Shivanna (1976a)
	Nicotiana	Pandey (1977)
Placental pollination	*Petunia*	Rangaswamy and Shivanna (1967, 1971a), Niimi (1970, 1976), Wagner and Hess (1973)
Interspecific incompatibility:		
Application of growth substances	*Lilium*	Emsweller and Stuart (1948)
Stump pollination	*Solanum*	Swaminathan (1955)
	Nicotiana	Swaminathan and Murty (1957)
Use of recognition/mentor pollen	*Populus* *deltoides* × *P. alba*	Stettler (1968), Knox et al. 1972b)
	Nicotiana *forgetiana* × *N. langsdorffii*	Pandey (1977)
Intra-ovarian Pollination	*Argemone* *mexicana* × *A. ochroleuca*	Kanta and Maheshwari (1963)
Placental pollination	*Melandrium* *album* × *M. rubrum*	Zenkteler (1967, 1970)
	M. album × *Silene schafta*	Zenkteler et al. (1975)
Treatment of stigma/pollen with organic solvents	*Populus*	Willing and Pryor (1976)

[a] The list includes only those investigations in which seed-set has been achieved. Although many other treatments have been reported to inactivate incompatibility reaction, to different degrees, their efficacy in inducing seed-set has not yet been demonstrated

nificant increase in seed-set (as much as 22% of the control). The application of crude ether extract of bee-collected pollen of *Brassica napus*, largely containing pollen-coat material, on to the stigma of Brussels sprouts (*B. oleracea* L. var *gemmifera*) greatly increased selfed seed-set, to as much as was obtained with bud pollination (Roggen 1975). (*B. napus* is compatible with *B. oleracea*). The application of recognition-pollen along with self-pollen has also been effective in increasing the selfed seed-set in *Petunia hybrida* (Sastri and Shivanna 1976 a) and many varieties of apple (Dayton 1974).

Recently, Pandey (1977) also reported success in overcoming both intra- and interspecific incompatibility by using mentor pollen in some species of *Nicotiana*. Detailed analysis of his results suggested that the role of mentor-pollen in overcoming incompatibility may be that of providing extra free pollen growth promoting substance rather than providing recognition factors.

These treatments are basically aimed at manipulating the pollen proteins involved in incompatibility. Comparable manipulations can also be carried out with the stigma-surface proteins, the other partners involved in incompatibility. In interspecific hybridisation of *Populus*, Willing and Pryor (1976) showed that the treatment of stigma with many organic solvents, such as anhydrous hexane and ethyl acetate, before pollination was remarkably effective in overcoming interspecific incompatibility. In some crosses the seed-set following incompatible pollination was almost as good as compatible pollination.

Intra-Ovarian Pollination and Test-Tube Fertilization

All the techniques, described so far, are aimed at overcoming incompatibility retaining the zone of inhibition in the pistil (i.e. the stigma and style). Theoretically, any attempt which aims at eliminating the stigma and style altogether, thus bringing the pollen in direct contact with the ovules and achieving fertilization and seed development, is the most effective technique. Significant success has been achieved on these lines at the Department of Botany, University of Delhi (see Rangaswamy 1977).

Initial attempts were made on members of Papavaraceae, as they are most suitable for such studies. The procedure of intraovarian pollination was the first technique standardised, and involves injecting pollen grain (suspended in a suitable medium) directly into the ovary, achieving pollen germination, pollen tube entry into ovule, and fertilization. Viable seeds following intra-ovarian pollination have been obtained in *Papaver somniferum* (Fig. 12 A, B), *P. rhoeas*, *Argemone mexicana*, and *A. ochroleuca*. This technique has also been applied to achieve interspecific hybridisation between *A. mexicana* and *A. ochroleuca* (Fig. 12 C) (Kanta and Maheshwari 1963; Maheshwari and Kanta 1961).

There are many limitations in extending the technique of intra-ovarian pollination to other taxa. It is not suitable to the taxa in which there is not enough space in the ovary to inject pollen suspension. Also, in taxa in which sugar is required for pollen germination, injection of pollen suspension containing sugar makes the ovary prone for bacterial and fungal infection.

These limitations have been overcome in techniques involving aseptic culture of whole pistils or ovules to achieve test-tube fertilization. Cultured pistils of *Nicotiana* (Dulieu 1963, 1966; Rao 1965), *Petunia* (Shivanna 1965), and *Antirrhinum* (Usha 1965), pollinated in vitro, have been successfully grown to obtain mature, viable seeds. As expected, the culture of whole pistils and pollination on the stigma

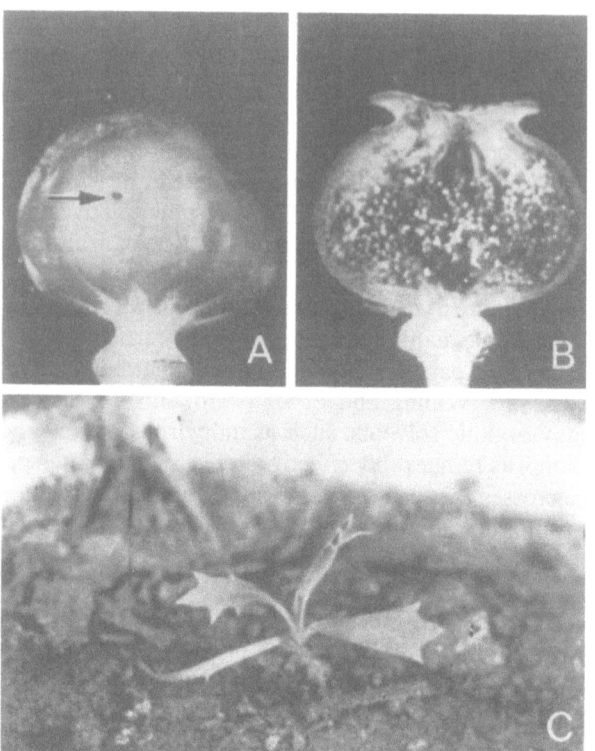

Fig. 12 A–C. Intra-ovarian pollination. **A, B** 27-day-old fruits of *Papaver somniferum* developed as a result of intra-ovarian pollination. *Arrow* in **A** shows point of injection of pollen suspension. In **B** vertical half of the fruit is shown; note the seeds. **C** 6-week-old hybrid seedling *(Argemone mexicana × A. ochroleuca)* raised from seed obtained through intra-ovarian pollination. (After Kanta and Maheshwari 1963)

was not effective in overcoming self-incompatibility (Shivanna 1965). This is because the inhibitory zone (the stigma and style) remains intact in the pistil. However, the culture of isolated ovules (bringing them in direct contact with pollen grains) has been very successful.

Kanta et al. (1962) cultured excised unpollinated ovules of *Papaver somniferum* on a nutrient medium and dusted pollen grains on and around the ovules (Fig. 13 A). Following ovule pollination, pollen grains germinated normally, pollen tubes entered the ovules, and effected double fertilization. Interestingly, fertilized ovules developed into mature seeds on the same medium (Fig. 13 B, C). Subsequently, the technique of test-tube fertilization was extended to *Argemone mexicana, Eschscholzia californica, Nicotiana tabacum* (Kanta and Maheshwari 1963; Maheshwari and Kanta 1964), *Dianthus caryophyllus* (Zenkteler 1965), and *Dicranostigma franchetianum* (Rangaswamy and Shivanna 1969). All these taxa are self-compatible, and fertilization achieved in vitro was through compatible pollen.

Rangswamy and Shivanna (1967) attempted to overcome self-incompatibility in *Petunia axillaris* by using the technique of test-tube fertilization. The culture of isolated ovules or groups of ovules together with the pollen grains did not result in fertilization or seed development, even with compatible pollen, although pollen germination was abundant. They modified the technique: instead of pollination of isolated ovules, the entire ovule mass of the ovary intact on the placentae to-

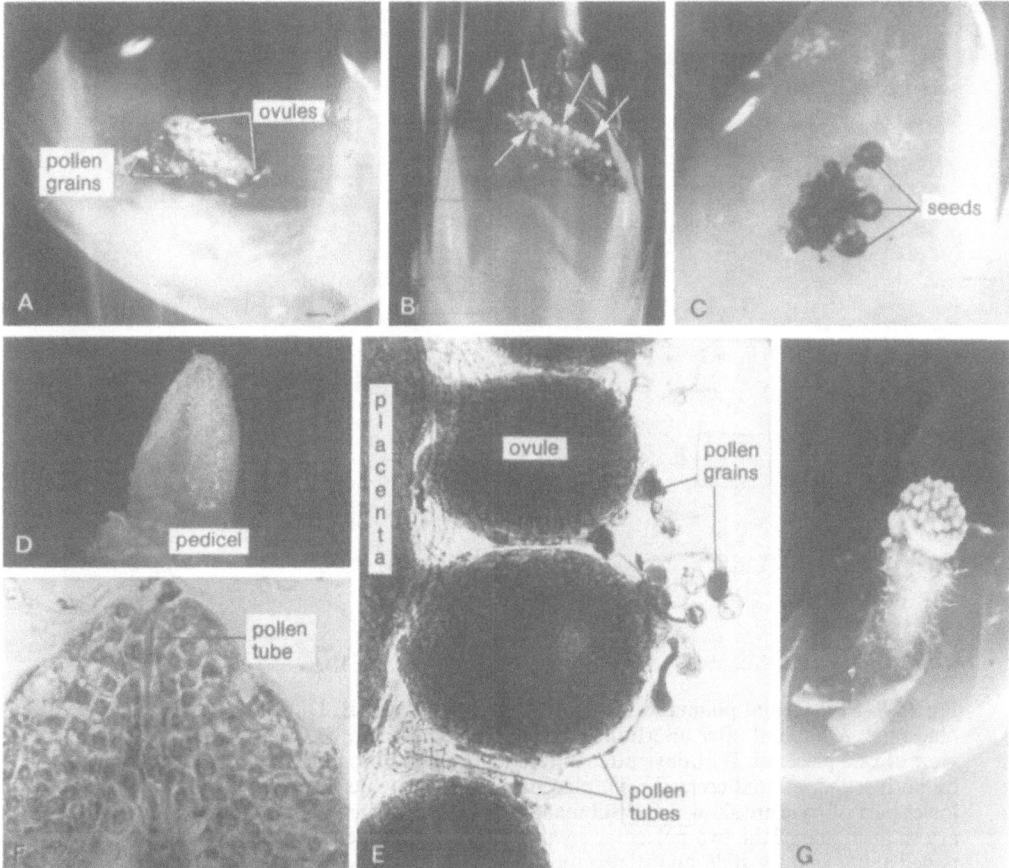

Fig. 13 A–G. Test tube fertilization. **A, B** *Papaver somniferum.* **C** *Argemone mexicana.* **D–G** *Petunia axillaris.* **A** Portion of ovules and placenta cultured on nutrient medium and pollinated with pollen. **B** 7-day-old culture showing many developing seeds *white bodies.* **C** About 4-week-old seeds developed on nutrient medium following ovule pollination. **D** Both placentae of an ovary with its entire mass of ovules dusted with pollen ready for culture. **E** Free-hand transection through self-pollinated placentae 24 h after culture; note marginal portion of one placenta, three ovules and many pollen grains. Also note pollen germination and profuse growth of pollen tubes. **F** Longisection of micropylar part of ovule, 2 days after selfing, to show entry of pollen tube. **G** 24 days after placental self-pollination; note mature seeds. (**A, B** after Kanta and Maheshwari 1963; **C** Maheshwari and Kanta 1964; **D–F** Rangaswamy and Shivanna 1971 a)

gether with a short length of pedicel was cultured on the medium and the ovules dusted with pollen (Fig. 13 D). This refined technique, termed placental pollination, helped to bring the pollen in direct contact with the ovules without disturbing their original arrangement and, thus, preventing any injury to them. Following placental pollination, pollen grains germinated readily on the ovules and placentae (Fig. 13 E), pollen tubes entered the ovules (Fig. 13 F), and effected fertilization. The fertilized ovules showed normal development of the embryo and endosperm,

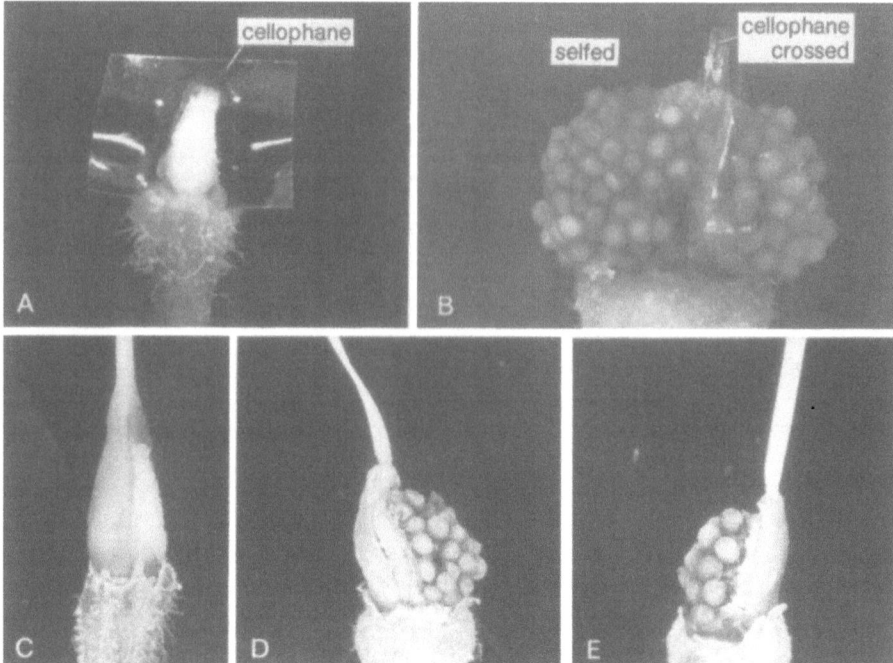

Fig. 14A–E. Placental pollination in *Petunia axillaris* **A** and **B.** Differential pollination of placentae. **A** Explant after insertion of a cellophane partition between the placentae, face view of one placenta. **B** 21 days after differential pollination; note the formation of seeds on both selfed *left* and crossed *right* placentae. **C–E** Two-site pollination; only ovary and lower part of style are shown. **C** Pistil made ready for two-site pollination by removing ovary wall on one of the two palcentae *right*. Ovary wall on the other placentae *left* was retained along with the style, and stigma (not shown). **D** 24-day-old culture in which both exposed placenta *right* and stigma were self-pollinated. **E** 21-day-old culture in which exposed placenta *left* was cross-pollinated, and the stigma self-pollinated. Only exposed placentae showed seed-set in both. (After Rangaswamy and Shivanna, **A, B**, 1971 a; **C–E**, 1971 b)

and mature viable seeds were obtained in 3 weeks after fertilization (Fig. 13 G). The same period is required for the seeds to mature in vivo. Unlike stigmatic pollination in which self-pollination is invariably a failure, in placental pollination both self- and cross-pollinations are equally effective in inducing seed-set.

The technique of placental pollination (in *Petunia*) was further modified to treat the ovules on the two placentae, of the same ovary, differently (Shivanna 1971; Rangaswamy and Shivanna 1971a). The two placentae were separated, mechanically, by introducing a piece of cellophane between them (Fig. 14 A). When one of the placentae was maintained as control, and the other pollinated with self- or cross-pollen grains, the ovules on the control invariably shrivelled, whereas many of the ovules on the pollinated placenta developed into seeds, irrespective of self- or cross-pollination. When one of the placentae was selfed and the other crossed, seeds developed on both the placentae equally well (Fig. 14 B), thus demonstrating conclusively the equal efficacy of both self- and cross-pollination.

These experiments demonstrated that once the stigma and style are eliminated, the ovules do not show any preferential receptivity to crossed pollen, and they led to further modification of the technique to study the interaction between stigmatic pollination and placental pollination. The technique of two-site pollination was devised in *P. axillaris* (Rangaswamy and Shivanna 1971 b). The ovary wall was carefully peeled to expose one of the placentae, retaining the ovary wall on the other placenta, and the style and stigma intact (Fig. 14 C). Thus, the pollination could be carried out both on the stigma and on the exposed placenta in the same pistil. A series of experiments using two-site pollination demonstrated that the stigmatic self-pollination does not affect the response of placental self-pollination, and vice-versa (Fig. 14 D, E).

The success of test-tube fertilization attracted the attention of many investigators, and attempts have been made to use the technique in many fundamental and applied aspects. It has also been successfully used to overcome self-incompatibility in *Petunia hybrida* (Niimi 1970, 1976; Wagner and Hess 1973). Balatkova and Tupý (1968) could achieve in vitro fertilization and seed formation in *Nicotiana tabacum*, by using already germinated pollen grains. They also showed that the pollen tubes in which gamete formation had occurred, when deposited on the unpollinated ovules, could effect successful fertilization. These studies demonstrate the feasibility of treating either the male gametes, or female gametes, without affecting each other. The technique of two-site pollination has been used (Wagner and Hess 1973) in *Petunia hybrida* to test the relative fertilization competence of pollen grains deposited on the stigma and on the ovule surface. The pollen deposited on the stigma was found to have a better chance for effecting fertilization, despite the longer way the pollen tube has to traverse, than the pollen deposited on the ovule.

Pollination with killed pollen, irradiated pollen, and pollen of alien species occasionally stimulate unfertilized egg to develop parthenogenetically. Logically, such pollinations carried out directly on the ovules would be more effective in inducing parthenogenesis, than those carried out on the stigma, as the factors which stimulate the egg are brought much closer to the embryo sac. The feasibility of this has been demonstrated by the report of Hess and Wagner (1974), who successfully induced parthenogenesis by using the technique of test-tube pollination. Their attempts to obtain androgenic haploids by anther culture technique in *Mimulus luteus* were unsuccessful. However, when its ovule mass was pollinated in vitro with the pollen of *Torenia fournieri*, 1% of the ovules developed parthenogenetically and gave rise to haploid plantlets.

Zenkteler (1967, 1970), and his associates (Zenkteler et al. 1975), successfully applied the technique of test-tube fertilization to raise interspecific and even intergeneric hybrids (Fig. 15 A–F). They could successfully obtain hybrid progeny following the cross *Melandrium album* (♀) × *Melandrium rubrum* (♂), as well as *M. album* (♀) × *Silene schafta* (♂). They also attempted to obtain hybrids between *M. album* (♀) and many other taxa belonging to Caryophyllaceae, Cruciferae, Solanaceae, and Campanulaceae (Fig. 15). Following such crosses, they reported many pre-fertilization as well as post-fertilization abnormalities. In many of them there was normal fertilization, and initiation of embryo and endosperm but, later, the embryo degenerated. Isolation of young embryos, and culturing them on a suitable nutrient medium, would probably enable the production of hybrids in these

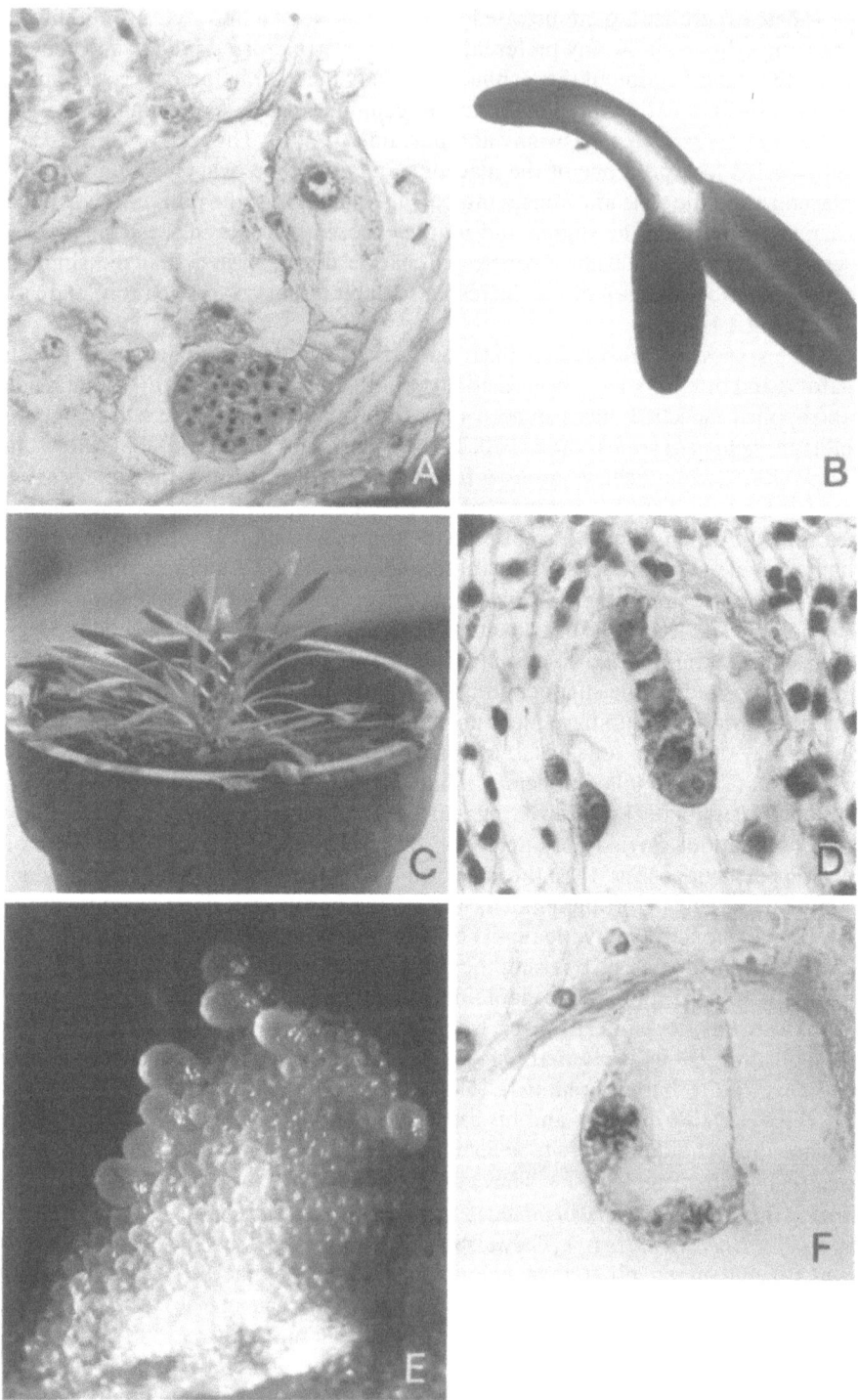

crosses also. Thus, the technique of test-tube fertilization offers immense possibilities not only to overcome intra- and interspecific incompatibility, but also in many fundamental studies concerning fertilization.

Concluding Remarks

From the foregoing discussion it is obvious that the progress in our understanding of the biology of pollen-pistil interaction and fertilization, although impressive, is far from complete. Detailed investigations, whether structural, physiological, or biochemical, have so far been confined only to a few selected taxa. In addition to the need for intensification of studies on the established taxa, it is important to extend them to other systems.

Much more attention must be paid to studying pollen and stigma, as their interaction during initial stages of pollination determines the subsequent events in the pistil. Localization of pollen-wall proteins and stigma-surface proteins, and their implication in pollen recognition, have been important achievements in recent years. Although the details of the origin of pollen-wall proteins, and their incorporation, have been followed in a few systems, the details of the stigma-surface proteins, particularly in dry types, are yet to be investigated. Attempts must also be made to characterize the heterogenous components of stigmatic surface and pollen wall, and in identifying the role of individual components in pollen physiology, stigma receptivity, pollen recognition, and eventual acceptance or rejection of the pollen tube. The investigations of Knox and his associates on *Gladiolus* are significant in this direction.

Extensive physiology and biochemical studies have been carried out on the growth of pollen tube in the pistil. Yet, our knowledge of many of the important aspects of pollen tube growth is totally inadequate. Some of these aspects are: the nature and origin of chemotropic substance, the nature of the stimulus which precedes the growing pollen tubes, and the mechanism of the release of male gametes and of fertilization.

Our knowledge of details of interspecific incompatibility is very inadequate, and extensive studies need to be conducted.

Recent investigations on various aspects of self-incompatibility have been rewarding. Most of the studies are confined only to homomorphic systems, largely to members of Cruciferae, Compositae, Liliaceae, and Solanaceae. The evidences in implicating exine proteins in controlling sporophytic incompatibility are more direct and conclusive, but implication of intine proteins in gametophytic incom-

---

Fig. 15 A–F. Interspecific hybridization through test tube fertilization. **A** *Melandrium album × Viscaria vulgaris*, hybrid embryo and endosperm, 5 days after test tube pollination. **B** *M. album × M. rubrum*, wholemount of embryo 12 days after pollination. **C** *M. album × V. vulgaris*, hybrid plants raised from seeds obtained in test tubes. **D** *Nicotiana tabacum × Petunia hybrida*, hybrid, linear embryo 4 days after pollination. **E** *N. tabacum × Hyoscyamus niger*, development of seeds 4 days after pollination. **F** *M. album × P. hybrida*, male gamete has entered the egg; male gamete and egg nucleus are undergoing mitotic division. (A–F through kind courtesy of Dr. M. Zenkteler, Poznan, Poland)

patibility needs conclusive evidences. There is an urgent need to extend these studies to other taxa, including heteromorphic systems. The ultimate aim of these investigations would be to isolate and characterize the proteins involved in pollen recognition and rejection.

There has not been much progress in understanding the mechanism of incompatibility. Many hypotheses have been put forward to explain the gametophytic incompatibility, on the basis of interaction of identical pollen and pistil proteins. However, the similarity of S-specific proteins from the pollen and pistil has been immunologically shown only in *Petunia*. Further immunological and electrophoretic investigations, similar to those carried out on *Brassica* by Nasrallah and his associates, should be conducted on other systems.

The efficacy of a large number of techniques have been demonstrated to overcome intraspecific incompatibility. Often, more than one technique works in a given system. It is, therefore, necessary to select the best technique suitable for the particular requirement. Attempts to devise and/or standardize simpler and more effective techniques should be continued. The techniques of recognition-pollen, and in vitro fertilization have great potential not only in overcoming self-incompatibility, but also in achieving interspecific and intergeneric crosses. Besides its practical application, the technique of in vitro pollination and fertilization offers great advantages in basic studies concerned with pollen-pistil interaction and fertilization. This is because of its suitability in controlling the environmental factors, and in carrying out experimental treatments.

Studies on the receptive surface of stigma, and the path of pollen tubes in the pistil have been extended to many other systems. Apart from esterases, acid phosphatases are also present on both dry (Ghosh and Shivanna 1980a) and wet stigma (Herrero and Dickinson 1979). Electron microscopic evidence indicates that the exudate is secreted by ER (Dumas at al. 1978); golgi vesicles do not seem to have any role (Kristen et al. 1979). Even in taxa characterized by wet stigma, the younger stigmas are comparable to the dry stigma, with a cuticle-pellicle layer (Shivanna and Sastri 1981). In solid-styled systems such as *Petunia* and *Nicotiana*, the pellicle is disrupted during the secretion of exudate. In hollow-styled systems such as *Amaryllis* and *Crinum*, the exudate is secreted by the tissues of stylar canal, and the pellicle-cuticle layer is not disrupted (Shivanna and Sastri 1981). The stigma of marine angiosperms (*Enhalus*, *Halophila* and *Thalassia*) is dry with a pellicle-cuticle layer, and is comparable to that in terrestrial plants (Pettitt 1980). In watermelon (Sedgley and Scholefield 1980) and *Acacia* (Kenrick and Knox 1981) pollination stimulates a second phase of stigmatic secretion which results in accumulation of copious exudate on the stigma.

The structure of transmitting tissue has been investigated in *Persea* (Sedgley and Buttrose 1978), *Vitis* (Considine and Knox 1979), *Petunia* (Herrero and Dickinson 1979), *Actinidia* (Hopping and Jerram 1979), members of Gramineae (J. Heslop-Harrison 1979a, J. Heslop-Harrison and Y. Heslop-Harrson 1980), and *Primula* (Y. Heslop-Harrison et al. 1981). Besides the presence of extracellular proteins on the surface of stigma, proteins are also present in intercellular spaces of the transmitting tissue (Herrero and Dickinson 1979, J. Heslop-Harrison and Y. Heslop-Harrison 1980, Y. Heslop-Harrison et al. 1981). Considerable significance is attached to the presence of extracellular proteins on the stigma, and in the path of the pollen tube, because of the possibility of its involvement in pollen recognition and incompatibility responses.

Investigations on pollen-pistil interaction highlight the importance of pollen adhesion and pollen hydration (J. Heslop-Harrison 1979b, Clarke et al. 1979, Stead et al. 1980, Woittiez and Willemse 1979). Adhesion of pollen is largely determined by the extent of wetness of the stigma, and on the sculpture of pollen wall (Woittiez and Willemse 1979). Wet stigma supports adhesion of both powdery and sticky pollen. This is largely mechanical, and does not seem to involve any specificity. The adhesion of pollen on dry stigma is more critical, and depends on the extent and composition of the pellicle and amount of surface-coat substances on pollen. Often the adhesion may involve morphological complementation between the pollen and stigma and, in many species, acts as a barrier for incompatible pollen. A theoretical consideration of pollen hydration on the stigma is provided by J. Heslop-Harrison (1979b). The rapidity of pollen hydration depends upon the nature of stigma. If the stigma is of dry type, hydration is gradual, and controlled by the water potential of stigma. When the stigma is covered with an aqueous exudate, the hydration is more rapid. There are evidences to indicate that, in desiccated pollen, the plasmalemma is in a dissociated condition, but its integrity is restored during hydration (J. Heslop-Harrison 1979b; Shivanna and J. Heslop-Harrison 1981).

The role of stigma-exudate in pollen germination varies from species to species (Shivanna and Sastri 1980). In species characterized by solid style, the exudate does not seem to contain factors involved in pollen germination. In species with hollow style, however, the exudate seems to contain pollen germination factors.

Observations on pollen-pistil interaction in a seagrass, *Amphibolis* (where pollination occurs under submerged condition), has revealed many interesting features (Pettitt et al. 1980). Pollen grains, soon after coming in contact with the stigma, are held tenaciously by a meniscus of adhesive material formed from the surface-coatings of both pollen and stigma. The adhesive binding, in contrast to terrestrial plants, is water-proof. In *Amphibolis* the pollen grains are filiform, and lack a preformed aperture. The aperture is formed by the focal autolysis through which the tube emerges. Anderson (1980) reported an unusual phenomenon in some members of Malphighiaceae with produce both chasmogamous and cleistogamous flowers. In cleistogamous flowers pollen grains germinate inside indehiscent anthers, pollen tubes grow through the filament of anther into the receptacle and, eventually, reach the carpel and the ovules. These observations need further study.

There has been significant progress in our knowledge on heteromorphic incompatibility. The presence of extracellular proteins, comparable to homomorphic systems, has been demonstrated in *Linum* (Ghosh and Shivanna 1980a), and *Primula* (Y. Heslop-Harrison et al. 1981). In *Primula* there is no basic difference in the receptive surface of stigma of the two morphs (Y. Heslop-Harrison et al. 1981); in *Linum* the stigma of long-styled form is dry, that of short-styled form wet (Ghosh and Shivanna 1980a). Also, the surface proteins of the two morphs of *Linum* show qualitative and quantitative differences which may have a role in incompatibility responses.

Unlike the earlier presumption which implied that incompatibility in heteromorphic systems was controlled by the absence of morphological comple-mentation between the stigma and pollen, and differences in the osmotic pressure of pollen grains and styles of the two morphs (see Nettancourt 1977) investigations on *Linum* (Ghosh and Shivanna 1980 b) have demonstrated the operation of phys-iological mechanisms – similar to those in homomorphic systems – in controlling incompatibility. Reports on *Primula vulgaris* (Shivanna et al. 1981) indicate that inhibition of incompatible pollen takes place at many levels – pollen hydration, germination, pollen tube entry into stigma and its growth through the stigma and style. The macromolecular components in the intercellular spaces of transmitting tissue of pistil appear to be involved in inhibiting the growth of incompatible tubes (Shivanna et al. 1981).

The cytology of pollen-pistil interaction (following compatible and incompat-ible pollinations) has been followed in several homomorphic taxa: members of Gramineae (Sastri and Shivanna 1979, J. Heslop-Harrison 1979 a, J. Heslop-Har-rison and Y. Heslop-Harrison 1980), *Petunia* (Herrero and Dickinson 1979, 1980 a, b, 1981, Cresti et al. 1979), and members of Commelinaceae (Herd and Beadle 1980, Owens 1981). S-allele specific proteins have been identified in the pistil of *Nicotiana alata*, through isoelectric focussing (Bredemeijer and Blass 1981). Ferrari et al. (1981) isolated and purified S-allele specific glycoprotein in *Brassica oleracea* var. *capitata*. It eluted as a single peak from Sephadex column, appeared as a single band (which stained with coomassie blue and periodic acid Schiff reagent) after polyacrylamide-gel electrophoresis and had protein to carbo-hydrate ratio of 1.3. In vitro pretreatment of pollen with the glycoprotein obtained from self-stigma prevented pollen germinating even on compatible stigma. Pre-treatment of pollen from the glycoprotein obtained from compatible stigma did not affect germination.

In homomorphic incompatibility also, contrary to the earlier concept which considered self-incompatibility as a one-step reaction, it has been convincingly proved that incompatibility barriers operate at many levels within a system, par-ticularly in sporophytic systems. In *Brassica* Stead et al. (1979, 1980) and Roberts et al. (1980) reported significant differences in the adhesion and hydration of pol-len, and the mobility of pollen wall components following compatible and incom-patible pollinations. Earlier data, although not emphasized, had brought out the differences in the proportion of pollen grains germinating on the stigma, and pollen tubes entering the stigma in selfed and crossed pistils. Incompatibility barrier, at least in sporophytic systems, therefore, operate at all levels of pollen–pistil interac-tion – pollen adhesion, pollen hydration, pollen germination, pollen tube entry into the stigma, and pollen tube growth through the style. Thus the details shown in Figs 4 and 8 require modifications. The number of pollen grains completing succes-sive post-pollination stages become progressivley reduced at each level. Such a re-alization raises serious doubts about the validity of the hypotheses on the oper-ation of incompatibility based on a single mechanism such as the inhibition of pro-tein synthesis in the pollen/pollen tube. It would be appropriate to consider self-incompatibility as a system of many superimposed mechanisms controlled by different genes. Such a basic system has undergone modifications in a number of taxa due to inactivation of one or more of the mechanisms and, thus, altered the phenotypic manifestations of self-incompatibility.

Recent genetic studies also indicate that self-incompatibility controlled by many alleles is more common (than was thought earlier), and covers both gametophytic and sporophytic systems (see Lewis 1979). Our understanding of the operation of intraspecific incompatibility is likely to change radically in the coming years.

References

Abdalla MMF (1974) Unilateral incompatibility in plant species: Analysis and implications. Egypt J Genet Cytol 3:133–154

Ahloowalia BS (1973) Self- and cross-incompatibility of tetraploid Italian and perennial rye grass *Lolium* spp. Incompatibility Newslett No 3:51–52

Anderson WR (1980) Cryptic self-fertilization in the Malphighiaceae. Science 207:892–893

Annerstedt I, Lundquist A (1967) Genetics of self-incompatibility in *Tradescantia paludosa* (Commelinaceae). Hereditas 58:13–30

Arasu NT (1968a) Self-incompatibility in angiosperms: A review. Genetica 39:1–24

Arasu NT (1968b) Overcoming self-incompatibility by irradiation. Rep E Malling Res Stn (1967):109–112

Ascher PD (1966) A gene action model to explain gametophytic self-incompatibility. Euphytica 15:179–183

Ascher PD (1977) Localization of the self- and the interspecific incompatibility reactions in style sections of *Lilium longiflorum*. Plant Sci Lett 10:199–203

Ascher PD, Peloquin SJ (1966a) Effect of floral ageing on the growth of compatible and incompatible pollen tubes in *Lilium longiflorum*. Am J Bot 53:99–102

Ascher PD, Peloquin SJ (1966b) Influence of temperature on incompatible and compatible pollen tube growth in *Lilium longiflorum*. Can J Genet Cytol 8:661–664

Ascher PD, Peloquin SJ (1968) Pollen tube growth and incompatibility following intra- and inter-specific pollination in *Lilium longiflorum*. Am J Bot 55:1230–1234

Ascher PD, Peloquin SJ (1970) Temperature and the self-incompatibility reaction in *Lilium longiflorum* Thunb. J Am Soc Hort Sci 95:586–588

Attia MS (1950) The nature of incompatibility in cabbage. Proc Am Soc Hortic Sci 56:369–371

Balatkova V, Tupý J (1968) Test-tube fertilization in *Nicotiana tabacum* by means of an artificial pollen tube culture. Biol Plant 10:266–270

Bali PN (1963) Some experimental studies on self-incompatibility in *Oenothera rhombipetala* Nutt. Φton (Argentina) 20:97–103

Bell J, Hicks G (1976) Transmitting tissue in the pistil of tobacco: Light and electron microscopic observations. Planta 131:187–200

Bredemeijer GMM, Blass J (1981) S-specific proteins in styles of self-incompatible *Nicotiana alata*. Theor Appl Genet 59:185–190

Brewbaker JL (1957) Pollen cytology and self-incompatibility systems in plants. J Hered 48:271–277

Brewbaker JL, Natarajan AT (1960) Centric fragments and pollen-part mutations of incompatibility alleles in *Petunia*. Genetics 45:699–704

Campbell RJ, Ascher PD (1972) High temperature removal of self-incompatibility in *Nemesia strumosa*. Incompatibility Newslett No 1:3–5

Carter AL, McNeilly T (1973) The effect of humidity on self-incompatibility in mature Brussels sprout flowers. Incompatibility Newslett 2:19–21

Chao CH (1971) A periodic acid-Schiff's substance related to the directional growth of pollen tubes. Am J Bot 58:649–654

Chao CH (1977) Further cytological studies of a periodic acid-Schiff's substance in the ovules of *Paspalum orbiculare*. Am J Bot 64:920–930

Christ B (1959) Entwicklungsgeschichtliche und physiologische Untersuchungen über die Selbsterilität von *Cardamine pratensis* L. Z Bot 47:88–112

Clarke A, Gleeson P, Harrison S, Knox RB (1979) Pollen-stigma interactions: Identification and characterization of surface components with recognition potential. Proc Natl Acad Sci USA 76:3358–3362

Coe GC (1954) Distribution of C_{14} in ovules of *Zephyranthes drummondii*. Bot Gaz 45:342–346

Considine JA, Knox RB (1979) Development and histochemistry of pistil of the grape, *Vitis vinifera*. Ann Bot 43:11–22

Crane MB, Lawrence WJ (1929) Genetical and cytological aspects of incompatibility and sterility in cultivated fruits. J Pomol Hortic Sci 7:276–301

Crane MB, Lewis D (1942) Genetical studies in pears. III. Incompatibility and sterility. J Genet 43:31–38

Cresti M, Van Went JL, Pacini E, Willemse MTM (1976) Ultrastructure of transmitting tissue of *Lycopersicum peruvianum* style: Development and histochemistry. Planta 132:305–312

Cresti M, Ciampolini F, Pacini E (1977) Ultrastructural aspects of pollen tube growth inhibition after gamma irradiation in *Lycopersicum peruvianum*. Theor Appl Genet 49:297–303

Cresti M, Ciampolini F, Pacini E, Sarfatti G, van Went JL, Willemse MTM (1979) Ultrastructural differences between compatible and incompatible pollen tubes in the stylar transmitting tissue of *Petunia hybrida*. J Submicros Cytol 11:209–210

Dashek WV, Thomas HR, Rosen WG (1971) Secretory cells of lily pistils. II. Electron microscope cytochemistry of canal cells. Am J Bot 58:909–920

Dayton DE (1974) Overcoming self-incompatibility in apple with killed compatible pollen. J Am Soc Hortic Sci 99:190–192

Denward T (1963) The function of the incompatibility allele in red clover (*Trifolium pratense* L.). Hereditas 49:189–334

Dickinson HG, Lewis D (1973a) Cytochemical and ultrastructural differences between intraspecific compatible and incompatible pollinations in *Raphanus*. Proc R Soc London Ser B 183:21–38

Dickinson HG, Lewis D (1973b) The formation of the tryphine coating the pollen grains of *Raphanus* and its properties relating to the self-incompatibility. Proc R Soc London Ser B 184:149–165

van der Donk JAWM (1974) Differential synthesis of RNA in self- and cross-pollinated styles of *Petunia hybrida* L. Mol Gen Genet 131:1–8

van der Donk JAWM (1975) Recognition and gene expression during incompatibility reaction in *Petunia hybrida* L. Mol Gen Genet 141:305–317

Dulieu HL (1963) Sur la fécondation in vitro chez le *Nicotiana tabacum* L. CR Acad Sci Paris 256:3344–3346

Dulieu HL (1966) Pollination of excised ovaries and culture of ovules of *Nicotiana tabacum* L. Phytomorphology 16:69–75

Dumas C (1974) Some aspects of stigmatic secretion in *Forsythia*. In: Linskens HF (ed) Fertilization in higher plants. North-Holland Publ, Amsterdam, pp 119–126

Dumas C, Rougier M, Zandonella P, Ciampolini M, Cresti M, Pacini E (1978) The secretory stigma of *Lycopersicum peruvianum* Mill.: Ontogenesis and glandular activity. Protoplasma 96:173–187

East EM (1934) Norms of pollen tube growth in incompatible matings of self-sterile plants. Proc Natl Acad Sci USA 20:225–230

Emsweller SL, Stuart NW (1948) Use of growth regulating substances to overcome incompatibilities in *Lilium*. Proc Am Soc Hortic Sci 51:581–589

Emsweller SL, Uhring J, Stuart NW (1960) The role of naphthalene acetamide and potassium gibberellate in overcoming self-incompatibility in *Lilium longiflorum*. Proc Am Soc Hortic Sci 75:720–725

Eyster WH (1941) The induction of fertility in genetically self-sterile plants. Science 94:144–145

Ferrari TE, Wallace DH (1976) Pollen protein synthesis and control of incompatibility in *Brassica*. Theor Appl Genet 48:243–249

Ferrari TE, Wallace DH (1977) A model for self-recognition and regulation of the incompatible response of pollen. Theor Appl Genet 50:211–225

Ferrari TE, Bruns D, Wallace D (1981) Isolation of a plant glycoprotein involved in control of intercellular recognition. Plant Physiol 67:270–277

Fett WF, Paxton JD, Dickinson DB (1976) Studies on self-incompatibility response in *Lilium longiflorum*. Am J Bot 63:1104–1108

Gastel AJG van, Nettancourt D de (1975) The effects of different mutagens on self-incompatibility in *Nicotiana alata* Link and Otto II. Acute irradiations with X-rays and fast neutrons. Heredity 34:381–392

Gerassimova H (1933) Fertilization in *Crepis capillaris*. Cellule 42:103–148

Ghosh S, Shivanna KR (1980a) Pollen-pistil interaction in *Linum grandiflorum*: Scanning electron microscopic observations and proteins of stigma surface. Planta 149:257–261

Ghosh S, Shivanna KR (1980b) Pollen-pistil interaction in *Linum grandiflorum*: Stigma-surface proteins and stigma receptivity. Proc Indian Natl Sci Acad B46:177–183

Gladding RW, Paxton JD (1975) Pollen "recognition" in *Lilium longiflorum*. Incompatibility Newslett No 6:24–28

Glendinning DR (1960) Selfing of self-incompatible cocoa. Nature (London) 187:170

Glenk HO, Schimmer D, Wagner W (1970) Die Calcium-Verteilung in *Oenothera*-pflanzen und ihr möglicher Einfluß auf den Chenotropisms der Pollenschläuche und auf die Befruchtung. Phyton (Austria) 14:97–111

Hecht A (1960) Growth of pollen tubes of *Oenothera organensis* through otherwise incompatible styles. Am J Bot 47:32–36

Hecht A (1961) Partial reduction of an incompatibility substance in the styles of *Oenothera organensis*. Genetica 46:869

Herd YR, Beadle DJ (1980) The site of self-incompatibility mechanism in *Tradescantia pallida*. Ann Bot 45:251–256

Herrero M, Dickinson HG (1979) Pollen-pistil incompatibility in *Petunia hybrida*: Changes in the pistil following compatible and incompatible intraspecific crosses. J Cell Sci 36:1–18

Herrero M, Dickinson HG (1980a) Ultrastructural and physiological differences between buds and mature flowers of *Petunia hybrida* prior to and following pollination. Planta 148:138–145

Herrero M, Dickinson HG (1980b) Pollen tube growth following compatible and incompatible intraspecific pollinations in *Petunia hybrida*. Planta 148:217–221

Herrero M, Dickinson HG (1981) Pollen tube development in *Petunia hybrida* following compatible and incompatible intraspecific matings. J cell Sci 47:365–383

Heslop-Harrison J (1975a) Physiology of the pollen grain surface. The Croonian Lecture, 1974. Proc R Soc London 190B:275–299

Heslop-Harrison J (1975b) Incompatibility and the pollen-stigma interaction. Annu Rev Plant Physiol 26:403–425

Heslop-Harrison J (1975c) Male gametophyte selection and the pollen-stigma interaction. In: Mulcahy DL (ed) Gamete competition in plants and animals. North Holland Publ, Amsterdam, pp 177–190

Heslop-Harrison J (1976) A new look at pollination. Amos Memorial Lecture. Rep E Malling Res Stn 1975:141–157

Heslop-Harrison J (1979a) Pollen-stigma interaction in grasses: A brief review. New Zealand J Bot 17:537–546

Heslop-Harrison J (1979b) An interpretation of the hydrodynamics of pollen. Am J Bot 66:734–743

Heslop-Harrison J, Heslop-Harrison Y (1975) Enzyme removal of the proteinaceous pellicle of the stigma papilla prevents pollen tube entry in the Caryophyllaceae. Ann Bot 39:163–165

Heslop-Harrison J, Heslop-Harrison Y (1980) The pollen-stigma interaction in the grasses. 1. Fine-structure and cytochemistry of the stigma of *Hordeum* and *Secale*. Acta bot Neerl 29:261–276

Heslop-Harrison J, Heslop-Harrison Y, Knox RB, Howlett B (1973) Pollen wall proteins: gametophytic and sporophytic fractions in the pollen walls of the Malvaceae. Ann Bot 37:403–412

Heslop-Harrison J, Knox RB, Heslop-Harrison Y (1974) Pollen-wall proteins: Exine held fractions associated with the incompatibility response in Cruciferae. Theor Appl Genet 44:133–137

Heslop-Harrison J, Knox RB, Heslop-Harrison Y, Mattsson O (1975a) Pollen wall proteins. Emission and role in incompatibility responses. In: Duckett JG, Racey PA (eds) The biology of the male gamete, vol VII. Biol J Linn Soc, pp 189–202

Heslop-Harrison J, Heslop-Harrison Y, Barber J (1975b) The stigma surface in incompatibility responses. Proc R Soc London Ser B 188:287–297

Heslop-Harrison Y, Shivanna KR (1977) The receptive surface of the angiosperm stigma. Ann Bot 41:1233–1258

Heslop-Harrison Y, Heslop-Harrison J, Shivanna KR (1981) Heterostyly in *Primula*. 1. Fine structural and cytochemical features of the stigma and style in *Primula vulgaris* Huds. Protoplasma 107:171–188

Hess D, Wagner G (1974) Induction of haploid parthenogenesis in *Mimulus luteus* by in vitro pollination by foreign pollen. Z Pflanzenphysiol 72:466–468

Hogenboom NG (1972) Breaking breeding barriers in *Lycopersicon*. Breakdown of self-incompatibility in *L.peruvianum* (L.). Mill. Euphytica 21:228–243

Hogenboom NG (1975) Incompatibility and incongruity: Two different mechanisms for the non-functioning of intimate partner relationships. Proc R Soc London Ser B 188:361–375

Hopper JE, Peloquin SJ (1968) X-ray inactivation of the stylar component of the self-incompatibility reaction in *Lilium longiflorum*. Can J Genet Cytol 10:941–944

Hopping ME, Jerram EM (1979) Pollination of Kivi fruit (*Actinidia chinensis* Planch.): Stigma-style structure and pollen tube growth. New Zealand J Bot 17:233–240

Howlett BJ, Knox RB, Paxton JH, Heslop-Harrison J (1975) Pollen-wall proteins: Physicochemical characterization and role in self-incompatibility in *Cosmos bipinnatus*. Proc R Soc London Ser B 188:167–182

Iwanami Y (1959) Physiological studies on pollen. J Yokohama Municipal Univ C-55 Biol No 175 24:1–31

Jensen WA (1973) Fertilization in flowering plants. Bioscience 23:21–27

Jensen WA, Fischer DB (1968) Cotton embryogenesis: The entrance and discharge of the pollen tube in the embryo sac. Planta 78:158–183

Jensen WA, Fischer DB (1969) Cotton embryogenesis: The tissues of the stigma and style and their relation to the pollen tube. Planta 84:97–121

Kakizaki Y (1930) Studies on the genetics and physiology of self- and cross-incompatibility in the common cabbage. Jpn J Bot 5:133–208

Kanta K, Maheshwari P (1963) Intraovarian pollination in some Papaveraceae. Phytomorphology 13:215–229

Kanta K, Rangaswamy NS, Maheshwari P (1962) Test-tube fertilization in a flowering plant. Nature (London) 194:1214–1217

Kapil RN, Bhatnagar AK (1975) A fresh look at the process of double fertilization in angiosperms. Phytomorphology 25:334–368

Kendall WA, Taylor NL (1969) Effects of temperature on pseudo-self-incompatibility in *Trifolium pratense* L. Theor Appl Genet 39:123–126

Kenrick J, Knox RB (1981) Post-pollination exudate from stigmas of *Acacia* Mimosaceae). Ann Bot 48:103–106

Knox RB (1971) Pollen-wall proteins: Localization, enzymic, and antigenic activity during development in *Gladiolus* (Iridaceae). J Cell Sci 9:209–237

Knox RB (1973) Pollen-wall proteins: Pollen-stigma interaction in ragweed and *Cosmos* (Compositae). J Cell Sci 12:421–443

Knox RB, Heslop-Harrison J (1969) Cytochemical localization of enzymes in the wall of the pollen grain. Nature (London) 223:92–94

Knox RB, Heslop-Harrison J (1970) Pollen wall proteins: Localization and enzymic activity. J Cell Sci 6:1–27

Knox RB, Heslop-Harrison J (1971 a) Pollen-wall proteins: Electron microscopic localization of acid phosphatase in the intine of *Crocus vernus*. J Cell Sci 8:727–733

Knox RB, Heslop-Harrison J (1971 b) Pollen-wall proteins: The fate of intine-held antigens on the stigma in compatible and incompatible pollination of *Phalaris tuberosa*. J Cell Sci 9:239–251

Knox RB, Heslop-Harrison J, Reed C (1970) Localization of antigens associated with the pollen grain wall by immunofluorescence. Nature (London) 225:1066–1068

Knox RB, Willing R, Ashford AE (1972 a) Role of pollen-wall proteins as recognition substances in interspecific incompatibility in poplars. Nature (London) 237:381–383

Knox RB, Willing R, Pryor LD (1972 b) Interspecific hybridization in poplars using recognition pollen. Silvae Genet 21:65–69

Knox RB, Clarke A, Harrison S, Smith P, Marchalonis JJ (1976) Cell recognition in plants: Determinants of the stigma surface and their pollen interactions. Proc Natl Acad Sci USA 73:2788–2792

Konar RN, Linskens HF (1966 a) The morphology and anatomy of stigma of *Petunia hybrida*. Planta 71:356–371

Konar RN, Linskens HF (1966 b) Physiology and biochemistry of the stigmatic fluid in *Petunia hybrida*. Planta 71:372–387

Kristen U, Biedermann M, Liebezeit G, Dawson R, Böhm L (1979) The composition of stigmatic exudate and the ultrastructure of the stigma papillae in *Aptenia cardifolia*. European J Cell Biol 19:281–287

Kroh M (1966) Reaction of pollen after transfer from one stigma to another. Contribution to the character of the incompatibility mechanism in Cruciferae. Züchter 36:185–189

Kroh M, Helsper JPFG (1974) Transmitting tissue and pollen tube growth. In: Linskens HF (ed) Fertilization in higher plants. North-Holland Publ, Amsterdam, pp 167–175

Kroh M, Miki-Hirosige H, Rosen W, Loewus F (1970) Incorporation of label into pollen tube walls from myoinositol-labelled *Lilium longiflorum* pistils. Plant Physiol 45:92–94

Kumar S, Hecht A (1970) Studies on growth and utilization of stylar carbohydrate by pollen tubes and callose development in self-incompatible *Oenothera organensis*. Biol Plant 12:41–46

Kwack B (1969) Chemotropic growth of *Clivia* and *Crinum* pollen toward pistils as influenced by calcium action. Korean J Hortic Sci 6:81–84 (cited from Mascarenhas 1975)

Labarca C, Loewus F (1973) The nutritional role of pistil exudate in pollen tube wall formation in *Lilium longiflorum*. II. Production and utilization of exudate from stigma and stylar canal. Plant Physiol 52:87–92

Labarca C, Kroh M, Loewus F (1970) The composition of the stigmatic exudate from *Lilium longiflorum*. Labelling studies with myo-inositol, D-glucose, and L-proline. Plant Physiol 46:150–156

Laffel RC (1963) Pseudo-self-compatibility and segregation of gametophytic self-incompatibility alleles in red clover *Trifolium pratense* L. Crop Sci 3:377–380

Larsen K (1977) Self-incompatibility in *Beta vulgaris* L. I. Four gametophytic, complementary S-loci in sugar beet. Hereditas 85:227–248

Lawson J, Dickinson HG (1975) The site of the incompatibility reaction in *Lilium henryi*. Incompatibility Newslett 6:18–21

Lewis D (1942) The physiology of incompatibility in plants. I. The effect of temperature. Proc R Soc London Ser B 131:13–26

Lewis D (1949) Incompatibility in flowering plants. Biol Rev 24:472–496

Lewis D (1952) Serological reactions of pollen incompatibility substances. Proc R Soc London Ser B 140:127–135

Lewis D (1954) Comparative incompatibility in angiosperms and fungi. Adv Genet 6:235–285

Lewis D (1956) Incompatibility and plant breeding. Brookhaven Symp Biol 9:89–100

Lewis D (1960) Genetic control of specificity and activity of the S-antigen in plants. Proc R Soc London Ser B 151:468–477

Lewis D (1965) A protein dimer hypothesis on incompatibility. In: Geerts SD (ed) Genetics today, vol III. Proc XI Int Congr Genet, The Hague, pp 657–663

Lewis D (1977) Sporophytic incompatibility with 2 and 3 genes. Proc R Soc London Ser B 196:161–170

Lewis D (1979) Genetic variability of incompatibility in plants. New Zealand J Bot 17:637–644

Lewis D, Crowe LK (1953) Theory of reversible mutation. Nature (London) 171:561

Lewis D, Crowe LK (1958) Unilateral interspecific incompatibility in flowering plants. Heredity 12:233–256

Lewis D, Burrage S, Walls D (1967) Immunological reactions of single pollen grains: Electrophoresis and enzymology of pollen protein exudates. J Exp Bot 18:371–378

Linskens HF (1960) Zur Frage der Entstehung der Abwehrkörper bei der Inkompatibilitätsreaktion von *Petunia*. III. Mitt. Serologische Teste mit Leitgewebs- und Pollenextrakten. Z Bot 48:126–135

Linskens HF (1965) Biochemistry of incompatibility. In: Geerts SD (ed) Genetics today, vol III. Proc XI th Int Congr Genet, The Hague, pp 629–636

Linskens HF (1975) Incompatibility in *Petunia*. Proc R Soc London Ser B 188:299–311

Linskens HF, Kroh M (1967) Incompatibilität der Phanerogamen. Encycl Plant Physiol 18:506–530

Linskens HF, Schrauwen JAM (1966) Measurement of oxygen tension changes in the styles during pollen tube growth. Planta 71:98–106

Linskens HF, Spanjers W (1973) Changes of the electric potential in the transmitting tissue of *Petunia* styles after cross- and self-pollination. Incompatibility Newslett 3:81–85

Linskens HF, Schrauwen JAM, van der Donk M (1960) Überwindung der Selbstinkompatibilität durch Röntgenbestrahlung des Griffels. Naturwissenschaften 46:547

Lundquist A (1975) Complex self-incompatibility systems in angiosperms. Proc R Soc London Ser B 188:235-245

Maheshwari P (1950) An introduction to the embryology of the angiosperms. McGraw-Hill, New York

Maheshwari P, Kanta K (1961) Intraovarian pollination in *Eschscholzia californica* Cham., *Argemone mexicana* L., and *Argemone ochroleuca* Sweet. Nature (London) 191:304

Maheshwari P, Kanta K (1964) Control of fertilization. In: Linskens HF (ed) Pollen physiology and fertilization. North-Holland Publ, Amsterdam, pp 187–194

Maheshwari P, Rangaswamy NS (1965) Embryology in relation to physiology and genetics. In: Preston RD (ed) Advances in botanical research, vol II. Academic Press, London New York, pp 219–312

Martin FW (1969) Compounds of the stigma of 10 species. Am J Bot 56:1023–1027

Martin FW (1970) Compounds of the stigmatic surface of *Zea mays* L. Ann Bot 34:835–842

Martin FW (1972) In vitro measurement of pollen tube growth inhibition. Plant Physiol 49:924–925

Martin FW, Ruberte R (1972) Inhibition of pollen germination and tube growth by stigmatic substances. Phyton 30:119–126

Mascarenhas JP (1966) The distribution of ionic calcium of *Antirrhinum majus*. Protoplasma 62:53–58

Mascarenhas JP (1975) The biochemistry of the angiosperm pollen development. Bot Rev 41:259–314

Mascarenhas JP, Machlis L (1962) The pollen tube chemotropic factor from *Antirrhinum majus*: Bioassay, extraction and partial purification. Am J Bot 49:482–489

Mascarenhas JP, Machlis L (1964) Chemotropic response of the pollen tubes of *Antirrhinum majus* to calcium. Plant Physiol 39:70–77

Matsubara S (1973) Overcoming self-incompatibility by cytokinin treatment on *Lilium longiflorum*. Bot Mag Tokyo 86:43–46

Mattsson O, Knox RB, Heslop-Harrison J, Heslop-Harrison Y (1974) Protein pellicle as a probable recognition site in incompatibility reactions. Nature (London) 213:703–704

Muraba AIM El (1957) The effect of high temperature on incompatibility in radish. Euphytica 6:268–270

Nasrallah ME (1974) Genetic control of quantitative variation in self-incompatibility proteins detected by immunodiffusion. Genetics 76:49–50

Nasrallah ME, Wallace DH (1967) Immunogenetics of self-incompatibility in *Brassica oleracea*. Heredity 22:519–527

Nasrallah ME, Barber JT, Wallace DH (1970) Self-incompatibility proteins in plants: Detection, genetics, and possible mode of action. Heredity 25:23–27

Nasrallah ME, Wallace DH, Savo RM (1972) Genotype, protein, phenotype relationships in self-incompatibility of *Brassica*. Genet Res 20:151–160

Nettancourt D de (1972) Self-incompatibility in basic and applied researches with higher plants. Genet Agrar 26:163–216

Nettancourt D de (1977) Incompatibility in angiosperms. Springer, Berlin Heidelberg New York

Nettancourt D de, Ecochard R (1968) Effects of chronic irradiation upon a self-incompatible clone of *Lycopersicum peruvianum*. Theor Appl Genet 38:289–293

Nettancourt D de, Ecochard R, Perquin MDG, van der Driff T, Westerhof M (1971) The generation of new S-alleles at the incompatibility locus of *Lycopersicum peruvianum* Mill. Theor Appl Genet 41:120–129

Nettancourt D de, Devreux M, Bozzini A, Cresti M, Pacini E, Sarfatti G (1973a) Ultrastructural aspects of the self-incompatibility mechanism in *Lycopersicum peruvianum* Mill. J Cell Sci 12:403–419

Nettancourt D de, Devreux M, Laneri U, Pacini E, Cresti M, Sarfatti G (1973b) Ultrastructural aspects of unilateral interspecific incompatibility between *Lycopersicum peruvianum* and *L. esculentum*. Caryologia 25:207–217

Nettancourt D de, Devreux M, Laneri U, Cresti M, Pacini E, Sarfatti G (1974) Genetical and ultrastructural aspects of self- and cross-incompatibility in interspecific hybrids between self-compatible *Lycopersicum esculentum* and self-incompatible *L. peruvianum*. Theor Appl Genet 44:278–288

Nettancourt D de, Devereux M, Carluccio F, Laneri U, Cresti M, Pacini E, Sarfatti G, van Gastel AJG (1975) Facts and hypothesis on origin of S-mutations and on the function of the S-gene in *Nicotiana alata* and *Lycopersicum peruvianum*. Proc R Soc London Ser B 188:345–360

Niimi Y (1970) In vitro fertilization in the self-incompatible plant *Petunia hybrida*. J Jpn Soc Hortic Sci 39:345–352

Niimi Y (1976) Effect of "stylar pollination" on in vitro seed setting of *Petunia hybrida*. J Jpn Soc Hortic Sci 45:168–172

Nishio T, Hinata K (1977) Analysis of S-specific proteins in stigma of *Brassica oleracea* L. by isoelectric focusing. Heredity 38:391–396

Ockendon DJ (1978) Effect of hexane and humidity on self-incompatibility in *Brassica oleracea*. Theor Appl Genet 52:113–117

Odland ML, Noll CJ (1950) The utilization of cross-incompatibility and self-incompatibility in the production of F_1-hybrid cabbage. Proc Am Soc Hortic Sci 55:391–398

Opeke LK, Jacob VJ (1969) Studies on methods of overcoming incompatibility in *Theobroma cacao* L. Proc II Int Cacao Res Conf, Bahia, pp 356–358

Owens SJ (1981) Self-incompatibility in the Commelinaceae. Ann Bot 47:567–581

Pandey KK (1956) Mutations of self-incompatibility alleles in *Trifolium pratense* and *T. repens*. Genetics 41:327–343

Pandey KK (1960) Incompatibility in *Abutilon hybridum*. Am J Bot 47:877–883

Pandey KK (1963) Stigmatic secretion and bud pollinations in self- and cross-incompatible plants. Naturwissenschaften 50:408–409

Pandey KK (1965) Centric chromosome fragments and pollen part mutation of the incompatibility gene in *Nicotiana alata*. Nature (London) 206:792–795

Pandey KK (1968) Compatibility relationships in flowering plants: Role of the S-gene complex. Am Nat 102:475–489

Pandey KK (1975) Model for incompatibility determination in flowering plants. Incompatibility Newslett 6:70–73

Pandey KK (1977) Mentor pollen: Possible role of wall-held pollen growth promoting substances in overcoming intra- and interspecific incompatibility. Genetica 47:219–229

Pettitt JM (1980) Reproduction in sea-grasses: Nature of the pollen and receptive surface of the stigma in the Hydrocharitaceae. Ann Bot 45:257–271

Pettitt JM, McConchie CA, Ducker SC, Knox RB (1980) Unique adaptations for submarine pollination in seagrasses. Nature (London) 286:487–489

van der Pluijm JE, Linskens HF (1966) Die Feinstruktur der Pollenschalanche im Griffel von *Petunia*. Genet Breed Res 36:220–222

Raff J, Knox RB (1977) Self-incompatibility in sweet cherry, *Prunus avium*. Incompatibility Newslett 8:36–39

Raghavan V (1976) Experimental embryogenesis in vascular plants. Academic Press, London New York

Rangaswamy NS (1964) Control of fertilization and embryo development. In: Maheshwari P (ed) Recent advances in the embryology of angiosperms. Int Soc Plant Morphol, Univ Delhi, Delhi, pp 327–353

Rangaswamy NS (1977) Applications of in vitro pollination and in vitro fertilization. In: Reinert J, Bajaj YPS (eds) Applied and fundamental aspects of plant cell, tissue and organ culture. Springer, Berlin Heidelberg New York, pp 412–425

Rangaswamy NS, Shivanna KR (1967) Induction of gametic compatibility and seed formation in axenic cultures of a diploid self-incompatible species of *Petunia*. Nature (London) 216:937–939

Rangaswamy NS, Shivanna KR (1969) Test-tube fertilization in *Dicranostigma franchetianum* (Prain) Fedde. Curr Sci 38:257–259

Rangaswamy NS, Shivanna KR (1971a) Overcoming self-incompatibility in *Petunia axillaris* (Lam.) BSP. II. Placental pollination in vitro. J Indian Bot Soc 50A:286–296

Rangaswamy NS, Shivanna KR (1971b) Overcoming self-incompatibility in *Petunia axillaris*. III. Two-site pollinations in vitro. Phytomorphology 21:284–289

Rao PS (1965) The in vitro fertilization and seed formation in *Nicotiana rustica* L. Phyton 22:165–167

Roberts IN, Stead D, Ockendon DJ, Dickinson HG (1980) Pollen stigma interactions in *Brassica oleracea*. Theor Appl Genet 58:241–246

Roggen HPJR (1967) Changes in enzyme activities during the progame-phase in *Petunia hybrida*. Acta Bot Neerl 16:1–31

Roggen HPJR (1975) Stigma application of an extract from rape pollen (*Brassica napus* L.) effects self-incompatibility in Brussels sprouts (*Brassica oleracea* L. var. *gemmifera* DC.). Incompatibility Newslett No 6:80–86

Roggen HPJR, Linskens HF (1967) Pollen tube growth and respiration in incompatible intergeneric crosses. Naturwissenschaften 54:542–543

Roggen HPJR, van Dijk AJ (1972) Breaking incompatibility in *Brassica oleracea* L. by steel brush pollination. Euphytica 21:424–425

Roggen H, van Dijk AJ (1976) Thermally aided pollinations: A new method of breaking self-incompatibility in *Brassica oleracea* L. Euphytica 25:643–646

Roggen HPJR, van Dijk AJ, Dorsman C (1972) Electric aided pollination: A method of breaking incompatibility in *Brassica oleracea*. Euphytica 21:181–184

Rosen WG (1961) Studies on pollen tube chemotropism. Am J Bot 48:889–895

Rosen WG (1971) Pollen-pistil interactions in *Lilium*. In: Heslop-Harrison J (ed) Pollen: Development and physiology. Butterworths, London, pp 239–254

Rosen WG, Thomas HR (1970) Secretory cells of lily pistils. I. Fine structure and function. Am J Bot 57:1108–1114

Rosenfield C, Loewus FA (1975) Carbohydrate interconversion in pollen-pistil interaction in lily. In: Mulcahy DL (ed) Gamete competition in plants and animals. North-Holland Publ, Amsterdam, pp 151–160

Rudramuniyappa CK, Panchaksharappa MG (1974) Histochemistry of pollen tube growth in vivo in *Triticum durum* Desf. Cytologia 39:665–671

Sassen MMA (1974) The stylar transmitting tissue. Acta Bot Neerl 23:99–108

Sastri DC, Shivanna KR (1976a) Recognition pollen alters incompatibility in *Petunia*. Incompatibility Newslett No 7:22–24

Sastri DC, Shivanna KR (1976b) Attempts to overcome interspecific incompatibility in *Sesamum* using recognition pollen. Ann Bot 41:891–893

Sastri DC, Shivanna KR (1979) Role of pollen-wall proteins in intraspecific incompatibility in *Saccharum bengalens*. Phytomorphology 29:324–330

Sears ER (1937) Cytological phenomena connected with self-incompatibility in flowering plants. Genetics 22:130–181

Sedgley M (1975) Flavanoids in pollen and stigma of *Brassica oleracea* and their effect on pollen germination in vitro. Ann Bot 39:1091–1095

Sedgley M (1976) Control by the embryo sac over pollen tube growth in the avocado (*Persea americana* Mill.). New Phytol 77:149–152

Sedgley M, Buttrose MS (1978) Structure of the stigma and style of the avocado. Aust J Bot 26:663–682

Sedgley M, Scholefield BP (1980) Stigma secretion in the water melon before and after pollination. Bot Gaz 141:428–434

Shivanna KR (1965) In vitro fertilization and seed formation in *Petunia violacea* Lindl. Phytomorphology 15:183–185

Shivanna KR (1971) Overcoming self-incompatibility in *Petunia*. Differential treatment in vitro of the whole placentae. Experientia 27:864–865

Shivanna KR (1977) Pollen-stigma interaction – recognition, acceptance and rejection. Symposium on basic sciences and agriculture. Indian Natl Sci Acad, New Delhi, pp 53–61

Shivanna KR (1979) Recognition and rejection phenomenon during pollen-pistil interaction. Proc Indian Acad Sci B88:115–141

Shivanna KR, Heslop-Harrison J (1981) Membrane state and pollen viability. Ann Bot 47:759–770

Shivanna KR, Rangaswamy NS (1969) Overcoming self-incompatibility in *Petunia axillaris*. I. Delayed pollination, pollination with stored pollen, and bud pollination. Phytomorphology 19:372–380

Shivanna KR, Sastri DC (1976) Stigma-surface proteins and bud pollination in *Petunia*: A correlative study. Incompatibility Newslett No 7:16–21

Shivanna KR, Sastri DC (1981) Stigma-surface proteins and stigma receptivity in some taxa characterized by wet stigma. Ann Bot 47:53–64

Shivanna KR, Heslop-Harrison Y, Heslop-Harrison J (1978) Pollen stigma interaction: Bud pollination in Cruciferae. Acta Bot Neerl 27:107–119

Shivanna KR, Heslop-Harrison J, Heslop-Harrison Y (1981) Heterostyly in *Primula*. 2. Sites of pollen inhibition and effects of pistil constituents on compatible and incompatible pollen tube growth. Protoplasma 107:319–338

Shivanna KR, Sastri DC, Jagadish MN (1976) Intraspecific incompatibility in grasses: rejection phenomenon and role of intine proteins. Abstr IV Int Palynol Conf, Lucknow India, p 166

Stanley RG, Linskens HF (1967) Oxygen tension as a control mechanism in pollen tube rupture. Science 157:833–834

Stead AD, Roberts IN, Dickinson HG (1979) Pollen-pistil interaction in *Brassica oleracea*. Events prior to pollen germination. Planta 146:211–216

Stead AD, Roberts IN, Dickinson HG (1980) Pollen-stigma interaction in *Brassica oleracea*. The role of stigmatic proteins in pollen grain adhesion. J cell Sci 42:417–423

Stettler RF (1968) Irradiated mentor pollen: its use in remote hybridization in black cotton wood. Nature (London) 219:746–747

Stout AB, Chandler C (1941) Change from self-incompatibility accompanying the change from diploid to tetraploid. Science 94:114

Swaminathan MS (1955) Overcoming cross-incompatibility among some Mexican diploid species of *Solanum*. Nature (London) 176:887–888

Swaminathan MS, Murty BR (1957) One-way incompatibility in some species crosses in the genus *Nicotiana*. Indian J Genet Plant Breed 17:23–26

Takahashi H (1974) Genetical and physiological analysis of intraspecific incompatibility between *Nicotiana alata* and *N.langsdorffii*. Jpn J Genet 49:247–256

Tara CP, Namboodiri AN (1976) Association between defective stigmatic exudate and sterility in *Impatiens*: Chromatographic evidences. Indian J Exp Biol 14:354–355

Townsend CE (1965) Seasonal and temperature effects on self-incompatibility in tetraploid alsike clover, *Trifolium hybridum* L. Crop Sci 5:329–332

Townsend CE (1966) Self-incompatibility response to temperature and the segregation of S-alleles in diploid alsike clover. *Trifolium hybridum* L. Crop Sci 10:558–567

Townsend CE (1971) Advances in the study of incompatibility. In: Heslop-Harrison J (ed) Pollen: Development and physiology. Butterworths, London, pp 281–309

Tupý J (1961) Investigation of free amino acids in cross-, self-, and non-pollinated pistils of *Nicotiana alata*. Biol Plant 3:47–64

Usha SV (1965) In vitro pollination in *Antirrhinum majus* L. Curr Sci 34:511–513

Vasil IK (1974) The histology and physiology of pollen germination and pollen tube growth on the stigma and in the style. In: Linskens HF (ed) Fertilization in higher plants. North-Holland Publ, Amsterdam, pp 105–118

Visser DL (1977) The effect of alternating temperatures on the self-incompatibility of some clones of Brussels sprouts (*Brassica oleracea* L. var *gemmifera* DC Schultz). Euphytica 26:273–277

Wagner G, Hess D (1973) In vitro-Befruchtungen bei *Petunia hybrida*. Z. Pflanzenphysiol 69:262–269

Welk sr M, Millington WF, Rosen WG (1965) Chemotropic activity and the pathway of the pollen tube in lily. Am J Bot 52:774–781

Willing RR, Pryor L (1976) Interspecific hybridization in poplar. Theor Appl Genet 47:141–151

Woittiez RD, Willemse MTM (1979) Sticking of pollen on stigmas: The factors and a model. Phytomorphology 29:57–63

Wricke G (1974) Seed-set in rye after selfing under high temperature conditions. Incompatibility Newslett No 4:23–27

Yasuda S (1934) Physiological researches on self-incompatibility in *Petunia*. Bull Imp Coll Agric For Morioka 20:1–95

Zenkteler M (1965) Test-tube fertilization in *Dianthus caryophyllus* Linn. Naturwissenschaften 23:645–646

Zenkteler M (1967) Test-tube fertilization of ovules in *Melandrium album* Mill. with pollen grains of several species of the Caryophyllaceae family. Experientia 23:775–777

Zenkteler M (1970) Test-tube fertilization of ovules in *Melandrium album* Mill. with pollen grains of *Datura stramonium* L. Experientia 26:661–662

Zenkteler M, Misiura E, Guzowska I (1975) Studies on obtaining hybrid embryos in test tubes. In: Mohan Ram HY, Shah JJ, Shah CK (eds) Form, structure, and function in plants. BM Johri Comm Volume. Sarita Prakashan, Meerut, pp 180–187

8. Endosperm Culture

P.S. SRIVASTAVA

In gymnosperms the nutritive tissue (female gametophyte) necessary for the growth and development of the embryo is already present at the time of fertilization. In flowering plants, on the other hand, the development of nutritive tissue is postponed until after fertilization (triple fusion). The endosperm develops as a result of repeated divisions of the primary endosperm nucleus, and is unique in genetic constitution.

The embryo, in angiosperms as in other plants, is a diploid product of fertilization, but the endosperm is usually triploid. The endosperm is situated between the preceding and succeeding sporophytic tissues. It is the main source of reserve food for the developing embryo and has a dynamic influence on its differentiation. Failure of proper development of endosperm results in the abortion of embryo.

Generally, the endosperm is a short-lived tissue, and is consumed during the development of the embryo (exalbuminous seed). In some plants, as castor (Euphorbiaceae), it persists as a massive tissue even in the mature seed (albuminous).

The endosperm is an excellent experimental system for morphogenic studies, since it lacks any differentiation and consists mostly of parenchymatous cells.

The exceptions where the endosperm does not develop are the families Trapaceae, Podostemonaceae, and Orchidaceae. In these taxa the embryo has alternative sources of nutrition.

Lampe and Mills (1933) were the first to report the proliferation of immature endosperm tissue of maize, grown on a medium containing extract of potato or young corn kernels. LaRue (1947) observed that in Nature, in some maize kernels, the pericarp ruptured and the endosperm exhibited a white tissue mass. Subsequently, he (LaRue 1949) demonstrated the potentiality of maize endosperm tissue for unlimited growth. It was reported that "... in *Zea mays*, a few specimens, less than one in a thousand developed root, and one formed a shoot-root axis and miniature leaves." Since then several workers have cultured maize endosperm but invariably failed to obtain organogenesis (Fig. 1 A, B; see Sehgal 1969). It may be noted that only the sugary variety of maize yields tissue cultures.

Culture of Immature Endosperm

Subsequent to LaRue's report, immature endosperm of many other species has been successfully cultured (Lampton 1952, Straus and LaRue 1954; Sternheimer 1954; Norstog 1956; Nakajima 1962; Nakano et al. 1975; Mu et al. 1977).

LaRue (1947) cultured maize endosperm on media supplemented with extracts of tomato (TJ), coconut milk (CM), and yeast extract (YE). Of these, 20% TJ proved most effective and stimulated 3- to 18-fold increase in fresh weight within

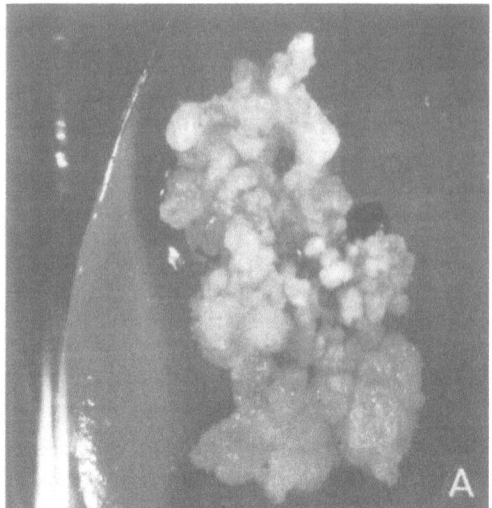

Fig. 1 A, B. *Zea mays.* A 6-week-old endosperm culture on WM + YE (0.5%) showing numerous nodular structures. B Section of 4-week-old callus; note compactly arranged cells in the nodules. (After Sehgal 1969)

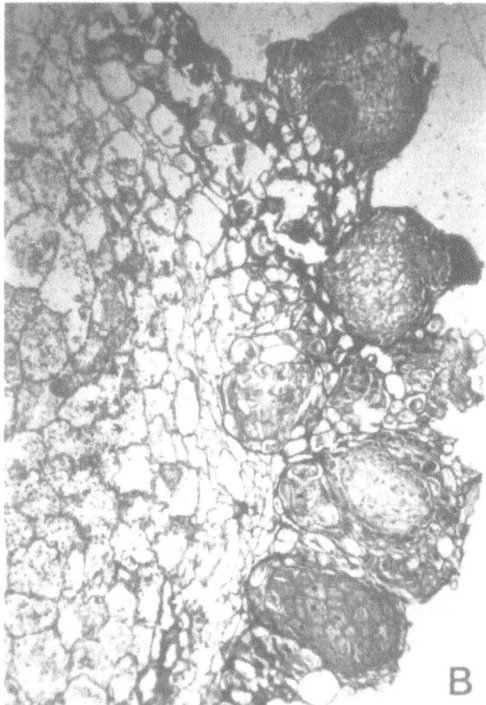

3 weeks, and the callus could be repeatedly subcultured. YE could substitute TJ to a considerable extent (Tamaoki and Ullstrup 1958). Straus (1960a) pointed out that the growth factor present in YE is not the minerals, but a substance soluble in 85% ethanol. He (Straus 1960b) also discovered a compound similar to asparagine in TJ. Asparagine supported better growth than YE, TJ, or casein hydrolysate (CH). Glutamic acid and aspartic acid also stimulated growth appreciably. The ad-

dition of aspartic acid to a medium containing asparagine was further stimulatory for the growth of maize endosperm tissue.

Among sugars, Straus and LaRue (1954) reported optimal growth of maize endosperm on sucrose, glucose, and fructose-fortified medium. Lactose, arabinose, galactose, and rhamnose proved inhibitory. For *Asimina* endosperm, even starch supports growth (Lampton 1952). Nakajima (1962) demonstrated that satisfactory growth of the immature endosperm of *Cucumis sativus* occurs on a medium containing indoleacetic acid (IAA) + 1,3-diphenylurea (DPU) + CH. These reports indicate that the endosperm tissue requires, in addition to minerals, an additional source of nitrogen, and CH or YE seems essential.

Earlier investigators emphasised the significance of the age of endosperm explant for proper in vitro growth. The endosperm tissue of maize younger than 8 or older than 12 days after pollination did not grow in cultures. The endosperm tissue of *Lolium* (Norstog 1956) and *Cucumis sativus* (Nakajima 1962) could be grown only when excised 7–10 days after pollination. In rice, Nakano et al. (1975) observed that only cellular endosperm containing starch grains excised 4–7 days after pollination responded to the culture medium. Coenocytic (free-nuclear) endosperm which lacked starch did not produce any callus. Perhaps callus formation is dependent on the level of organization of cells in the developing endosperm. In maize, the endosperm becomes cellular 3–5 days after pollination, and for 8 days the cells remain meristematic. Later, the mitoses are restricted to only the outermost layer.

Culture of Mature Endosperm

The earlier contention that the endosperm of only a certain age responds favourably to culture treatments has been disproved during the last decade. The mature endosperm of several taxa has now been grown successfully. Rangaswamy and Rao (1963) established tissue cultures from the mature endosperm of *Santalum album*, and Satsangi and Mohan Ram (1965) from *Ricinus communis*. Since then mature endosperm of several taxa has yielded a continuously growing tissue (see Johri 1971): *Exocarpus cupressiformis* and *Scurrula pulverulenta* (Bhojwani 1968); *Phoradendron tomentosum* (Bajaj 1970); *Dendrophthoe falcata, Taxillus cuneatus, T. vestitus*, and *Leptomeria acida* (all semi-parasitic angiosperms, Nag 1970). However, only a few autotrophic taxa of angiosperms have given equally good results: *Croton bonplandianum* (Bhojwani 1966); *Jatropha panduraefolia, Putranjiva roxburghii, Ricinus communis*, and *Sapium sebiferum* (Srivastava 1971 a); *Coffea arabica* (Monaco et al. 1977), and *Petroselinum hortense* (Masuda et al. 1978). While the tissue generally exhibited unlimited growth, organ formation was achieved in all the above taxa except *C. arabica, R. communis*, and *S. album*.

Mostly, the mature endosperm of parasitic taxa shows optimal growth on a medium containing either only a cytokinin or a cytokinin + an auxin. However, in autotrophic members, in addition to an auxin and cytokinin, CH or YE is also essential. In *C. arabica* malt extract (ME) and YE serve the purpose (see Monaco et al. 1977).

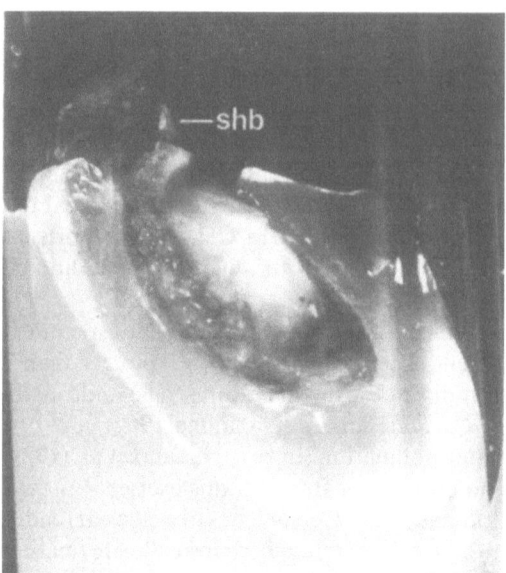

Fig. 2. *Putranjiva roxburghii*. 6-week-old culture of endosperm-half (without embryo) pre-treated with GA$_3$, on WM+IAA+KN+CH; there is scanty callusing of endosperm along the margin. Note the formation of a shoot bud *shb* (After Srivastava 1971a)

Rangaswamy and Rao (1963) obtained a continuously growing endosperm culture of *Santalum album* on White's medium (WM) containing 2,4-dichlorophenoxyacetic acid (2,4-D), kinetin (KN), and YE. The same medium also supported optimal growth of the endosperm tissues of *Croton bonplandianum* (Bhojwani and Johri 1971), *Jatropha panduraefolia* (Srivastava 1971c; Srivastava and Johri 1974), and *Ricinus communis* (Srivastava 1971 b; Johri and Srivastava 1972). However, for *Putranjiva roxburghii*, neither 2,4-D nor YE proved promotive, and a continuously growing callus was obtained only on WM+IAA+KN+CH (Srivastava 1973). The mature endosperm of *Dendrophthoe falcata*, *Taxillus vestitus*, *T. cuneatus*, and *Leptomeria acida* proliferated on WM+IBA (Indolebutyric acid)+CH+KN. Endosperm cultures of *D. falcata* started callusing on WM+IAA+KN after 6 weeks (Nag and Johri 1971).

Embryo Factor

The mature endosperm tissue fails to proliferate if cultured without the embryo. Various investigators have reported that gibberellin-like substances are released from the embryo during its germination (Paleg 1960; Ingle and Hageman 1965; see also Brown et al. 1970). When pieces of endosperm were grown without the embryo, these did not proliferate even when the medium was supplied with gibberellic acid.

Bhojwani (1968) and Srivastava (1971a) conducted some experiments to replace this "embryo factor." They presoaked the endosperm pieces (before implantation on the medium) in different concentrations of GA$_3$, IAA, or KN solution for different periods of time. The explants soaked in IAA (2.0 ppm) enlarged considerably in cultures, but did not proliferate. A similar, and somewhat better, response, was observed with the endosperm of *Putranjiva* and *Ricinus*. The endosperm pieces of *Putranjiva* soaked in GA$_3$ (1.0 and 2.0 ppm) for 36 h and cultured on WM + IAA + KN + CH or WM + 2,4-D + KN + YE callused along the margin and, after 4 weeks, formed small green nodules. Later, shoot buds differentiated from the callused margin (Fig. 2). Thus, the embryo factor can be replaced by treatment of endosperm with GA$_3$.

Interestingly, proliferation in endosperm cultures of parsley *(Petroselinum hortense)* was observed without the embryo (Masuda et al. 1978). It appears that in this taxon the endosperm may itself have a system of growth factor(s) which induce proliferation.

The mature endosperm of *Putranjiva*, cultured on WM + KN, developed numerous peripheral nodules; within 15 days these nodules proliferated profusely upon transfer to WM containing IAA, KN, and CH (Srivastava 1971 a). In semiparasitic angiosperms the cultured endosperm shows a tendency for direct differentiation of organs with slight marginal callusing or without prior callusing (Nag and Johri 1971).

Histological and Cytological Studies

The mature endosperm of *Santalum album* proliferated in 21–25 days after culture (Rangaswamy and Rao 1963); the endosperm tissue of euphorbiaceous members took lesser time. *Putranjiva* required 21 days (Srivastava 1971 a), *Jatropha* 10–12 days (Srivastava 1971 c; Srivastava and Johri 1974), and *Ricinus* just 10 days (Srivastava 1971 a; Johri and Srivastava 1972). This period (in *Ricinus*) could be reduced to 7 days by presoaking the explants in GA$_3$ (2.0 ppm). Brown et al. (1970) pointed out that maximal proliferation of endosperm could be achieved if it is cultured 2 days after the germination of the seed.

In *Putranjiva*, after 4 weeks of culture, numerous nodules, or growth centers, are discernible. Mostly, proliferation starts with the laying down of a cambium-like layer or by internal divisions, followed by cell enlargement. The initial growth pattern is similar to that of carrot root segments. In endosperm cells the mitoses are more frequent in the peripheral cells, and begin only after a lag phase. In *Osyris* (Bhojwani 1968) localized peripheral meristems lead to the formation of superficial outgrowths in which they differentiate large patches of "tracheidal" cells. This is also true of *Jatropha*, *Putranjiva*, and *Ricinus* (Srivastava 1971 a). In endosperm cultures of *Dendrophthoe* and *Taxillus* the epidermal cells are the first to divide (Nag 1970).

In maize the initial growth of cultured endosperm occurs as a result of divisions in the aleurone cells and the underlying layers (see Straus and LaRue 1954). The cells of the outermost layer in the endosperm, excised 8–12 days after pollination, continue to divide in cultures, forming localised meristematic zones which give rise to nodules (Sehgal 1969).

For the purpose of studying caffeine synthesis in endosperm tissue, *Coffea arabica* endosperm has been cultured on Murashige and Skoog's (MS) medium. The explants proliferated, and there was profuse callusing on malt extract (ME)- and CH-fortified medium (see Monaco et al. 1977).

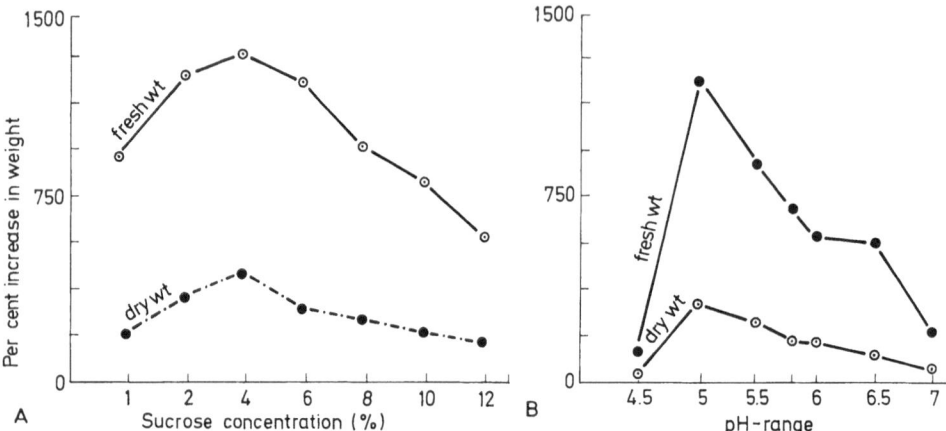

Fig. 3 A, B. *Ricinus communis.* Medium: WM + 2,4-D + KN + YE. **A** Effect of sucrose on the growth of endosperm callus. **B** Effect of pH on the growth of endosperm callus. (After Johri and Srivastava 1972)

Anatomical studies of callused endosperm of *Jatropha, Putranjiva,* and *Ricinus* revealed that the cotyledons and radicle, of the associated embryo, also proliferated slightly. Soon after, however, the proliferated radicle and cotyledons showed signs of degeneration. To avoid any contamination of the endosperm callus from that of cotyledons and radicle, the endosperm callus was transferred to a fresh medium. In squash preparations and microtome sections, 4-week-old callus showed parenchymatous cells. Differentiation of tracheidal cells was observed in sections of 6-week-old callus.

The disposition of thick-walled cells in *Jatropha, Putranjiva,* and *Ricinus* was at first irregular and later became restricted to the cambium-like zones in the callus. Trachiedal cells differentiated in 12-week-old callus. The organization of trachiedal cells and cambium-like layer(s) resembled vascular bundles in 20-week-old cultures of *Ricinus* endosperm. In both squash preparations and free-hand sections, the phloroglucinol-HCl test for lignin was positive (Srivastava 1971 a).

Optimal growth of callus (Fig. 3 A) and differentiation of trachiedal cells in the endosperm callus of *Ricinus* was produced with 2%–4% sucrose (Johri and Srivastava 1972). A gradual decrease in the growth of callus and formation of trachiedal cells was observed with increasing concentration of sucrose to beyond 4%. With arabinose, cellobiose, lactose, mannose, and sorbose the growth of *Ricinus* endosperm callus was poor and trachiedal cells did not differentiate. There was neither callusing nor trachiedal differentiation without sugar. Thickening of cells of the callus tissue and differentiation and organization of trachiedal cells varied a great deal on different media. Cell groups with thickened walls were observed in 9-week-old callus of *Jatropha* (Srivastava and Johri 1974). After 12 weeks these groups developed reticulate thickenings. Mention may be made of the ovule culture of *Gossypium hirsutum,* in which the endosperm occasionally developed thick walls and contained amoeboid nuclei (Joshi and Johri 1972). In 96-day-old cultures on WM + CH (250 ppm) the endosperm cells at the chalazal end showed pitted walls.

In squash preparations the endosperm callus of *Jatropha* and *Ricinus* on YE-supplemented medium showed actively dividing cells, as well as multicellular "filaments" and cell clusters with hypertrophied nuclei and fragmented nucleoli. Occasionally, cell assemblages simulating stages of embryogeny were also observed (Srivastava 1971 b). Differentiation of trachiedal cells in *Jatropha* and *Putranjiva* accompanied organogenesis (Srivastava 1971 a).

As a result of meristematic activity in the epidermal and hypodermal cells of endosperm, differentiation of shoot buds occurred in *Scurrula pulverulenta* (Bhojwani and Johri 1970). In *Exocarpus* (Johri and Bhojwani 1965) these buds developed from the peripheral cells. In *Dendrophthoe* (Johri and Nag 1968) and *Taxillus* (Johri and Nag 1970) also proliferation of the endosperm started from the cells in direct contact with the medium and later extended to other regions. The peripheral cells divided first. Division of cells in a row was occasionally observed in *Leptomeria* (Nag and Johri 1971). The differentiation of shoot buds in endosperm cultures of loranthaceous and santalaceous plants was achieved either by direct divisions of epidermal cells, or after the formation of callus. Nag (1970) often noticed anticlinal divisions in the epidermal cells.

The initiation of proliferation in *Jatropha*, *Putranjiva*, and *Ricinus* (all belonging to Euphorbiaceae) occurred as a result of laying down of a cambium-like layer in peripheral zone (Srivastava 1971 a). In *Putranjiva* a layer of periderm was observed in 24-week-old callus (Srivastava 1973). The shoot buds pierced through the multilayered periderm and organised into a leafy shoot. Roots developed from the nodules of endosperm callus, and the root primordia exhibited calyptrogen and other layers in *Croton* (Bhojwani and Johri 1971), *Putranjiva* and *Jatropha* (Srivastava 1971 a). In *Croton* a mature root had a root cap, and in transverse sections the epidermal layer was followed by cortex; the pericycle and endodermis were also distinct (Bhojwani and Johri 1971).

In 10- to 12-month-old cultures the endosperm callus, like other plant tissue cultures, exhibited cells of different ploidy. Occurrence of aneuploid and polyploid cells was observed in maize endosperm by Straus and LaRue (1954), and in *Lolium* by Norstog (1956), Norstog et al. (1969). In the endosperm cultures of *Croton* (Bhojwani 1968), and *Jatropha* (Srivastava 1971 c) also, the tissue exhibited cells of different ploidy. However, organogenesis did not ever take place from the cells of higher ploidy. In *Dendrophthoe* and *Taxillus* spp., even in 2-year-old cultures, neither there was a change in chromosome number, nor a loss in organ-forming capacity of the callus (Nag and Johri 1971).

Johri and Nag (1974) reported that in the diploid (from embryonal callus) and triploid (from endosperm callus) leaf cells of *Dendrophthoe* there was no difference in the size of chromosomes. Nag (1970) observed that there was no effect of either the nutrients or the position of explant in the medium on the chromosomal constitution of the cells. This may be because, in this species *(Dendrophthoe falcata)*, "... the natural mechanism leading to polyploidy is either absent or much less manifest ..." (see Partanen 1965).

The organ-forming capacity of the endosperm callus of *Putranjiva* was not lost even after 24 months of continuous culture on the same medium. This may be attributed to the fact that the callus retained its constancy in chromosome number, and thus maintained cytologic stability.

Growth Requirements

Physical Factors

Besides the earlier work on the role of physical factors by Lampton (1952) on *Asimina triloba*, and Straus and LaRue (1954) on maize, Nag (1970) and Srivastava (1971 a) studied the effect of pH, temperature, and light on the growth response of endosperm in a few other members.

The pH of the nutrient media is usually maintained between 4.5 and 6.3. Straus and LaRue (1954) observed that on fresh weight basis pH 7.0 supported the best growth of maize endosperm tissue, whereas on dry weight basis pH 6.1 appeared optimal. An increase of 100% in fresh weight of endosperm callus of *Asimina* occurred at pH 4.0, and 95% increase at pH 5.0. Maximal growth of endosperm callus of *Ricinus* occurred at pH 5.0 (Fig. 3 B) and at pH 5.6 in *Jatropha* and *Putranjiva* (Srivastava 1971 c, 1973).

In maize there was at least 50% decrease in growth at 30 °C, and a fourfold increase at 25 °C as compared to that at 20 °C (Straus and LaRue 1954). In *Jatropha* and *Ricinus* endosperm cultures too, 24°–26 °C supported best growth.

The growth of maize endosperm was satisfactory only when maintained in the dark (Straus and LaRue 1954). In *Ricinus* best growth occurred under continuous light (1,500 lux). Norstog (1956) did not find any significant effect of light on the growth of the endosperm callus of *Lolium*. The endosperm of *Coffea arabica* thrived best under 12 h light/dark conditions, at a temperature of 27 °C (see Monaco et al. 1977).

Chemical Factors and Differentiation

Organ formation has so far been demonstrated in *Dendrophthoe, Exocarpus, Leptomeria, Scurrula*, and *Taxillus* (see Johri 1971), all parasitic endosperms; and in *Croton* (Bhojwani 1966), *Jatropha* (Srivastava 1971 c), *Putranjiva* (Srivastava 1973), *Oryza* (Nakano et al. 1975), apple (Mu et al. 1977), *Petroselinum* (Masuda et al. 1978), *Citrus* (Wang and Chang 1978), and *Codiaeum* (Chikkannaiah and Gayatri 1974; Gayatri 1978), all autotrophs. In almost all the semi-parasitic members differentiation of shoot buds from the mature endosperm occurred without callusing, whereas in the autotrophic members the endosperm usually formed a callus mass followed by the differentiation of shoot buds or roots.

The only reports on the differentiation of roots and plantlets from the callus of the immature endosperm are those of Nakano et al. (1975) in rice, and Mu et al. (1977) in apple. Immature endosperm of rice excised 4–7 days after pollination proliferated on Linsmaier and Skoog's medium (LS) supplemented with YE. Proliferation and callus formation was also observed on LS + 2,4-D + YE, LS + 2,4-D + CM, and modified WM + YE + 2,4-D. Differentiation of rootlets also occurred on WM + YE, with or without KN or IAA. Plantlets were formed on WM supplemented with YE and IAA, or YE + KN.

The mature endosperm tissue of *Exocarpus cupressiformis* showed organogenesis on WM enriched with IAA, KN, and CH. The embryo formed a normal seed-

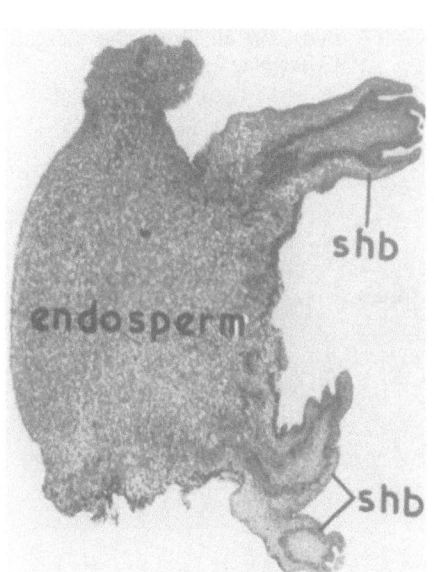
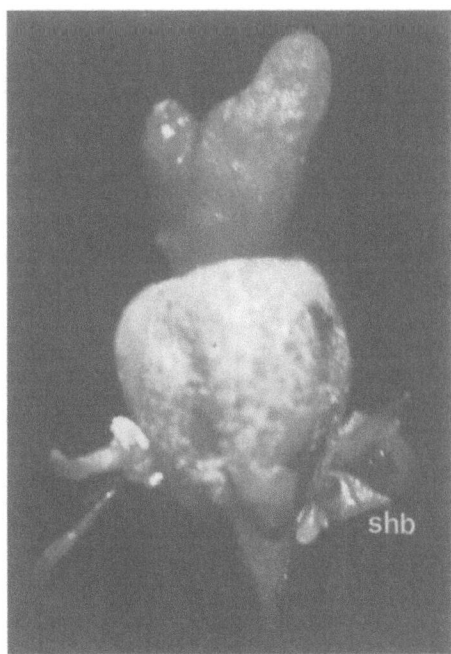

Fig. 4. *Exocarpus cupressiformis.* Section of endosperm with differentiated shoot buds *shb*, on WM + IAA + KN + CH. (After Johri and Bhojwani 1965)

Fig. 5. *Scurrula pulverulenta.* 16-week-old culture showing chlorophyllous shoot buds *shb* differentiated from endosperm, on WM + zeatin. (After Bhojwani and Johri 1970)

ling, while the endosperm proliferated slightly. Shoot buds differentiated from the endosperm (Fig. 4) after 5 weeks. Johri and Bhojwani (1965) observed that with the omission of CH from the medium, the percentage of cultures showing shoot bud differentiation increased from 13 to 26. In later experiments, therefore, only IAA and KN were added to the medium.

The shoot buds which differentiated from the mature endosperm of *Exocarpus* showed limited growth. When these buds were excised and planted on fresh medium of same composition as the one on which they differentiated, the cut ends of the shoot proliferated and, after 11 weeks, shoot buds and haustoria differentiated from the callus (Bhojwani 1968). The leaves which developed from the shoot buds, however, did not bear the characteristic trichomes.

In the "seed" (endosperm with embryo) cultures of *Scurrula pulverulenta*, on WM + Zeatin, differentiation of chlorophyllous buds from the endosperm was observed (Fig. 5). Interestingly, when subcultured on fresh medium, haustoria differentiated from these buds which is a unique feature (Bhojwani and Johri 1970). In this species differentiation of buds did not occur without an exogenous supply of cytokinin. While 6-(γ,γ-dimethylallylamino) purine was most effective and triacanthine the least, SD 8339 (1.3×10^{-5} M) also favoured the differentiation of buds.

The differentiation of shoot buds from the mature endosperm tissue of *Taxillus vestitus* and *T. cuneatus* has also been studied. In *T. vestitus* the orientation of the

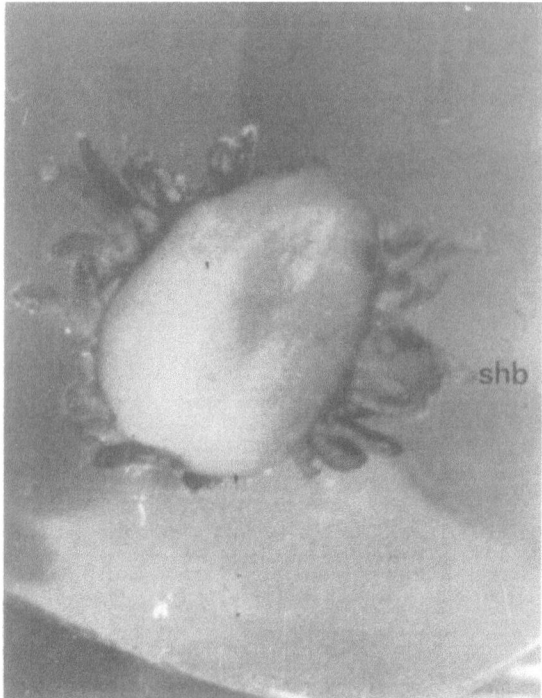

Fig. 6. *Taxillus vestitus.* 10-week-old culture of endosperm, on WM + 6-benzylaminopurine; note the differentiation of shoot buds *shb* all along the margin. (After Nag 1970)

explant in relation to the agar medium played a very significant role. If the half-split endosperm was implanted with the cut surface in contact with the medium (WM + 6-benzylaminopurine or KN), 100% cultures developed 12–18 shoot buds per culture, after 10 weeks (Fig. 6). Moreover, the response was also dependent on the cytokinin added to the medium. With 5.0 ppm KN shoot buds differentiated after 9 weeks, whereas 10.0 ppm hastened the process by 3 weeks. Amongst the substituted aminopurines tested, 6-(γ,γ-dimethylallylamino) purine proved most effective in the differentiation of shoot buds. Adenine (AD), 3-(γ,γ-dimethylallylamino) purine, 6-methylaminopurine, or 6-benzylamino-9-(2-tetrahydropyrenyl) purine failed to bring about any response (Fig. 7). The majority of buds were confined to the injured portion of the endosperm in direct contact with the medium. The half-split endosperm implanted vertically or horizontally always developed the first two buds from the epidermal cells near the cut-end. The half-split endosperm pieces soaked in 0.025 ppm KN for 4 h and cultured on WM alone developed buds all over the surface (Johri and Nag 1970). A similar response was also noticed with transverse segments of endosperm (with the embryo intact) (Nag 1970). In *T. cuneatus* callusing invariably preceded bud differentiation. A few haustoria also developed at the base of buds. The buds obtained from the endosperm of this species were comparatively much longer (12 mm) than those of *T. vestitus* (5 mm).

In *D. falcata* Nag (1970) observed that the distal end of the mature endosperm (with embryo) callused in 6 weeks along the surface in contact with the medium (WM + IAA + KN + CH). After another week, 3 or 4 shoot buds appeared on the

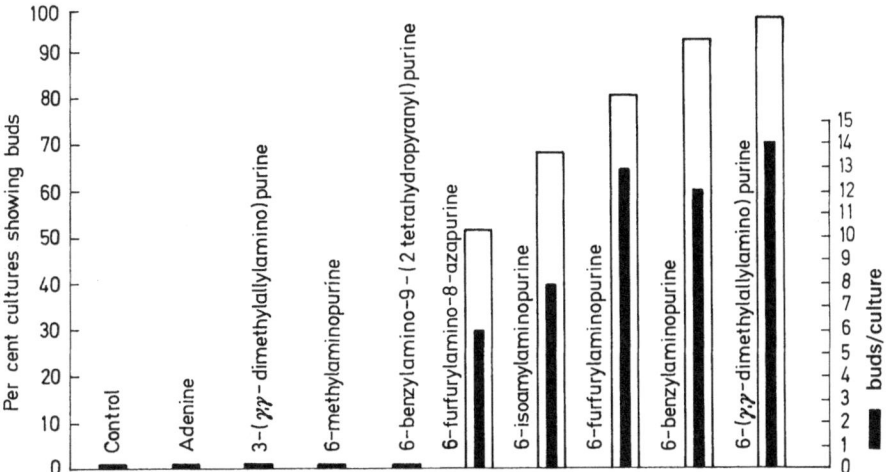

Fig. 7. *Taxillus vestitus.* Effect of some substituted aminopurines, and adenine at 2×10^{-5} M, used with WM (control), on shoot bud formation from endosperm; 8-week-old cultures. (After Johri and Nag 1970)

uncallused surface and, about 6 weeks later, these buds grew into well-developed shoots with leaves. Differentiation of both shoot and haustoria was observed in in vitro-formed leaf callus cultures (Fig. 8A–D).

Like *Dendrophthoe*, in *Leptomeria acida* also the endosperm callused and buds differentiated when both IAA and KN were added to WM (Nag 1970). In *D.falcata* IBA induced profuse callusing, while IAA proved more effective in inducing buds.

Incorporation of IAA or IBA and a cytokinin in WM, decreased the potentiality for bud formation in *Taxillus* spp. This effect, however, could be annulled by increasing the concentration of cytokinin (Johri and Nag 1970). In contrast to *Exocarpus* (Johri and Bhojwani 1965), the endosperm of *Dendrophthoe* showed an enhancement in the percentage of cultures producing shoot buds on CH-fortified medium (Johri and Nag 1968; Nag and Johri 1971). At 2,000 ppm CH, 46% of cultures showed shoot bud differentiation in comparison to 32% without it. Buds were mostly restricted to the distal end of the endosperm. Although indolepropionic acid (IPA) induced shoot buds in 35% of cultures, the buds were rather deformed. Other auxins, namely naphthaleneacetic acid (NAA), 2,4-D, and 2,4,5-trichlorophenoxyacetic acid (2,4,5-T) could not induce bud formation. Although IBA promotes callusing, only 5% of cultures showed bud formation. On WM + IAA + KN + CH, the endosperm of *T.cuneatus* formed shoot buds in 85% and haustoria in 40% of cultures after 10 weeks. When IAA was replaced by IBA, buds were observed in 55%, and haustoria in 60% of cultures. Unlike *Dendrophthoe* and *T.cuneatus*, in *T.vestitus* IBA in combination with KN and CH in WM proved ineffective (see Nag and Johri 1971).

The endosperm cultures of *Leptomeria acida* (Nag 1970) did not show significant response to CH alone. But when CH was used along with IAA, or IBA + a cytokinin, promotion of shoot bud formation was observed. This medium, when

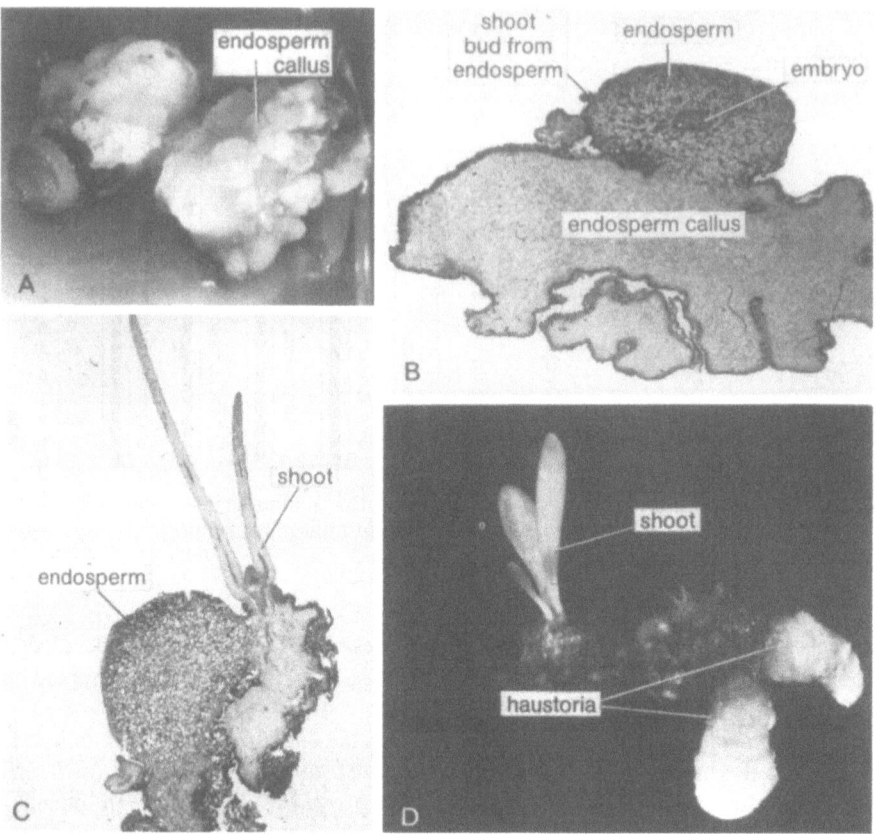

Fig. 8 A–D. *Dendrophthoe falcata.* **A** 6-week-old cultures of "seed" on WM + IAA + KN + CH; note the callusing of endosperm. **B** Transection of 12-week-old culture of "seed." **C** 18-week-old culture in transection; note a shoot with well-developed leaves. **D** 16-week-old culture of in vitro-formed leaf (from endosperm) callus showing both shoot and haustoria. (After Nag and Johri 1971)

it contained IBA, was also most suitable for a rapid growth of the endosperm callus. When the endosperm callus was grown on the above medium containing IAA instead of IBA, buds differentiated in almost all the cultures after 7 weeks; these buds developed into shoots in another 3 weeks.

As stated earlier, the mature endosperm of only a few autotrophic taxa (*Zea mays, Lolium, Santalum, Ricinus, Coffea arabica, Croton, Jatropha, Putranjiva, Petroselinum hortense*, and apple) have been induced to form callus, and only the last five of these have so far yielded organ differentiation. The callus tissue in some of these taxa became compact upon continuous subculture on WM + 2,4-D + KN + YE, but did not undergo organogenesis on this medium. The compact callus of *Jatropha* obtained from endosperm (cultured with embryo) on WM + 2,4-D + KN + YE in 6–7 weeks, when transferred to WM + KN + CH became friable after 5 weeks. After another 2 weeks small, green buds (Fig. 9 A–C) developed in 22%

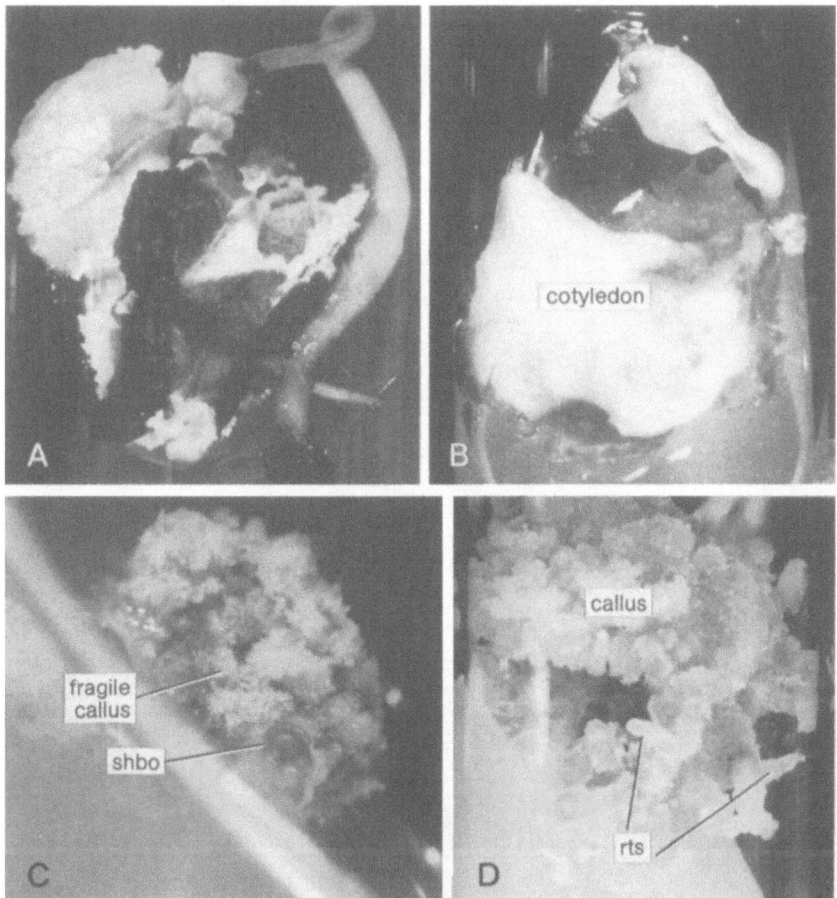

Fig. 9 A–D. *Jatropha panduraefolia.* **A** 7-week-old culture on WM + 2,4-D + KN + YE. Note the proliferated endosperm, and seedling. **B** Proliferated endosperm and enlarged cotyledons in 6-week-old culture, on WM + 2,4-D + KN + YE. **C** 7-week-old culture on WM + KN + CH with shoot bud *shbo*, transferred from WM + 2,4-D + KN + YE after 18 weeks. **D** 8-week-old callus, on WM + IAA + CH, with roots *rts*. (After Srivastava and Johri 1974)

of cultures (Srivastava and Johri 1974). These buds attained a length of up to 2 cm, but did not develop into normal leaves. Addition of an auxin failed to influence bud differentiation.

Buds differentiated in 80% of cultures from the nodulated callus obtained on WM + IAA + KN + CH in 8-week-old endosperm cultures of *Putranjiva*. In 11 weeks the buds developed into shoots. These shoots reached a length of 4 cm and bore 3–4 pairs of leaves (Fig. 10 A–D). The response of mature endosperm of *Putranjiva* (Srivastava 1973), to different auxins, was similar to that of other endosperm tissue. IPA proved much more effective than IBA. Next to IAA, IPA brought about maximum percentage of bud differentiation. Numerous shoot primordia differentiated on WM + IPA + KN + CH. On this medium the shoot buds

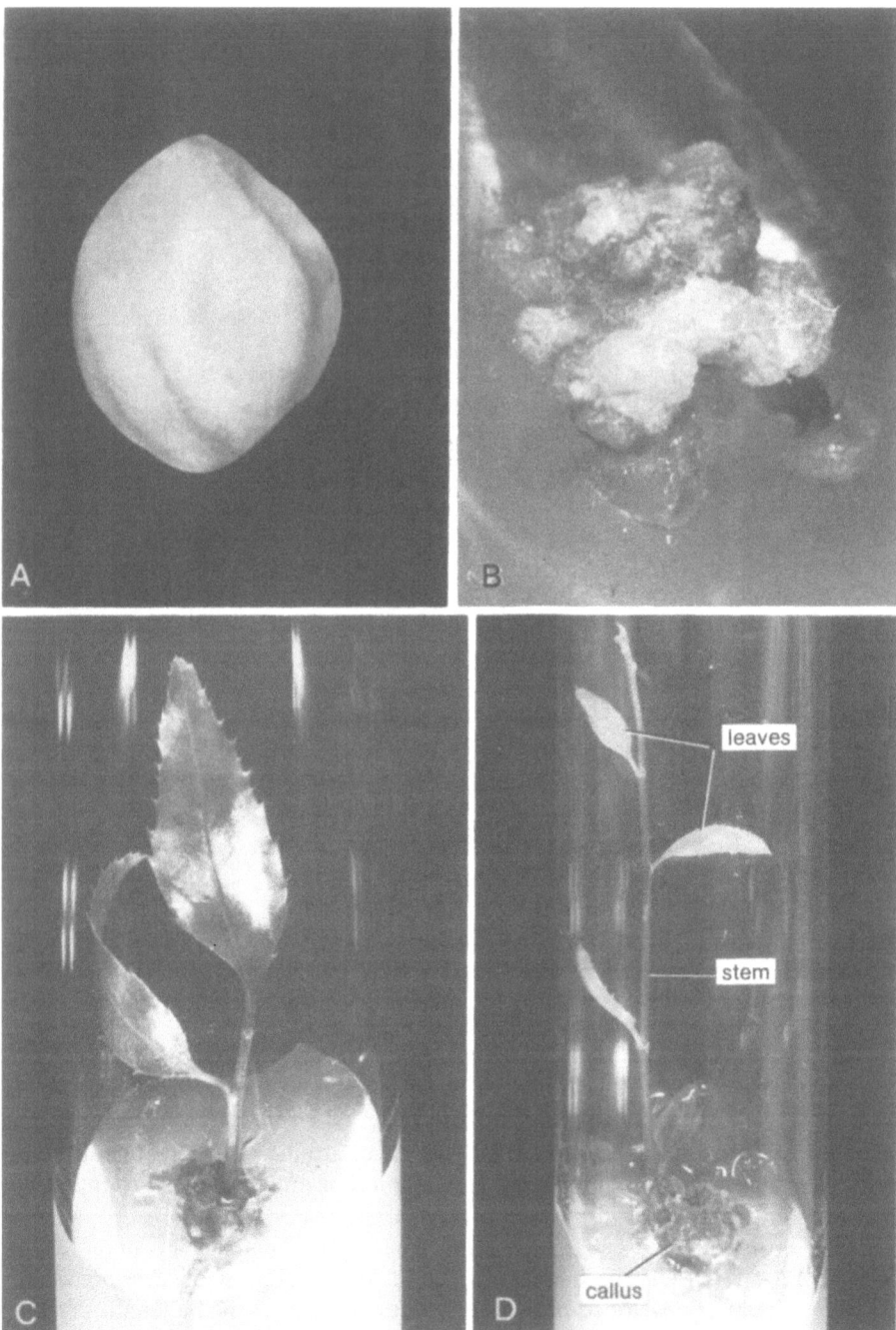

Fig. 10 A–D. *Putranjiva roxburghii.* **A** Endosperm with embryo intact (decoated seed) at culture. **B** 6-week-old nodulated callus on WM + IAA + KN + CH. **C** Differentiation of triploid shoot bearing leaves, on WM + IAA + KN + CH, 19-week-old culture; the leaves have a characteristic dentate margin. **D** 20-week-old culture showing a well-developed plantlet with root-shoot axis bearing leaves, on WM + IAA + KN + CH. (**A, B, D** after Srivastava 1971 a; **C** after Srivastava 1973)

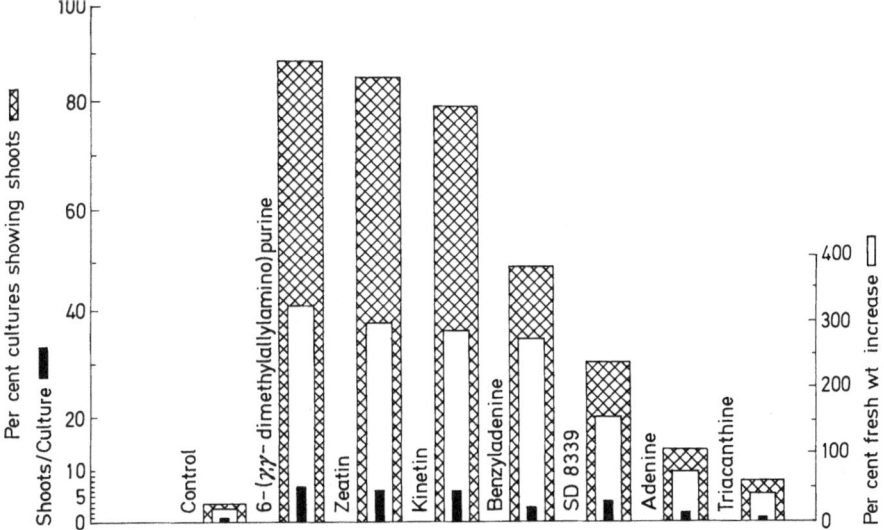

Fig. 11. *Putranjiva roxburghii.* Growth response of callus, after 12 weeks, to some substituted aminopurines, and adenine at 3×10^{-5} M, used with WM + IAA + CH (control). (After Srivastava 1973)

grew up to 2 cm into the agar medium and developed leaves. If left inside the medium, the stem and leaves became robust, and later showed signs of proliferation. In autotrophic taxa, 6-(γ,γ-dimethylallylamino) purine induced shoot bud formation in the maximal number of cultures. Adenine did not support as much callus growth as 6-(γ,γ-dimethylallylamino) purine (Fig. 11; see also Fig. 7). The same is true of *Exocarpus*. In *Putranjiva* CH was indispensable for the differentiation of shoot buds from the endosperm callus (Srivastava 1973). Contrarily, CH had inhibitory effect in inducing shoot buds of *Exocarpus* (Bhojwani 1968). Like *Dendrophthoe*, in *Putranjiva* also the growth of callus was much better with 2,000 ppm CH. With 3,000 ppm CH, the shoots did not grow beyond 0.8 cm, while 5,000 ppm CH caused profuse callusing.

In *Putranjiva* CH significantly promoted (87% of cultures) shoot bud differentiation. CH is a complex mixture of 13 amino acids and certain other substances. In an attempt to replace CH with one or more of the constituent amino acids, it was observed that when all the 13 amino acids were used together the growth of the callus was slow as compared to CH, and only 72% of cultures showed poorly developed shoots. Individually, only valine could replace the effect of CH to any great extent (Srivastava 1971 a; Srivastava and Johri 1978).

A noteworthy feature of *Petroselinum* is the occurrence of embryogenesis without any exogenous supply of growth substances. This seems to be the first report concerning embryogenesis in endosperm cultures (Masuda et al. 1978). However, this has to be accepted with caution. The embryos could just as well have differentiated from the proliferation of zygotic embryo.

Rooting has been rare in endosperm cultures. Bhojwani and Johri (1971) reported that the mature endosperm of *Croton* did not exhibit any organogenesis when grown on WM + CH. However, if the callus obtained on WM + 2,4-D +

KN+CH was transferred to WM, or WM+CH, several roots differentiated. Liquid or semi-solid medium did not in any way affect the morphogenic behaviour of endosperm callus. IBA proved most effective for the induction of roots. With the decreasing concentration of IBA, the number of cultures showing root formation increased.

The differentiation of roots also occurred in the endosperm callus of *Jatropha* (Srivastava 1971c; Srivastava and Johri 1974). The callus obtained on WM+2,4-D+KN+YE grew profusely upon transfer to WM+IAA+CH. This tissue turned green and in eight weeks showed differentiation of roots in 43% of the cultures (Fig. 9D). In *Putranjiva* roots appeared only in a few cultures, although the differentiation of shoots was very common (Srivastava 1973). Even the shoots which reached up to 15 cm failed to differentiate roots. However, in 20% of cultures, roots also differentiated more or less in continuation of the base of the shoot embedded in the callus. Eventually, complete plantlets with root-shoot axis were obtained (Fig. 10). When the shoots were excised from the callus and planted on a fresh medium, differentiation of roots was noticed in some cultures. These triploid plantlets were comparable to normal seedlings obtained on WM (Srivastava 1973). The embryoids from endosperm callus in *Citrus* (Wang and Chang 1978) produced roots and shoots developing into triploid plantlets.

The addition of KN to the medium seems to be indispensible for the differentiation of shoot buds or shoots. In semi-parasitic angiosperms the most significant feature is the formation of triploid haustoria (without the host stimulus) under in vitro conditions.

Concluding Remarks

When cultured, the endosperm usually develops into an unorganized mass of tissue, irrespective of its genetic constitution (which depends on the total number of nuclei taking part in the formation of primary endosperm nucleus). The formation of shoots, roots, haustoria, and plantlets establishes the totipotency of endosperm cells. While organogenesis has been observed in apple, *Citrus*, *Codiaeum*, *Croton*, *Dendrophthoe*, *Exocarpus*, *Jatropha*, *Leptomeria*, *Putranjiva*, *Scurrula*, and *Taxillus* spp., embryogenesis in *Petroselinum*, the endosperm of maize, *Ricinus*, *Nuytsia*, and *Santalum* has yielded only a continuously growing callus.

A pertinent question to ask is why, in some cases, organogenesis occurs directly from the endosperm tissue (especially in semi-parasitic taxa examined), while in others the endosperm first forms a callus followed by organogenesis. Correlations among explants callusing, and undergoing vascularization and organogenesis, remain to be explained. Also, how is it that, in spite of the occurrence of polyploid cells in the callus, differentiation of organs invariably takes place only from triploid cells.

A challenging problem concerning morphogenesis of endosperm tissue has been amply demonstrated, and the formation of triploid plantlets has been achieved. Further investigations are necessary to develop the technology of inducing triploids from endosperm tissue of many more taxa, and obtaining plantlets which could be utilized for horticultural and genetical studies.

The report of Lakshmi Sita et al. (1980) on the development of triploid plantlets in cultures of mature endosperm of *Santalum album* is of much significance. They succeeded in obtaining endosperm (cultured without embryo) callus on a variety of media, but MS supplemented with benzylaminopurine (BAP) and NAA gave better results than with 2,4-D. "Embryoid"-formation occurred on MS + GA (1.0–2.0 mg/l), MS + BAP (0.3–2.0 mg/l) + IAA (1.0 mg/l), and MS + KN (0.3–2.0 mg/l) + GA (1.0 mg/l). Extensive embryogenesis was observed on MS + BAP (0.3 mg/l) + IAA (1.0 mg/l), and MS + GA (1.0 mg/l) + KN (0.3 mg/l). The embryoids developed further upon transfer to White's medium.

These authors claim that embryogenesis in the endosperm cultures of sandalwood followed the same sequence as the embryos in situ (Nature). This does not seem convincing, since sequential developmental stages up to maturity have not been studied. Judging from the illustrations, the irregular structures can be considered as masses of callus which undergo regeneration, and produce triploid plants.

This is the first report of transplantation of cultured triploid plants to pots in garden soil, and then to experimental plots.

At present, healthy plants are growing in the forest nursery at Bangalore. *Personal information – BMJ*

References

Bajaj YPS (1970) Growth responses of excised embryos of some mistletoes. Z Pflanzenphysiol 63:408–415

Bhojwani SS (1966) Morphogenetic behaviour of mature endosperm of *Croton bonplandianum* in culture. Phytomorphology 16:349–353

Bhojwani SS (1968) Morphogenetic studies on cultured mature endosperm and embryo of some angiosperms. PhD Thesis, Univ Delhi

Bhojwani SS, Johri BM (1970) Cytokinin-induced shoot bud differentiation in mature endosperm of *Scurrula pulverulenta*. Z Pflanzenphysiol 63:269–275

Bhojwani SS, Johri BM (1971) Morphogenetic studies on cultured mature endosperm of *Croton bonplandianum*. New Phytol 70:761–766

Brown DJ, Canvin DT, Zilkey BF (1970) Growth and metabolism of *Ricinus communis* endosperm in tissue culture. Can J Bot 48:2323–2331

Chikkannaih PS, Gayatri MC (1974) Organogenesis in endosperm tissue culture of *Codiaeum variegatum* Blume. Curr Sci 43:23–24

Gayatri MC (1978) In vitro studies on *Codiaeum variegatum*: Growth and organogenesis in endosperm tissue. Phytomorphology 28:395–400

Ingle J, Hageman RH (1965) Metabolic changes associated with the germination of corn. III. Effect of gibberellic acid on endosperm metabolism. Plant Physiol 40:672–675

Johri BM (1971) Differentiation in plant tissue cultures. Presidential Address, Sect Botany. Proc 58th Indian Sci Congr Bangalore Part II:159–186

Johri BM, Bhojwani SS (1965) Growth responses of mature endosperm in cultures. Nature (London) 208:1345–1347

Johri BM, Nag KK (1968) Experimental induction of triploid shoots in vitro from endosperm of *Dendrophthoe falcata* (L.f.) Ettings. Curr Sci 37:606–607

Johri BM, Nag KK (1970) Endosperm of *Taxillus vestitus* Wall.: A system to study the effect of cytokinin in vitro in shoot bud formation. Curr Sci 39:177–179

Johri BM, Nag KK (1974) Cytology and morphogenesis in embryo and endosperm tissue in vitro of *Dendrophthoe* and *Taxillus*. Cytologia 39:801–813

Johri BM, Srivastava PS (1972) In vitro growth responses of mature endosperm of *Ricinus communis* L. In: Murty YS, Johri BM, Mohan Ram HY, Varghese TM (eds) Advances in plant morphology. Prof V Puri Comm Vol, Sarita Prakashan, Meerut, pp 339–358

Joshi PC, Johri BM (1972) In vitro growth of ovules of *Gossypium hirsutum*. Phytomorphology 22:195–209

Lakshmi Sita G, Raghava Ram NV, Vaidyanathan CS (1980) Triploid plants from endosperm cultures of sandalwood by experimental embryogenesis. Pl Sci Lett 20:63–69

Lampe L, Mills CO (1933) Growth and development of isolated endosperm and embryo of maize. Abstr Pap Bot Soc Boston

Lampton RK (1952) Developmental and experimental morphology of the ovule and seed of *Asimina triloba*. PhD Thesis, Univ Michigan, Ann Arbor

LaRue CD (1947) Growth and regeneration of the endosperm of maize in culture. Am J Bot 34:585–586

LaRue CD (1949) Cultures of the endosperm of maize. Abstr Am J Bot 36:798

Masuda K, Koda Y, Okazawa Y (1978) Callus formation and embryogenesis of endosperm tissue of parsley seed cultured on hormone-free medium. Physiol Plant 41:135–138

Monaco LC, Sondahl MR, Carvalho A, Crocomo OJ, Sharp WR (1977) Applications of tissue culture in the improvement of coffee. In: Reinert J, Bajaj YPS (eds) Applied and fundamental aspects of plant cell, tissue, and organ culture. Springer, Berlin Heidelberg New York, pp 109–126

Mu S, Liu S, Zhou Y, Qian N, Zhang P, Xie H, Zhang F, Yan Z (1977) Induction of callus from apple endosperm and differentiation of the endosperm plantlet. Sci Sinica 20:370–377

Nag KK (1970) Morphogenic studies on endosperm, embryo, and other sporophytic tissues of some parasitic angiosperms. PhD Thesis, Univ Delhi

Nag KK, Johri BM (1971) Morphogenic studies on endosperm of some parasitic angiosperms. Phytomorphology 21:202–218

Nakajima T (1962) Physiological studies of seed development, especially embryonic growth and endosperm development. Univ Osaka Prefect Ser B 13:13–48

Nakano H, Tashiro T, Maeda E (1975) Plant differentiation in callus tissue induced from immature endosperm of *Oryza sativa* L. Z Pflanzenphysiol 76:444–449

Norstog K (1956) Growth of rye-grass endosperm in vitro. Bot Gaz 117:253–259

Norstog K, Wall WE, Howland GP (1969) Cytological characteristics of ten-year-old rye-grass endosperm tissue cultures. Bot Gaz 130:83–86

Paleg LG (1960) Physiological effects of gibberellic acid. II. On starch hydrolyzing enzymes of barley endosperm. Plant Physiol 35:902–906

Partanen CR (1965) On the chromosomal basis for cellular differentiation. Am J Bot 52:204–209

Rangaswamy NS, Rao PS (1963) Experimental studies on *Santalum album:* Establishment of tissue culture of endosperm. Phytomorphology 13:450–454

Satsangi, Asha, Mohan Ram HY (1965) Continuously-growing tissue cultures from mature endosperm of *Ricinus communis* L. Phytomorphology 15:20–30

Sehgal CB (1969) Experimental studies on maize endosperm. Beitr Biol Pflanzen 46:233–238

Srivastava PS (1971 a) Morphogenic studies on mature endosperm of *Jatropha, Putranjiva*, and *Ricinus*. PhD Thesis, Univ Delhi

Srivastava PS (1971 b) In vitro growth requirement of mature endosperm of *Ricinus communis*. Curr Sci 40:337–339

Srivastava PS (1971 c) In vitro induction of triploid roots and shoots from mature endosperm of *Jatropha panduraefolia*. Z Pflanzenphysiol 66:93–96

Srivastava PS (1973) Formation of triploid plantlets in mature endosperm cultures of *Putranjiva roxburghii*. Z Pflanzenphysiol 69:270–273

Srivastava PS, Johri BM (1974) Morphogenesis in mature endosperm cultures of *Jatropha panduraefolia*. Beitr Biol Pflanzen 50:255–268

Srivastava PS, Johri BM (1978) Triploid plants of *Putranjiva roxburghii* from endosperm. Beitr Biol Pflanzen 54:381–397

Sternheimer Elizabeth P (1954) Method of culture and growth of maize endosperm in vitro. Bull Torrey Bot Club 81:111–113

Straus J (1960 a) Anthocyanin synthesis in corn endosperm. II. Effect of certain inhibitory and stimulatory agents. Plant Physiol 35:645–650

Straus J (1960b) Maize endosperm tissue grown in vitro. III. Development of a synthetic medium. Am J Bot 47:641–647

Straus J, LaRue CD (1954) Maize endosperm tissue grown in vitro. I. Culture requirements. Am J Bot 41:687–694

Tamaoki T, Ullstrup AJ (1958) Cultivation in vitro of excised endosperm and meristem tissues of corn. Bull Torrey Bot Club 85:260–272

Wang T, Chang C (1978) Triploid *Citrus* plantlet from endosperm culture. In: Proc Symp Plant Tissue Culture. Sci Press Peking, pp 463–468

9. Embryo Culture

V. RAGHAVAN and P.S. SRIVASTAVA

Several reviews on growth, differentiation, and morphogenesis of cultured embryos have been published during the past several years (Brink and Cooper 1947; Rappaport 1954; Sanders and Ziebur 1963; Narayanaswami and Norstog 1964; Maheshwari and Rangaswamy 1965; Wardlaw 1965; Degivry 1966; Maheshwari 1966; Raghavan 1966). Some of the general techniques employed in the culture of embryos have been described by Sanders and Ziebur (1963), Raghavan (1967), and Torrey (1973).

Depending upon the age of the embryo, the stimulus for its continued growth may be said to be present within its own cells, or in the cells of the surrounding tissue of the endosperm. The fertilized egg and the early division-phase embryos generally develop at the expense of the nutritional resources of the endosperm. They are, thus, heterotrophic and are provided with specialized nutritional substances, including amino acids, carbohydrates, purines, pyrimidines, perhaps vitamins, hormones, and other essential metabolites. In a typical dicotyledonous embryo, only in the late heart-shaped stage, with the beginning of cotyledonary development and the consequent attainment of internal differentiation, does the embryo become sufficiently independent and autotrophic to make it possible to culture it in vitro on a nutrient medium. This critical age varies in the embryos of different species.

Based on this trend of biochemical specialization of developing embryos, we will consider separately the culture of differentiated and mature embryos, and culture of proembryos. The former roughly corresponds to the autotrophic embryos, and the latter to the heterotrophic embryos. Some authors (for example, Rijven 1952) refer to them as post-germinal and pre-germinal embryos, respectively. Attention is also drawn to a comparative evaluation of growth of embryos in vivo and in vitro.

Culture of Differentiated and Mature Embryos

Current interest in the culture of embryos can be traced to the work of Hannig (1904), who grew under aseptic conditions relatively mature embryos (of different ages) of *Raphanus caudatus*, *R. landra*, *R. sativus*, and *Cochlearia danica* in a mineral salt medium supplemented with sugar, and obtained transplantable seedlings. Since this early work, cultivation of embryos as isolated systems has expanded rapidly in scope and depth, continuing to provide useful information on the physiology of their growth. Brown (1906) studied the relative efficiency of various organic nitrogen compounds on the growth of excised barley embryos cultured on a medium containing mineral salts and sucrose. The dependence of the embryo on nutrients from its own endosperm seemed less probable, as a general rule, from the finding that embryos of several Gramineae could grow well when grafted on each other's endosperm (Stingl 1907). In other studies the role of storage tissues of the seed in the growth of the embryo was explored by separating the latter from the endosperm, or by decotylating embryos and planting them in nutrient solutions (Dubard and Urbain 1913; Buckner and Kastle 1917). The role of scutellum of graminean embryo in the absorption of nutrients was implied in a study which showed that embryo devoid of this organ grew somewhat feebly (Andronescu 1919).

Knudson (1922) succeeded in growing orchid embryos into plantlets (in the absence of the symbiotic fungus) by culturing them on nutrient agar containing sugar. In the absence of sugar, embryos failed to develop beyond the protocorm stage. The specific stimulation of growth in the orchid embryo by a simple carbohydrate led to an appreciation of the role of mycorrhizal fungi in the conversion of starch and other complex polysaccharides into simpler forms. The asymbiotic propagation of orchids, now practised on an unparalleled commercial scale, had its inception in these experiments.

Based on the behaviour of cultured embryos of several plants, Dieterich (1924) pointed out two generalizations which have proved to be significant in the understanding of the physiology of growth of embryos. One was that the embryo grown in vitro usually skipped a rest period (observed when it is part of the intact seed) and germinated. In addition it was found that a solid medium with Knop's mineral salts and 2.5%–5% sucrose could support normal growth of embryos isolated from mature seeds, but in the same medium embryos from immature seeds tended to form malformed seedlings – a type of growth designated as "precocious germination." At about the same time Laibach (1925) envisioned the immense application of embryo culture technique in rearing viable seedlings from otherwise unsuccessful crosses.

This paved the way for a series of future investigations to surmount barriers to crossability in such crosses where embryos aborted before germination. It was possible to raise a second generation of plants by culturing the hybrid embryos of *Cerasus vulgaris* × *C. tomentosa* and *Ribes nigrum* × *Grossularia reclinata* (Kravtsov and Kas'yanova 1968). Partial success has also been achieved in rearing hybrid embryos of cotton *(Gossypium arboreum* × *G. hirsutum)* to maturity (Joshi and Pundir 1966).

During recent years breeding work with crop plants has been facilitated by embryo culture technique. Hybrid seedlings have now been raised by culturing embryos from crosses: *Hordeum sativum* × *H. bulbosum* (Konzak et al. 1951), *Lycopersicon esculentum* × *L. peruvianum* (Alexander 1956), *Melilotus officinalis* × *M. alba* (Schlosser-Szigat 1962), *Corchorus capsularis* × *C. olitorius* (Islam 1964), and in other interspecific crosses such as *Phaseolus* (Honma 1955), *Chrysanthemum* (Kaneko 1957), *Lilium* (Emsweller et al. 1962), *Medicago* (Fridriksson and Bolton 1963), *Lathyrus* (Pecket and Selim 1965), and intergeneric hybrids in *Hordeum jubatum* × *Secale cereale* (Brink et al. 1944), *Triticum durum abyssinicum* × *Secale cereale* (Rédei 1955), *Tripsacum* × *Zea* (Farquharson 1957), and *Triticum* × *Elymus* (Ivanovskaya 1962).

The ability of excised embryos to grow in culture has been demonstrated in other genera and species as well. Deviations in detail concern the nature of the me-

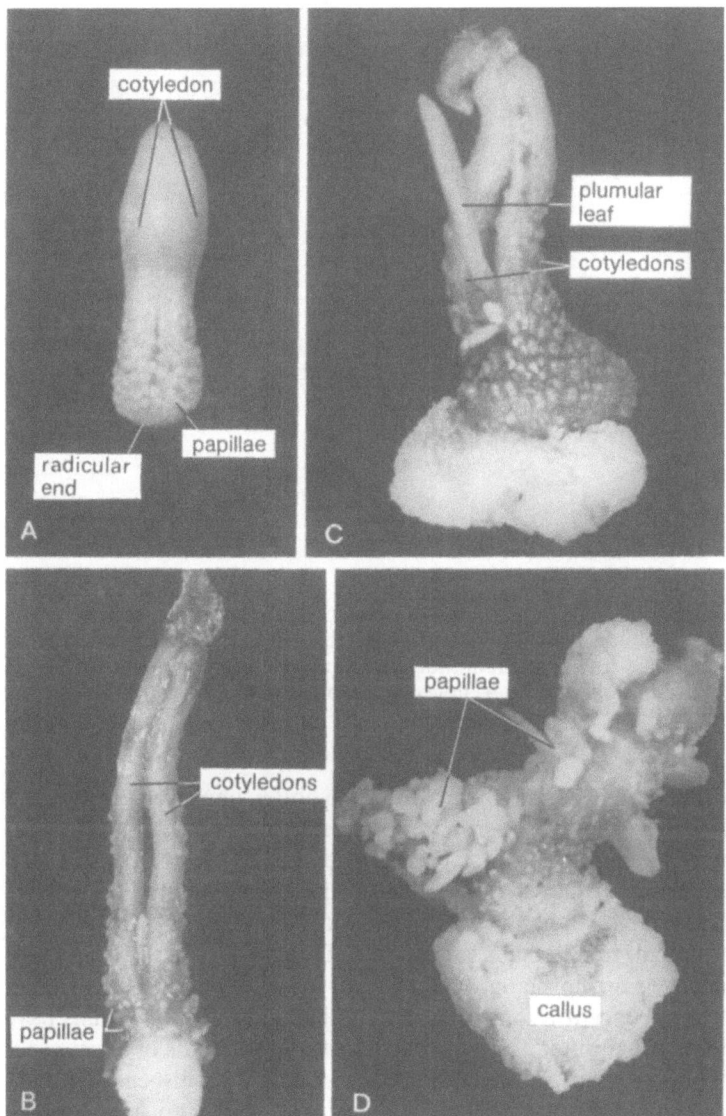

Fig. 1 A–D. *Amyema pendula.* **A** Mature embryo showing papillate outgrowths on the radicular end. **B** 3-week-old culture of embryo on White's medium (WM)+IAA+CH, showing proliferated tip of cotyledons, elongation of papillae, and formation of a holdfast. **C** 8-week-old seedling on WM+IAA+CH; note the plumular leaves and callused holdfast. **D** 15-week-old seedling showing extensive callusing and differentiation of papillae into shoots. (After Bajaj 1967)

dium and the type of growth that ensues. These early attempts utilized full-grown embryos excised from mature seeds, and provided evidence that mature embryos require an extremely simple medium. Although mineral salts, and a carbohydrate source in the form of sucrose, were the only major components of the medium,

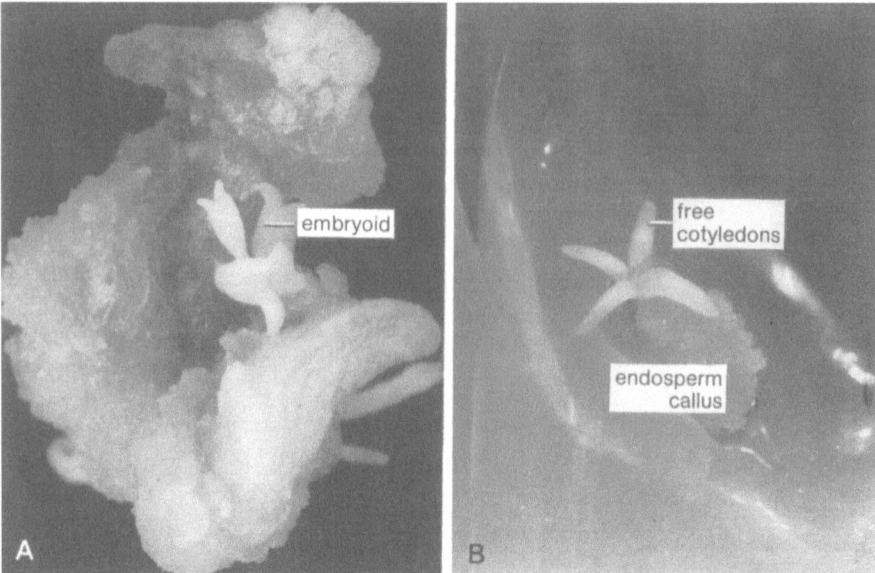

Fig. 2A, B. *Nuytsia floribunda.* **A** Embryoids with free cotyledons, differentiated from embryonal callus after subcultures on WM + indolebutyric acid (IBA) + adenine (AD) + CH for 6 weeks. **B** Mature embryo with a segment of endosperm cultured on WM + IBA + AD + CH (compare the embryo with embryoids in **A**; note endosperm callus). (After Nag 1970)

other compounds were often added to the medium to obtain improved growth, and for modifying the existing pattern of growth. As a result, incorporation of vitamins such as biotin, nicotinic acid, ascorbic acid, pantothenic acid, pyridoxine, and thiamine in the basal medium is now a part of the routine embryo culture technique. Several investigators have studied the effects of different organic nitrogen compounds, and amino acids in particular, and indicated the most suitable compounds and their optimal levels for normal growth of embryos of different species in culture. A wide range of concentrations of auxins, gibberellins, and cytokinins have been tested for their growth-modifying effects on cultured embryos of diverse species of plants. However, none of the added compounds are indispensable components of the medium, although many of them were shown to have specific growth-regulating effects on different parts of the embryo, ranging from growth promotion to inhibition and callus growth. Finally, there is the collective influence of the usual parameters of the environment such as radiation, temperature, pH, supply of gases, etc., which might affect the growth of embryos. These aspects of embryo culture have been the subject of detailed study, and frequent comment in past years (see Sanders and Ziebur 1963; Narayanaswami and Norstog 1964; Degivry 1966; Raghavan 1966). A suggestive feature of these investigations is that the mature embryos are autotrophic, and that their subsequent development is to a large extent under the control of factors inherent in their own cells.

The mature embryos of *Amyema miquelii* and *A. pendula* cultured on a medium containing indoleacetic acid (IAA) and casein hydrolysate (CH) germinated in 3–

5 weeks, producing a massive holdfast and a pair of plumular leaves (Bajaj 1967). The papillate outgrowths around the radicular end (hypocotyl), whose morphology is as yet unknown, developed into shoot buds (Fig. 1 A–D). Johri and Bajaj (1964) reported similar results with *Amylotheca dictyophleba*.

From the seeds (i.e., endosperm enclosing the embryo; testa is absent) of *Exocarpus cupressiformis* cultured on a medium containing IAA and CH, the radicle emerged after 3 weeks (Bhojwani 1968). Sometimes, the radicular end proliferated followed by the differentiation of both root and shoot. The mature embryos of *Dendrophthoe falcata* proliferated, and shoot buds and haustoria differentiated from the callus. Sometimes, shoot buds differentiated directly from the embryo (Johri and Bajaj 1963). A similar response was observed in *Scurrula pulverulenta*, *Taxillus cuneatus*, and *T. vestitus*. In *Nuytsia floribunda* (Nag 1970) embryoids, later shoot buds, differentiated only after the embryo had callused (Fig. 2A, B).

Culture of Proembryos

With refinements in tissue culture techniques, emphasis in embryo culture shifted from mature to comparatively young, immature embryos, mostly in the early and late heart-shaped and torpedo-shaped stages. The small embryos generally failed to survive on transfer to media which supported growth of mature embryos, or showed only a few additional divisions before turning into an undifferentiated callus. In succeeding attempts to culture heart-shaped and torpedo-shaped embryos, auxins, vitamins, amino acids, and complex organic additives such as yeast extract (YE) and CH, were tested with some degree of success. As early as 1932, White grew young heart-shaped embryos of *Portulaca oleracea* in a medium containing mineral salts, glucose, and fibrin digest, and obtained their growth to the size of nearly mature embryos complete with root primordium and cotyledons. In 1936 LaRue successfully grew embryos as small as 0.5 mm long isolated from a number of angiosperms and gymnosperms, in media containing inorganic salts with sucrose, IAA, and YE, and obtained transplantable seedlings. Even in the presence of organic additives, individually or in combination, it was not possible to culture heart-shaped embryos about 200–300 μm long, which at best expanded to a larger size but did not grow.

Young embryos of *Cuscuta reflexa*, when grown on a medium containing IAA and CH, proliferated and numerous embryoids differentiated from the callus. These embryoids again proliferated and differentiated, and repeated the cycle. However, on transfer to fresh medium the embryos formed normal shoots (Maheshwari and Baldev 1962). Johri and Bajaj (1963, 1965) obtained similar results with the globular proembryos of *Dendrophthoe falcata* (Fig. 3 A–F).

In view of the small size of embryo, the seeds of *Orobanche aegyptiaca* (Rangaswamy 1963, 1967), *Cistanche tubulosa* (Rangan and Rangaswamy 1968), and *Striga angustifolia* (Rangaswamy and Rangan 1969) were cultured to study embryo morphogenesis. In *O. aegyptiaca* and *C. tubulosa*, which are obligate parasites, the embryo remains undifferentiated so that it lacks the normal organisation of radicle, hypocotyl, cotyledons, and plumule. The radicular end of the embryo

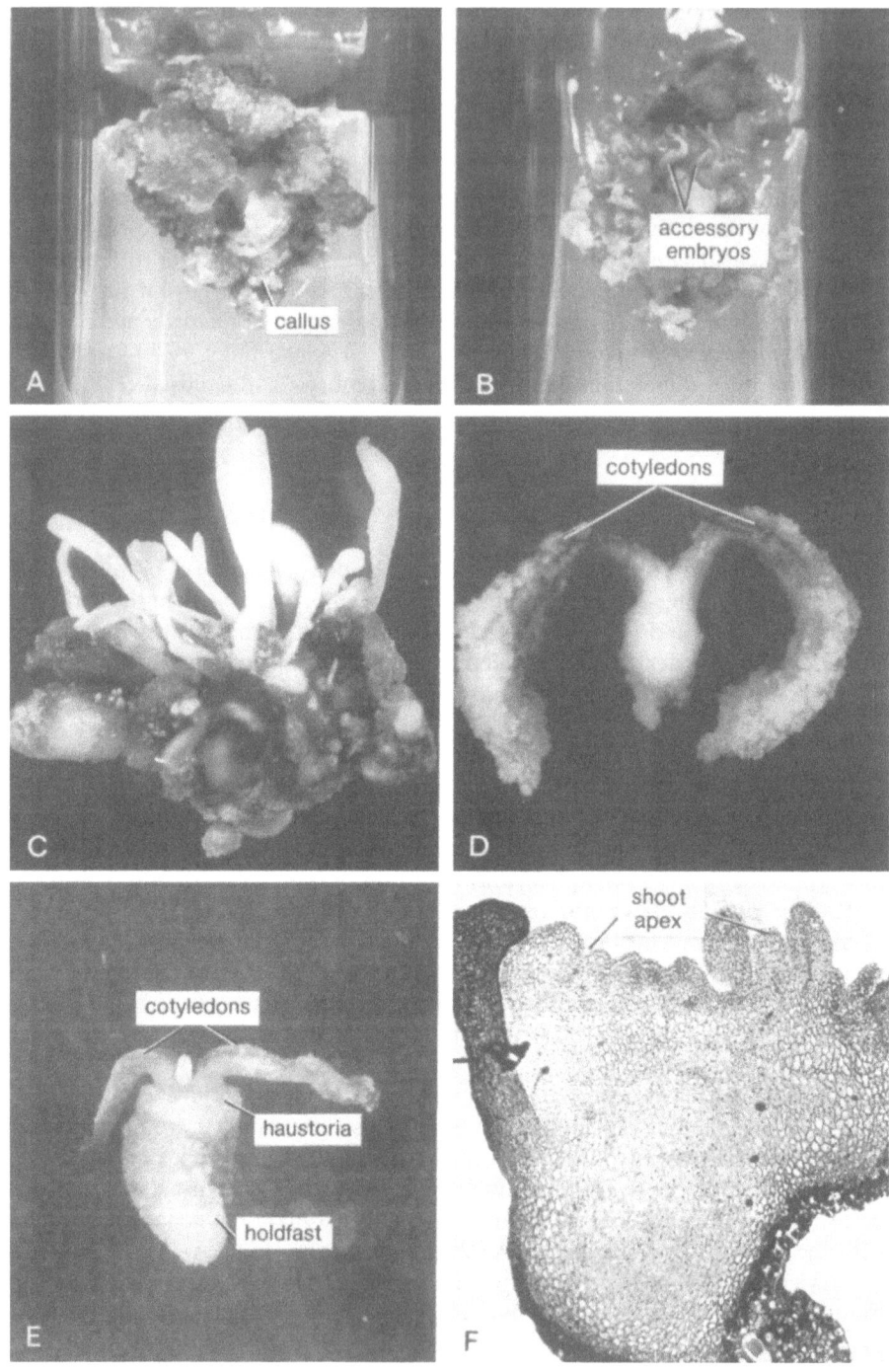

of *Orobanche* proliferated on CH-supplemented medium. In subcultures shoot buds differentiated from the callus. The seeds of *Cistanche* germinated only when coconut milk (CM) was added to the medium, and the radicular end of the embryo callussed (Fig. 4A–D). Differentiation of shoot buds occurred when this tissue was grown in dark on a complex medium. The seeds of *Striga* germinated after they had been washed in water for 48 h and treated with kinetin (KN). The embryo formed a callus tissue which differentiated only roots in subcultures on a medium containing CH and CM.

Role of Natural Plant Extracts

Coconut Milk. If, as mentioned above, studies of Hannig and others demonstrated the possibility of culturing mature embryos under aseptic conditions, it was the work of van Overbeek, Conklin, and Blakeslee (1941, 1942) which provided the lead that is widely recognised as one of the principal cornerstones in the culture of proembryos. They pointed out that whereas mature embryos of *Datura stramonium* were self-nourishing and grew into seedlings in a simple medium containing mineral salts and 1% of dextrose, torpedo-shaped and heart-shaped embryos required an organic addenda containing glycine, thiamine, ascorbic acid, nicotinic acid, pyridoxine, adenine, succinic acid, and pantothenic acid for normal growth in culture. In this enriched medium still smaller embryos failed to grow, or grew only feebly before turning into undifferentiated callus. Culture of such embryos was possible only when the medium was supplemented with non-autoclaved CM. It was observed, for example, that globular proembryos, initially about 200 to 500 μm long increased several times their original length in the course of a week in the CM medium without germinating precociously. In further studies (van Overbeek 1942; van Overbeek et al. 1944) CM was fractionated and, after eliminating the toxic principles, an active fraction designated as "embryo factor" was obtained in a relatively pure form. This preparation was still more dramatic in its effects, and induced growth in the cultured embryo when added in a dilution as low as 1:19,000 on a dry weight basis, compared with 1:110 for the crude milk.

The physiological effects of coconut milk vary in embryos of different species. For example, seedlings were reared from embryos of sugarcane initially 66 μm long by supplementing a standard embryo culture medium with CM (Warmke et al. 1946), but the extract had no beneficial effects on the growth of 10-day-old (more

◀——————————————————————————————

Fig. 3 A–F. *Dendrophthoe falcata.* A 4-week-old culture of embryonal callus on WM + IAA + CH. B Same, 10-week-old, showing differentiation of a large number of accessory embryos. C 30-week-old polyembryonal mass (from a globular proembryo) showing a large number of accessory embryos on WM + IAA + CH; note the germination of some of the embryos in situ. D 10-week-old subculture of an accessory embryo on WM + IAA + CH, showing callussed cotyledons. E Same, 20-week-old. Note the development of haustorial disc. F Longisection of a 15-week-old accessory embryo showing multiple shoot apices. (After Johri and Bajaj 1963)

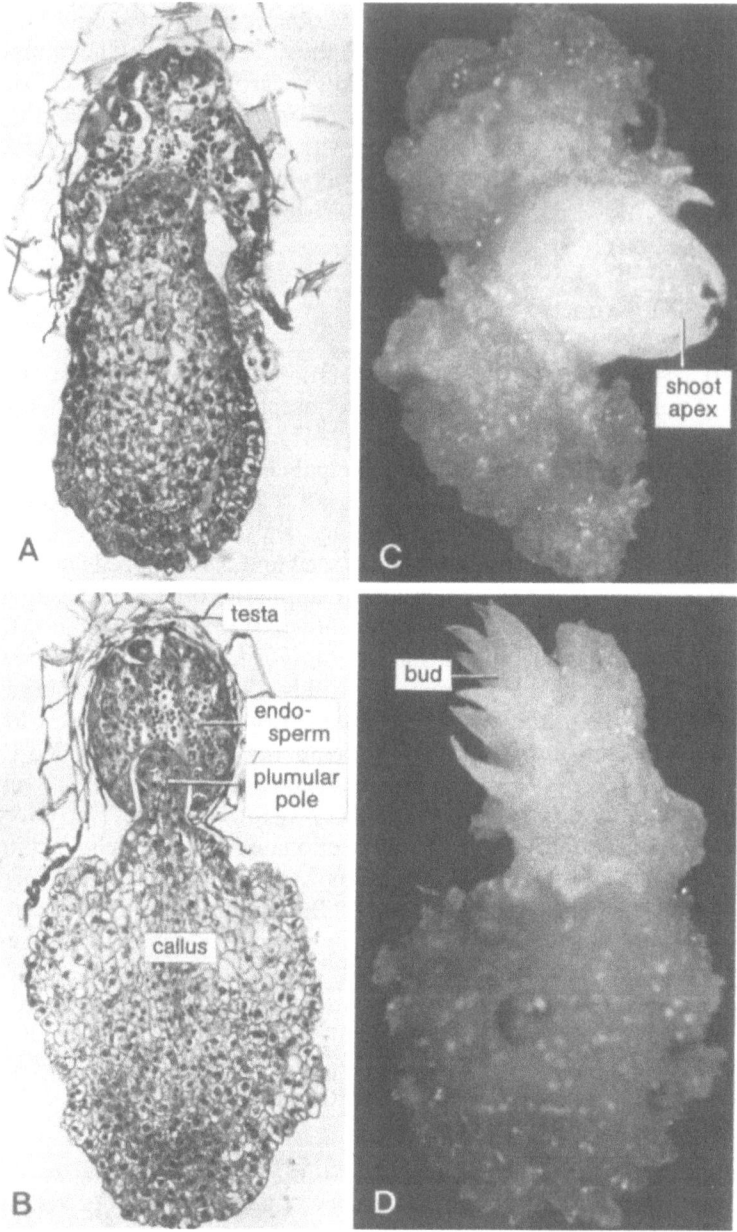

Fig. 4 A–D. *Cistanche tubulosa.* **A** Longisection of germinating seed cultured on WM +
CM + CH. Note the marked activity of the radicular end of embryo. The inactive plumular
pole is in situ. **B** Same, later stage. **C** Germinated seeds incubated for one week in dark on
WM + GA$_3$ + IAA + CM. Note the initiation of shoot bud. **D** Same, incubated for 4 weeks.
Note the differentiation of shoot buds from tubercle. (After Rangan and Rangaswamy
1968).

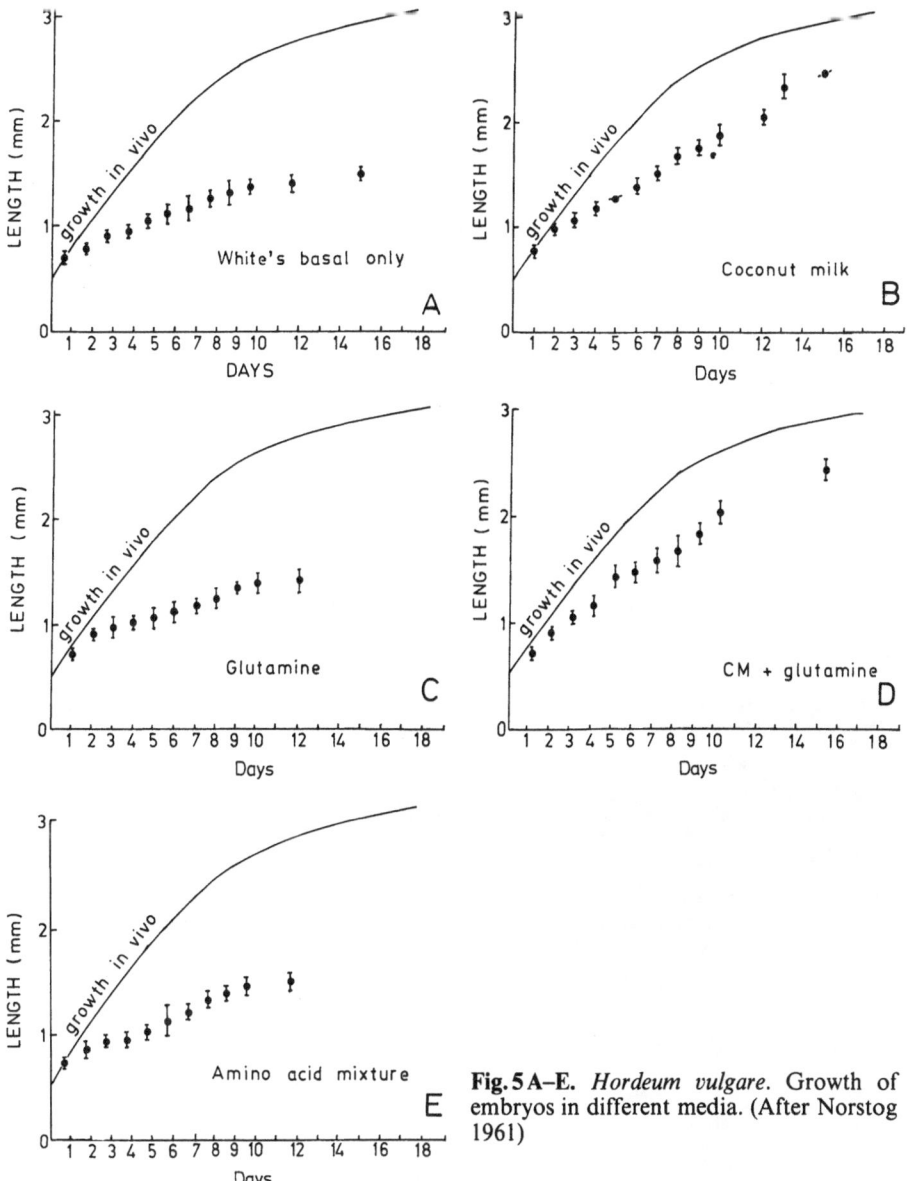

Fig. 5 A–E. *Hordeum vulgare.* Growth of embryos in different media. (After Norstog 1961)

than 300 µm long) corn embryos (Haagen-Smit et al. 1945). Coconut milk had also a depressing effect on the growth of shoot and root systems in 2- or 3-week-old corn embryos. When the extract was applied about 24 h after the transfer of embryos, part of the growth inhibition was relieved, but the overall growth of embryos was still less than in control (Uttaman 1949 a, b). These results indicate that substances necessary for growth induction in isolated corn embryos are different from those present in CM.

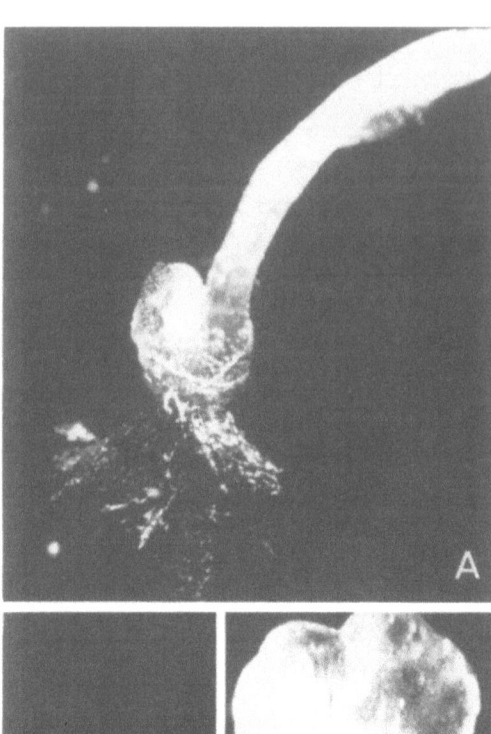

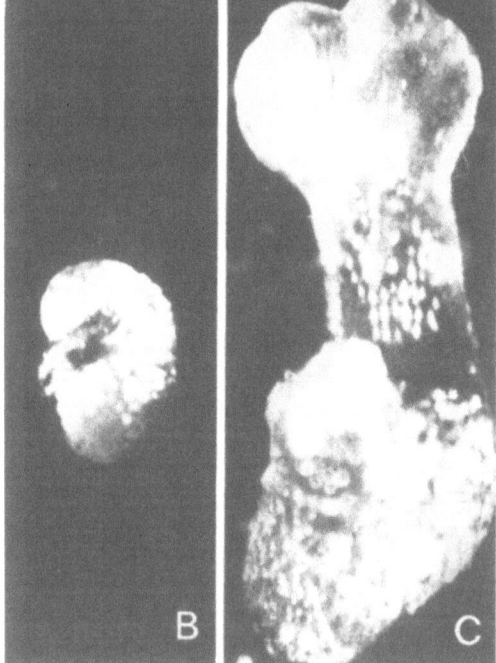

Fig. 6 A–C. *Todea barbara.* Effect of CM on embryo growth. **A** Normal sporophytic growth in an embryo excised 20 days after fertilization, and cultured in a mineral salt medium for 30 days. **B** Absence of adult organs in a 17-day-old embryo grown in the same medium. **C** Formation of normal sporophyte in a 17-day-old embryo grown in a medium supplemented with 10% of CM. (After De-Maggio and Wetmore 1961 b)

An embryo factor necessary for growth of 20-day-old tomato embryos is reportedly present in non-autoclaved CM, and a temporary maintenance of embryos in a medium containing 50% of CM is advocated for rearing healthy plants (Choudhury 1955). According to Zenkteler et al. (1961), irrespective of the presence of accessory substances in the basal medium, CM seemed to serve as a much-improved substrate for the survival and growth of embryos of *Daucus carota* excised at the very young cotyledon stage (450–600 µm). The specificity of this requirement may be questioned since growth was not improved by the addition of CM to a basal medium of a different composition.

Although earlier investigators did not observe any favourable effects of CM on the growth of immature barley embryos, later work (Norstog 1956a, 1961) showed that undifferentiated barley embryos which generally failed to survive in White's medium could be successfully cultured by supplementing it with 20% of CM (Fig. 5A–E). The smallest embryo, thus cultured, had probably no more than 100 cells (about 60 µm long). Its survival and subsequent growth in culture were enhanced by the addition of glutamine (400 mg/l), or a mixture of amino acids (total 585 mg/l-glutamic acid, leucine, asparagine, arginine, histidine, lysine, tyrosine, tryptophan, cysteine, methionine, alanine, valine, proline, and phenylalanine), to the medium. The interaction between CM and amino acids, or glutamine, in this system is not entirely clear.

The stimulatory effect of CM does not seem to be restricted to angiosperm embryos, as seen from the work of DeMaggio and Wetmore (1961 a, b) on the culture of embryos of the fern, *Todea barbara*. Given an adequate nutrition in the form of a simple mineral salt medium containing sucrose, embryos isolated 20 days after fertilization developed into normal sporophytes. When 17-day-old embryos were cultured on the same medium, they increased in size, but never reached a point where distinctive adult organs were formed. Addition of 10% of autoclaved CM to the medium resulted in normal growth of the sporophyte and, in many cases, leaf growth and maturation of cultured embryos exceeded that of their counterparts growing in a natural environment in the greenhouse (Fig. 6A–C).

Other Endosperm and Plant Extracts. The role of other natural plant extracts, most of which are of endospermic origin, in the culture of immature embryos has gradually emerged, during the last several years, to suggest that embryo factors must be of widespread occurrence. If small embryos of *Ginkgo biloba* were cultured in a medium containing an extract of *Ginkgo* "endosperm," their growth was enhanced considerably (Li 1934; Li and Shen 1934). Much the same results were obtained in inducing growth of immature corn embryos by an extract of corn kernel (Voss 1939). According to Kent and Brink (1947 a, b), water extracts of dates and bananas, wheat gluten hydrolysate, and tomato juice promoted growth of excised embryos of barley to the same extent as CH. In further work (Ziebur and Brink 1951), when immature embryos (0.3–1.1 mm long) were cultivated in a solidified medium containing 12.5% of sucrose, mineral salts, and CH, improved growth was obtained by placing aseptically excised barley endosperm around the embryos. The activity of barley endosperm in stimulating growth of embryos of *Raphanus sativus* and *Capsella bursa-pastoris* is an important, and as yet unexplained, phenomenon. Although this may indicate a lack of specificity of the endosperm in growth induction, conceivably the limiting factors in barley endosperm tissue might be related to the chemical control mechanism involved in initiating cell division in the embryos of the two latter species. Similarly, embryos excised from young caryopses of *Hordeum* and *Triticum* have been cultured with success on a medium containing the milky endosperm of corn, or an active fraction precipitated therefrom (Györffy et al. 1955).

In a study designed to discover natural plant extracts which could substitute for CM, Matsubara (1962, 1964) found embryo factor activity in alcohol diffusates from young seeds of *Datura stramonium* and *Lupinus luteus*, and old seeds of *Sechium edule*. The extract of *L. luteus* showed a wide spectrum of activity when tested against embryos of a number of species and was most effective for embryos of *Bidens biternata*, *Brassica campestris*, *Capsella bursa-pastoris*, *Datura tatula*, *Lupinus luteus*, *Pharbitis nil*, and of the hybrid *Brassica pekinensis* × *B. chinensis*; only slightly effective for embryos of *Antirrhinum majus*, *Astragalus sinicus*, *Stellaria media*, and *Triticum aestivum*; more or less ineffective for embryos of *Vicia faba* and *V. sativa*; and inhibitory for embryos of *Iris pseudoacorus*. Diffusates of young seeds and fruits of several other plants also showed embryo factor activity, and promoted growth of young embryos of *D. tatula*. It follows from the results that there is no species-specificity in the effects of diffusates in promoting growth of embryos, or in the response of embryos to the diffusates. Since promotion of embryonic growth by diverse plant extracts appears widespread, it is unfortunate that little information is available on the mechanism by which this growth is attained.

There is some evidence that even undifferentiated tissues originating from the endosperm and subcultured for prolonged periods may retain some of their growth promoting activity, and promote growth of isolated embryos placed in contact with them (Pieczur 1952; Norstog 1956 b). This leads to the inference that the composition of essential metabolites of the cultured tissue does not differ from that of the intact tissue from which it is derived, although the former is growing in the absence of restraint that applies normally. Another interesting finding is that undifferentiated tissues originating from mature embryos may impart some growth promoting properties for growth of an isolated proembryo placed in contact with it (Thomas 1972).

Chemical Factors of Endosperm Extracts. Some authors have approached the question of growth of embryos induced by endosperm extracts with a view to defining clearly the factors contained in them that control cellular growth and differentiation. One of the most persuasive piece of evidence for a link between chemical factors present in CM and embryo growth stems from the work of DeMaggio and Wetmore (1961 a, b). The success of these authors in inducing growth of 17-day-old embryos of *Todea barbara* by the addition of coconut milk to the medium is described earlier. When a mixture of the hexitol, *myo*-inositol, and the sugar alcohol sorbitol was substituted for CM, growth of embryos exceeded even that obtained by the addition of the most favourable concentration of CM. It seems that *myo*-inositol and sorbitol, known to be the chief constituents of the neutral fraction of coconut milk (Pollard et al. 1961), can substitute for whole CM in the normal growth of excised embryos.

The ability of the endosperm extracts of pumpkin (*Cucurbita maxima* and *C. moschata*) and cucumber *(Cucumis sativus)* to promote growth of their respective embryos is probably the result of the action of an auxin, a cytokinin, and several organic nitrogen compounds (Nakajima 1962). When embryos were grown in a medium containing IAA, growth of embryonic leaves lagged behind, probably due to lack of cell divisions. Upon addition of 1,3-diphenylurea, or KN, leaf growth was restored, although the overall growth of the embryo was poor. A medium containing IAA, 1,3-diphenylurea, and CH supported growth of embryos to the same extent as did a medium containing embryo factor. Coincidentally, the same set of factors that control growth of embryos was found to regulate the growth of endosperm in vitro. It is easy to visualize the interesting possibility that endosperm which controls the growth of the embryo by supplying it with specific nutrients is not capable of producing them within itself but receives them from the parent tissue. Attractive though this idea may be, there is minimal evidence to support it. More recently, Norstog and Smith (1963) concluded that a phosphate-enriched White's medium at an optimum pH of 4.9 fortified with glutamine (400 mg/l) and alanine (400 mg/l) as major nitrogen sources, and lesser amounts of leucine (20.0 mg/l), tyrosine, phenylalanine, cysteine, and tryptophan (all at 10.0 mg/l) can substitute for CM in the culture of barley embryos as small as 60–90 μm in length. In a further modification of this medium, it has been shown that survival of embryos in culture can be greatly enhanced by $5 \times$ to $10 \times$ increase in the concentration of KNO_3 and KCl in the medium (Norstog 1967). The conclusion seems inescapable that growth induction in cultured embryos by endosperm extracts is a measure of the effect of specific chemical components present in the endosperm.

Synthetic media containing known chemical substances capable of supporting growth of proembryos have only been achieved in a few cases. Only when the culture of proembryos is possible in a defined medium will our understanding of their physiology and metabolism be furthered.

Effect of Protein Preparations. Some investigators have shown that addition of malt extract, CH, and other commercial preparations containing amino acids to the medium is effective in promoting growth of immature embryos and, in some cases, the additives may actually substitute for the requirement for CM or other growth-promoting extracts. Work in this direction began with a report (Blakeslee and Satina 1944) that a cold-sterilized solution of powdered malt induced growth of small hybrid embryos of *Datura* (less than 50 μm long) as effectively as CM. Autoclaved malt extract was inhibitory, due to the presence of an inhibitor possibly formed by the breakdown of some precursors during autoclaving and which masked the activity of the embryo factor (Solomon 1950). Matsubara and Nakahira (1965) observed that among a number of compounds tested only CH, tryptone, and peptone approached the level of activity of *Lupinus luteus* extract in promoting growth of heart-shaped embryos of *D. tatula*. Addition of CH to the basal medium promotes growth of immature embryos of cotton (Lofland 1950; Mauney 1961), barley (Ziebur et al. 1950), and *Capsella bursa-pastoris* (Rijven 1952). In the culture of proembryos of *Ginkgo biloba* (Radforth 1936) and *Pinus nigra* var. *austriaca* (Radforth and Pegoraro 1955) best growth was obtained in a medium containing sucrose, or dextrose, and YE. According to Rangaswamy (1961), supplementation of White's basal medium with 400 mg/l CH enabled proembryos of *Citrus microcarpa*, even down to 28 μm in length, to grow and become fully organised. In contrast, most of the embryos cultivated in the basal medium alone showed little or no growth during prolonged periods in culture. Among a number of substances tested, only CH and YE proved suitable for rearing young embryos of *Gnetum ula* (Vasil 1963). It seems safe to assume that in all of the above cases it is the non-specific mixture of amino acids in CH and other products that contributes in some measure to the growth induction in embryos. A prevalent notion that growth-promoting effect of CH is due to its high osmotic pressure is discussed below.

Role of High Osmotic Concentration

The experimental evidence discussed in the preceding pages appeared to be a convincing lead to explain the mechanism that initiates growth in small embryos. Such evidence was, however, inconsistent with the hypothesis of a number of investigators on the possible role of high osmotic pressures in regulating embryo growth in culture. This view is clearly in harmony with the common observation that the amorphous liquid endosperm in which young embryos are constantly bathed has a high osmolarity (Ryczkowski 1960, 1961, 1965, 1969; Kerr and Anderson 1944; Mauney 1961; Smith 1973). Artificially increasing the osmotic concentration of the culture milieu by addition of sucrose, or mannitol, lent itself to the culture of embryos of a number of plants which did not grow previously even in the most complex media tried. A favourable osmotic concentration, besides preventing a possible osmotic shock to the embryo excised from an environment of high osmolarity, also inhibits cell elongation and precocious germination. By suppressing the germination potential of the embryo and switching the cells from a state of elongation to one of division, the high osmotic value of the sap exerts its apparent growth-promoting effect. This is illustrated in some early studies where precocious increase in length and other deleterious effects of an inorganic salt solution on the growth of embryos of several species were overcome by the addition of high concentrations of sucrose to the medium (Hannig 1904; Dieterich 1924). Similarly, an osmotic effect appears to be the basis for a requirement for increasing concentrations of sucrose for growth of embryos isolated at progressively earlier stages from deciduous trees (Tukey 1934, 1938; Lammerts 1942). While these reports are indirect in their manifestation, and can thus hardly be considered conclusive, Ziebur and Brink (1951; see also Kent and Brink 1947 a, b; Ziebur 1951) provided the first clear-cut evidence of a relationship between osmolarity of the medium and growth of small embryos. Their work was the outcome of attempts to control precocious germination of barley embryos excised and cultured 10–15 days after pollination. The germination-inhibiting effects of 1% of CH and the parallel effects of isotonic solutions of sucrose, or mannitol, on embryos led them to attribute the effects of CH to the high osmotic pressure it produced in the medium. It was found, for example, that immature embryos about 1.4–2.8 mm long, when planted in the basal medium, did not generally continue embryonic growth but germinated to form spindly seedlings. Normal embryonic growth was sustained by the addition of 1% of CH to the medium. After a growth period of one week, such embryos were slightly larger than mature seed embryos and had a high percentage of dry matter.

Amino acids, sodium chloride, and inorganic phosphates are the usual components of commercial preparations of CH. Additional evidence for the osmotic role of CH comes from experiments in which growth of embryos was studied with the individual components of this preparation added singly, or in combinations of two and all three together to the basal medium. It was observed that precocious germination was completely blocked in a medium containing both amino acid mixture and sodium chloride, with or without phosphate. Thus, the main function of CH in this system is probably not nutritional in the sense of supplying combustible substances for metabolism, but to provide the correct osmotic conditions for growth.

Datura provides another instance which illustrates the significance of the osmotic value of the medium in the growth of embryos. Mature embryos (4.5 mm or longer) of D. stramonium grow even in the absence of sucrose in the medium, while successively younger embryos require progressively higher concentrations of sucrose. In this system the role of sucrose as an osmoticum was confirmed by substituting nutritionally inert substances like mannitol, or glycerol, in different concentrations, in the medium. When the osmotic values were kept constant and sucrose concentrations varied, irrespective of the age of embryos, optimum growth was observed at 2% of sucrose (Rietsema et al. 1953). Similarly, in media of constant osmotic value secured by the addition of different amounts of mannitol and ammonium nitrate, maximum growth of embryos of D. tatula was obtained at 0.0075 M ammonium nitrate, and increase in concentration of ammonium nitrate did not further enhance growth (Matsubara 1964; Matsubara and Nakahira 1965). These results lead one to conclude that differences in carbon energy or nitrogen requirements of embryos of different ages can be due to differences in their osmotic requirement.

There are few other data as complete as those on barley and Datura on the relation between osmotic value of the medium and the growth of embryos of different ages. Several other investigators have also established the need to provide culture media of high osmotic values to induce growth of embryos which do not usually grow even in the most complex media tried, and it is to such studies that attention is now directed. Rijven (1952) and Veen (1961, 1962, 1963), for example, routinely used 12%–18% of sucrose in a liquid medium in attempts to culture torpedo-shaped and heart-shaped embryos of Capsella bursa-pastoris. Later work (Raghavan and Torrey 1963) has, however, shown that the need for high osmolarity can be dispensed with by growing embryos in a simple solidified medium containing the usual macro- and micronutrients, vitamins, and 2% of sucrose. When cultured in this medium, even heart-shaped embryos progressed through the normal embryonic stages of growth without showing signs of germination and gave rise to small plantlets. There is every reason to suppose that the solid versus liquid medium may account for these differences but, until experimental evidence for this is presented, the role of high osmotic pressure in growth induction in the embryos of Capsella must remain uncertain.

Rangaswamy (1961) reported that nucellar proembryos of Citrus microcarpa reared in a medium containing 5%–10% of sucrose remained healthy for as long as 20 days, while no growth, beyond swelling, was obtained in a lower concentration of sucrose. Following Ziebur and Brink (1951), Norstog (1961) used 12% of sucrose in his successful culture of small barley embryos in glutamine-CM medium. High osmotic pressure of the medium reportedly has an inhibitory effect on root elongation in rice embryos (Amemiya 1964). Finally, in long-term cultures of embryos, it is necessary to make provision for their changing osmotic requirements, and to the periodical transfer of embryos to media with progressively low osmotic values for the production of transplantable seedlings (Mauney 1961; Pecket and Selim 1965). These observations are an interesting and important corollary to the work demonstrating a role for high osmoticum in the initiation of growth in small embryos.

Against the background of experimental work discussed above, it seems permissible to suggest that growth induction in proembryos of several species in artificial culture is achieved by the establishment of high osmotic value in the medium, which probably allows for an effective flow of metabolites. Unfortunately, even in species which have been extensively studied, no data on the uptake and utilization of specific nutrients as a result of increased osmotic pressure are available.

Effect of Growth Hormones

Although there are ample data on the effects of growth hormones on the morphogenesis of embryos, reports of their use in the induction of growth of proembryos are limited and many of them are negative (LaRue 1936; Tukey 1938; Sterling 1949; Sanders 1950). Among the successful use of growth hormones in the culture of embryos may be mentioned the work of Loo and Wang (1943) who grew one- to several-celled embryos of *Pinus* and *Keteleeria* in a medium supplemented with auxin. In another case, fortification of the basal medium containing YE with IAA and gibberellic acid (GA) is reported to have led to successful culture of young embryos of *Corchorus olitorius* (Iyer et al. 1959).

Van Overbeek et al. (1942) observed a marked root growth inhibition in *Datura stramonium* embryos, perhaps because of the auxin fraction present in coconut milk. Excised embryos are reported to show inconsistent response to growth hormones. Inhibition of root growth with auxin was later confirmed by Rietsema et al. (1953). Low concentrations of auxin promote root growth. Kruyt (1952, 1954) showed that in pea embryos low concentrations of auxin stimulated root and shoot growth, whereas at high concentrations the growth rate declined. Auxin sensitivity of the different organs of the embryo of *Phaseolus vulgaris* was studied by Furuya and Soma (1957). The embryonic shoot of this plant had a higher auxin optimum than the root, and the intermediate regions had intermediate effect with respect to promotion or inhibition of growth. In the embryos of *Avena sativa* also an extremely low concentration of IAA stimulated root growth, and a higher range of IAA, although promoted coleoptile and mesocotyl growth, suppressed root elongation (Guttenberg and Wiedow 1952).

Significant increases in length of torpedo-shaped embryos of *Capsella bursa-pastoris* have been attributed to the addition of IAA and GA to a medium containing 12%–18% of sucrose, but neither compound provoked growth in still smaller embryos (Rijven 1952; Veen 1962, 1963). On the other hand, addition of KN seemed to increase the chances of small embryos to survive, probably by inducing a few rounds of cell divisions. As seen from Fig. 7A–I, in a manner not clearly understood, some of the embryos treated with KN grew abnormally and, unlike

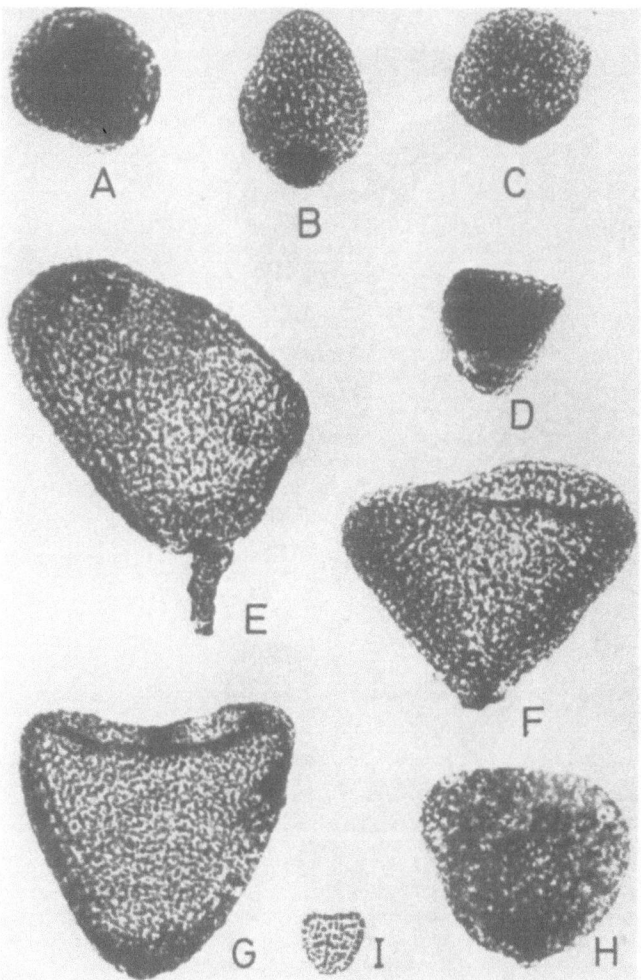

Fig. 7 A–I. *Capsella bursa-pastoris.* Growth of embryos cultured in basal medium with 18% of sucrose + varying concentrations of KN (g/ml); cotyledons failed to differentiate. **A, B** Growth period – 10 days; KN (10^{-8}). 60 µm long embryos grew to 187 µm **A**, and 68 µm to 196 µm **B. C, E, G, H** Growth period – 12 days; KN (10^{-8}). 68 µm long embryos grew to 153 µm **C**, 94 µm to 323 µm **E**, 119 µm to 340 µm **G**, and 85 µm to 221 µm **H. D** Embryo cultured at 85 µm enlarged to 136 µm after 10 days on KN (10^{-7}). **F** 10-day-old 323 µm long embryo cultured at 77 µm, on KN (10^{-9}). **I** 10-day-old 85 µm long embryo cultured on basal medium alone. There was no change in size. (After Veen 1963)

normal embryos, did not develop cotyledons. A partial resumption of growth in such embryos, manifest by an increased elongation of the hypocotyl (Fig. 8 A–E), was possible by transferring them to a medium containing 12% of sucrose and GA (Veen 1963).

By the use of a balanced mixture of IAA, KN, and adenine, some success has been achieved in inducing growth of proembryos of *Capsella bursa-pastoris* in an

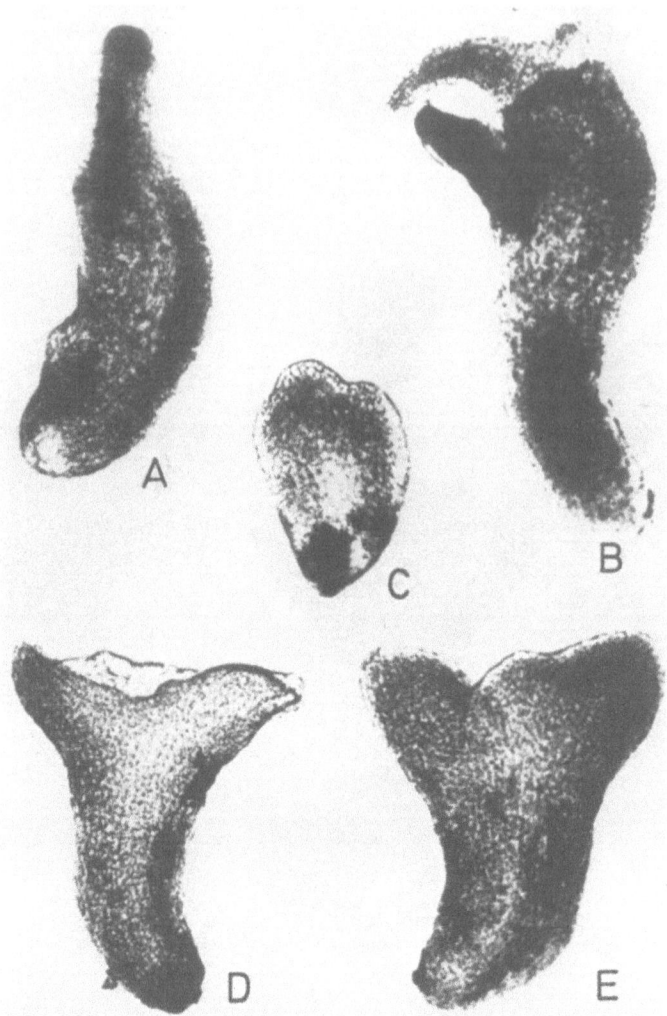

Fig. 8 A–E. *Capsella bursa-pastoris.* 12-day-old embryos, initially cultured on basal medium (18% sucrose) + KN (10^{-8} g/ml) for 8 days; embryos showed abnormal growth. The embryos resumed normal growth, partially, when transferred to basal medium (12% sucrose) + GA (10^{-5} g/ml). **A** 119 µm long embryo increased to 374 µm after 8 days, and 790 µm after 12 days. **B** Initial length 127 µm, after 8 days 416 µm, and 4 days later 884 µm. **C** At the time of culture 85 µm, after 8 days 297 µm, and after 12 days 374 µm. **D** 136 µm long embryos enlarged to 425 µm after 8 days, and 654 µm after 12 days. **E** Initial size 93 µm, after 8 days 433 µm, and after 12 days 697 µm. (After Veen 1963)

osmotically unadjusted medium in which slightly older embryos were grown (Raghavan and Torrey 1963). The development of an early globular embryo, initially 54 µm long, in the basal medium supplemented with IAA (0.1 mg/l), KN (0.001 mg/l), and adenine sulphate (0.001 mg/l), is illustrated in Fig. 9A–C. During the first 7 days in culture, the embryo increased in size as a sphere by irregular cell

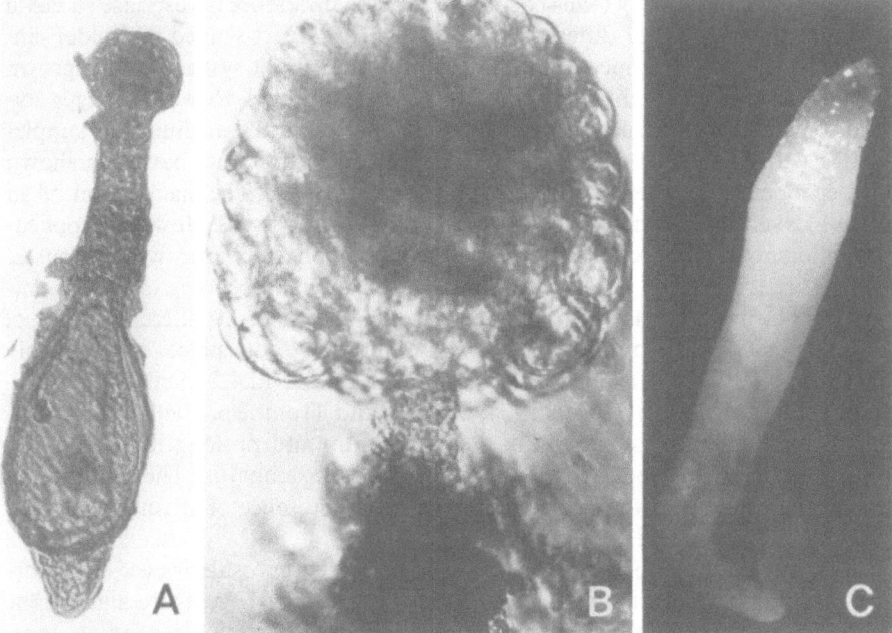

Fig. 9 A–C. *Capsella bursa-pastoris.* Stages in the development of the proembryo in a medium containing IAA (0.1 mg/l), KN (0.001 mg/l), and adenine sulfate (0.001 mg/l). **A** The embryo (54 μm) at the time of excision. **B** Formation of a mass of cells after 10 days in culture. **C** Differentiation of the cotyledons and root meristem after 6 weeks. (After Raghavan and Torrey 1963)

divisions and cell enlargement. After it had attained a few hundred cells, bilateral differentiation to form the cotyledons was observed. However, even this medium was inadequate to induce growth in embryos in the very early division stages. Securing a suitable range of growth hormones and other activating substances, the need for which seems highly probable in the light of general nutritional and metabolic studies, may be the crux of the problem in designing a medium for the growth of such embryos. The only other supplements to the medium which supported growth of embryos of comparable length were either a high concentration of sucrose or major salts.

The ease with which heart-shaped embryos of *Capsella bursa-pastoris* grow in a simple osmotically unadjusted mineral salt medium, and the induction of growth in still smaller embryos by growth hormones, or by increased concentrations of sucrose or major salts, pose a problem of considerable magnitude on the precise role of osmotic milieu of the medium and growth hormones in the regulation of embryo growth. Obviously, the physical control of a high osmotic pressure and the chemical control of hormones must be linked in some fashion. It is possible that the activity of one or more components of the balanced hormonal control system is in turn controlled by the high sucrose or high salt concentrations, perhaps through osmotic processes preventing cell elongation.

Raghavan and Torrey (1964) demonstrated the difference in response to auxin and light on embryos of different ages in *Capsella*. Heart-shaped and older embryos grown in dark formed long primary roots and shoot systems. When grown in 12 h dark and 12 h light cycle, root growth was inhibited. However, longer embryos (> 1,000 µm) did not show inhibition of growth if the medium was supplemented with low concentration of IAA. Guha and Johri (1966) have also shown age-dependent responses of *Allium cepa* embryos to IAA. The inactivation of an auxin-like substance is necessary for root initiation in embryos. However, application of auxins to embryos is also reported to cause non-specific growth responses. Rappaport et al. (1950) obtained an irregular twisting and bending of *Datura* embryos under the influence of naphthaleneacetic acid (NAA). Different grades of fusion of cotyledons of *Eranthis hiemalis* embryos took place when freshly collected seeds were soaked in 2,4-D, NAA, or 2,4,5-T (2,4,5-trichlorophenoxyacetic acid) for 12–14 h (Haccius 1955; Haccius and Trompeter 1960). Rabéchault (1962) could enhance the growth of the haustorium and prolong its survival by adding auxin to the long-term cultures of oil palm embryos. The embryos of several plants are known to produce callus in the presence of auxin (see Bulard 1967; Button et al. 1971).

IAA proved inhibitory for the growth of cotton embryos. However, the addition of GA to the medium could counteract the effect of IAA. Dure and Jensen (1957) reported that cotton embryos behaved differently to GA depending upon their age. In the relatively immature embryos, GA stimulated maturation of cotyledons and cell elongation, and in older embryos it accelerated cell division and elongation and axis growth. Stimulation of growth of embryos with GA has been observed in many other plants (see Skene 1969; van Staden et al. 1972; Kochba et al. 1974).

Apart from auxin and GA, cytokinins are also reported to play an important role in the growth and development of cultured embryos. However, like auxin and GA, the influence is subject to the age of the explant and cultural conditions. Raghavan and Torrey (1964) noticed suppression of root growth in the heart-shaped and older embryos of *Capsella*. Cytokinins seem to be inhibitory for the growth of orchid embryos. Kinetin was inhibitory for the growth of hybrid embryos of *Dendrobium* (Kano 1965). Kinetin seems to act by enhancing nitrate reductase activity, especially in excised embryos of *Agrostemma githago* (see Borriss 1967; Kende et al. 1971; Hirschberg et al. 1972).

There also exists an interaction between KN and other hormones for organ initiation in embryos. In isolated embryos of *Acer pseudoplatanus* Pinfield and Stobart (1972) obtained stimulation in elongation of the radicle by KN, but unrolling of the cotyledons was accelerated by GA and not by KN. This may prove to be an excellent system for studying the interaction of exogenous growth hormones on plant organs.

Comparative Growth of Embryos in Vivo and in Vitro

When viewed in its entirety, the changing patterns of growth of embryos in vivo provide a frame of reference for assessing the extent of growth and developmental

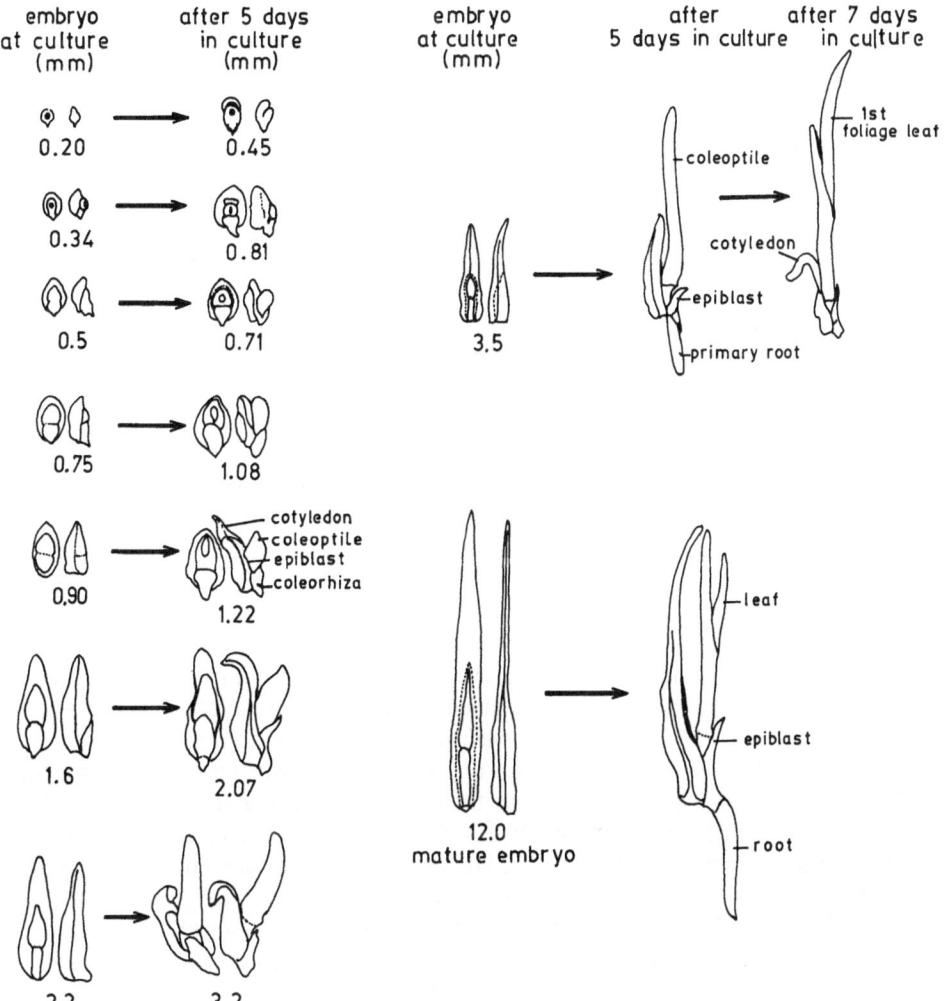

Fig. 10. *Zizania aquatica.* Note the size *mm* and outline of embryos at the time of excision, and response after 5 days; and in one case after 7 days also. (After LaRue and Avery 1938)

morphology they attain in culture. The ultimate aim in embryo culture studies is to raise seedlings closely comparable to those obtained from germinated seeds, but this goal is seldom achieved due to the prevalence of precocious germination in culture (Dieterich 1924; Yoshii 1925; LaRue 1936; Tukey 1938; Ziebur et al. 1950; McLane and Murneek 1952; Guzowska and Zenkteler 1969). The seedlings from immature embryos, if transplanted to soil and allowed to grow, are smaller and weaker than those raised from mature seeds.

When embryos of different ages are cultured, the final morphology attained by seedlings run a wide gamut of morphogenetic aberrations. In several varieties of pear *(Pyrus communis)* there was no development in culture of embryos excised prior to 46 days after anthesis. Embryos excised 66 days after anthesis showed greening of the cotyledons, while in 69-day-old embryos there was also elongation of the cotyledonary axis which terminated in a small rosette of leaves. Small, transplantable seedlings with slender long stems, stipule-like leaves, and elongate roots, were obtained from culture of 75-day-old embryos. Similar, but large plantlets with typical normal leaves and vigorously formed roots originated from culture of 81-day and 97-day embryos. Embryos of different ages of sour cherry, sweet cherry, apricot, plum, apple, and peach behaved in a similar manner with respect to the final morphology of seedlings (Tukey 1938).

Growth in culture of comparatively immature embryos of *Zizania aquatica* was generally characterised by precocious development of the shoot, and retarded growth of the primary root (LaRue and Avery 1938). In mature embryos, elongation of the primary root was normal. Thus, one can visualize a progressive retardation of growth in culture of increasingly younger embryos, as shown in Fig. 10. Similar observations have been made in the embryos of *Hordeum sativum* (Merry 1942), *Lychnis alba* (Devine 1949), and cotton (Lofland 1950).

Considerable variation exists in the development in culture of embryos of different ages of *Todea barbara* (DeMaggio and Wetmore 1961 b). The major difference is that while embryos isolated at advanced stages of development attained a normal proportion of growth in a suitable nutrient medium, those isolated after fertilization, before the formation of the first division wall, eventually developed into flat thalloid structures, more resembling gametophytes than sporophytes (see p. 205). An interesting correlation between the stages of embryonic development at excision and the extent of unorganized growth prevails in cultured embryos of *Cuscuta reflexa* (0.5–2.5 mm) (Maheshwari and Baldev 1962), and *C. gronovii* (Truscott 1966). Upon culture in an appropriate medium, relatively young embryos (less than 0.4 mm long) lose their original form and proliferate callus throughout their length. Slightly older, 0.4–1.2 mm long, embryos while retaining their original form in culture differentiate callus at the radicular end, but not at the plumular pole. Embryos 1.5 mm and longer form normal shoots in addition to a radicular callus. It, thus, seems that there is a stage in development at which an irreversible determination of the morphogenetic pattern of the embryo is reached. Accordingly, the formation of a callus would seem to be the simplest expression of growth of an isolated embryo capable of cell division. Only when it has acquired determination, can the embryo form a normal shoot without slipping back to callus growth.

Several investigators have commented upon the slow growth made by embryos in culture relative to their growth in ovulo. A comparative study of the growth of barley embryos in vitro and in ovulo (Chang 1963) showed that embryos about 0.4 mm long required 10–15 days to reach a length of 3.0 mm under natural conditions, whereas they reached only 1.8 mm in length after 30 days in culture. At the morphological level, the scutellum was poorly developed, and the coleoptile appeared anomalous in cultured embryos. The poor growth of embryos in culture seems to be due to the relatively simple medium employed. This was confirmed by Norstog (1965) who showed that embryos of the same size studied by Chang if grown in a complex medium containing several amino acids and increased salt concentrations exhibited a rate of growth nearly equal to that of natural embryos and, eventually, surpassed the latter in length. Growth of embryos of different ages of

Capsella bursa-pastoris was also slow in culture, although the development of cotyledons was characteristically out of proportion to the growth attained by other parts (Monnier 1968). Considering that the expansion of the cotyledons should be greatly restricted in ovulo, their aberrant behaviour in culture is probably due to the release from restraint.

The above examples have sought to emphasize the effect of age of embryos at excision on their final morphology in culture. From a developmental point of view one would like to know whether, irrespective of culture conditions, the preferred developmental pattern of embryos of a particular age group can be considered implicit in their age. Is it conceivable that as the embryo matures, the total milieu of the embryo sac exercises less and less influence on the final embryonic pattern attained in culture? Does the biochemical development of the embryo determine its morphological development in culture? An examination of these questions will serve to present in a new light the issues raised by the bewildering variations observed in the morphology of cultured embryos of different ages.

Culture of Embryonal Segments

The interdependence between different organs of the embryo is not clearly understood. The in vitro culture of various organs would yield valuable information.

The scutellum, an important tissue of the graminaceous embryo, influences the growth of embryos. The mature embryos of maize and other cereals cultured without the scutellum did not attain normal growth (in length) of root and shoot primordia (Andronescu 1919; Narayanaswami 1963). Smirnov and Pavlov (1964) could enhance the growth of shoot in the cultured embryonal axis of maize by using high concentrations of nitrate nitrogen. The potential morphogenic capacity of the scutellum and its relation with the meristematic regions of the embryo has been reported by LaRue (1952) and Narayanaswami (1959). Possibly, the scutellum helps in channeling the nutrients from the medium.

The role of cotyledons for the normal growth of root and shoot in embryos has also been demonstrated by surgical experiments. In decotylated embryos the growth is abnormal. In peach (Kester 1953) the embryos with a portion of cotyledons formed normal seedlings, but those without the cotyledons failed to grow in cultures. In pea the growth of root and shoot primordia was inhibited in the absence of cotyledons. Kruyt (1952) believed that, apparently, growth hormones present in the medium interact with the reserve substances of the cotyledons and promoted the growth of root and shoot. Embryos presoaked in auxin and cultured along with the intact cotyledons showed better growth (Kruyt 1954). The stimulation of growth depended upon the size of the cotyledon left with the cultured embryos. Similarly, in the embryos of *Vigna sesquipedalis* (Hotta 1957) decotylation resulted in the inhibition of development of shoot, and formation of lateral roots. Nuchowicz (1965) and Abou-Zeid (1972) also observed various types of growth inhibition in the embryos of cherry and peanut if cultured without the cotyledons.

The essentiality of cotyledons for normal growth of embryos is very clear in *Hevea brasiliensis*. Muzik (1956) observed that it was necessary to culture the embryonal axis with at least two-thirds of the cotyledons for normal growth.

Table 1. Growth response of decotylated embryo segments of *Cassytha filiformis* (Rangaswamy and Rangan 1971)

Portion of cotyledon(s) severed	Growth response
Plumular half of one cotyledon	Seedling well developed
Radicular half of one cotyledon	Seedling well developed
Plumular half of each cotyledon	Seedling well developed
One entire cotyledon	Hypocotyl and plumule both inhibited
One entire cotyledon and plumular half of other cotyledon	Plumule quiescent
One entire cotyledon and radicular half of other cotyledon	Plumule quiescent
Radicular half of each cotyledon	Plumule quiescent
Radicular half of one cotyledon and plumular half of other cotyledon	Growth of hypocotyl limited, plumule inhibited
Small portion from the equatorial region of each cotyledon	Seedling formation occasional
Entire cotyledons retaining only a belt around the plumular pole of embryonal axis	Growth inhibited, plumule quiescent
Both cotyledons leaving only the embryonal axis	Growth suppressed, plumule quiescent, rooting rare

Rangaswamy and Rangan (1963, 1971) studied the effects of decotylation on embryos of a parasitic angiosperm, *Cassytha filiformis*. These investigators report that there is a direct correlation between the partial or entire removal of the cotyledons and inhibition of morphogenesis of the shoot. The minimum intact cotyledonary tissue required for normal growth of shoot is the radicular halves of both cotyledons, or the radicular half of one cotyledon together with more than one half of the other cotyledon. The results of their experiments are summarized in Table 1.

The experimental results with decotylation of embryos suggest that hormonal and nutritional interaction affects the growth and normal development of seedling. Khan et al. (1972) presume that decotylation might affect the embryonal axis in a very specific way, by interfering with the synthesis of isoenzymes.

Other surgical experiments concern the longitudinal splitting of the entire embryo, or parts thereof, and growing the portions in culture to determine the initiation and development of a particular organ. Hanawa and Ishizaki (1953) reported that each portion of the longitudinally divided shoot apex and subjacent regions of the embryos of *Sesamum indicum* regenerated shoots with normal leaf arrangement. Similarly, the radicular end of embryos of *Dendrophthoe falcata*[1], split longitudinally, develops new meristems which differentiate holdfast[2] in both segments (Bajaj 1966). If the plumule is also involved, twin shoots are formed. These results provide evidence to indicate that embryonic root and shoot initials are not predetermined, and their fate depends on the position effect (see also Lee 1955; Furuya and Soma 1957; Sen and Verma 1959, 1963; Johri and Bajaj 1963; Nuchowicz 1965; Modrzejewski et al. 1970).

---

1 The embryo lacks a true radicle
2 Holdfast is a knob-like growth with which the semi-parasite attaches itself to the host

Embryo segments of *Cajanus cajan* containing unorganized axillary bud initials undergo significant regeneration (Kanta and Padmanabhan 1964). By culturing separately the plumule and two axillary bud primordia of the cotyledonary node, it was possible to obtain at least three seedlings from a single embryo. In *Azadirachta indica* the embryo exhibits a greater morphogenic plasticity. Rangaswamy and Promila (1972) obtained the differentiation of as many as 20 shoot buds from the decapitated plumule of embryos cultured even in distilled water. Ball (1946), working with pieces of the apical region of shoot of the mature embryo of *Lupinus albus*, indicated a high degree of autonomy of this tissue.

The role of suspensor in the developing embryos is now being established by the embryo culture technique. Cionini et al. (1976) studied the response of embryos of *Phaseolus coccineus* at different stages of development (0.5–5 mm length), to excision of the suspensor. The removal os suspensor had no effect on the growth of older embryos (5.0 mm), but development was retarded if the suspensor was removed from younger embryos.

The vital role of suspensor during embryogenesis was first demonstrated by Corsi (1972). In the young embryos of *P. coccineus*, Nagl (1974) reported that in most cases embryos deprived of suspensor succumbed unless GA was added to the medium. Cionini et al. (1976) also demonstrated that GA (10^{-8} to 10^{-6} M) replaced the function of suspensor in 0.5–1.5 mm long embryos of *P. coccineus*. However, the same concentration of GA reduced the development of larger embryos (2–3 mm in length). This indicates that, probably, the endogenous level of growth hormones increases with the increasing age of the embryo. *P. coccineus* embryos with cotyledons have a higher level (ten times more) of gibberellin than in the heart-shaped embryos (Alpi et al. 1975).

Since the suspensor completes its growth quite early during the embryogenic phase, the morphogenic studies with suspensor can be undertaken by culturing very young proembryos. According to Poddubnaya-Arnoldi (1959), growth of the suspensor in the ovule cultures of *Dendrobium nobile* at the zygote stage can be altered to a vesicular structure rather than the normal branched haustoria common in vivo.

The suspensor, no doubt, has a significant role in providing nutrition to the embryo, especially during the earlier stages. The earlier concept that the suspensor merely pushes the developing embryo deeper into the endosperm to obtain nutrition is no more valid.

Concluding Remarks

The main import of the discussion presented in this chapter bears upon the problems involved in the regulation of growth in culture of embryos of different ages. The results support the view that growth of embryos is continually open to external physical or chemical control mechanisms which vary with their age.

The successful culture of the isolated embryo was the starting point of a forward movement in the understanding of the physiology of their growth, as revealed by the documentation of the nutritional requirements of embryos of a considerable number of species. Even in the present state of our knowledge, the requirements for the culture of zygotes, and embryos in early division phases, that are so perfected in the embryo sac of each species, remain deeply mysterious. For handling this problem, the investigator will have to evolve probably very different methods, involving a great deal of manipulative skill and a great many delicate combinations of hormonal and chemical substances. If and when he is armed with the tools and tricks, the prospects of establishing a firm basis for the understanding of the complexities of embryonic differentiation will be distinctly in view. It should remain a challenging field in the years ahead.

Raghavan (1976, 1977, 1980) has published several reviews on embryo culture. Strong interest continues in the use of embryo culture techniques in raising hybrids from unsuccessful crosses. In pasture legumes barriers to interspecific hybridization are associated with endosperm failure leading to development of embryos precariously balanced between abortion or abnormal differentiation (White and Williams 1976; Williams and White 1976). In crosses between *Trifolium repens* × *T. ambiguum* and *T. ambiguum* × *T. hybridum*, hybrids have been obtained by culturing embryos dissected from young ovules shortly after interspecific pollination (Williams 1978, 1980). Hybridity of the surviving plants reared by embryo culture was established by karyotype mapping, and banding patterns of leaf isozymes. Embryo culture methods have also been used successfully in raising plants from interspecific or intervarietal crosses in *Lilium* (North 1975), *Allium* (Doležel et al. 1980), *Solanum* (Sharma et al. 1980) and cherry (Ivanička and Prečová 1980), and intergeneric hybrids between *Triticum* and *Aegilops* (Chueca et al. 1977), barley and rye (Pickering and Thomas 1979), and wheat and rye (Taira and Larter 1978). In some intergeneric crosses the initial development of embryo in the ovule may be so retarded that subsequent culture in vitro is usually ineffective. In such cases pretreatment of the female plants after pollination with specific chemicals restores the growth of embryos to a stage when they can be excised and cultured. According to Taira and Larter (1978), pretreatment of fertilized ovules with ε-amino-*n*-caproic acid, or its analogue lysine, removes physiological barriers to subsequent growth of embryos of wheat × rye crosses in a defined medium. In other cases culture of hybrid embryos resulted in the formation of a callus from which hybrid plants were regenerated by standard tissue culture techniques (Wang and Chang 1975, 1978; Cooper et al. 1978).

An alternative method to the direct culture of hybrid embryos in vitro is the nurse-culture technique in which the underdeveloped hybrid embryo enveloped in a nurse endosperm from a normally-developing ovule is cultured to the seedling stage. Such a system may prove to be independent of the complex media that are necessary for the direct culture of embryos. By the nurse-culture method, interspecific hybrids have been raised from the pasture legume genera *Trifolium*, *Lotus* and *Ornithopus* (de Lautour et al. 1978, Williams and de Lautour 1980). In an extension of embryo culture methods to raise seedlings from inviable crosses, fertilized ovules have been cultured. Subsequently, embryos excised from such ovules were transferred to a medium, of a different composition, to promote the final development of hybrid seedlings to a transplantable stage (Stewart and Hsu 1978; Inomata 1978; Matsuzawa 1978; Takeshita et al. 1980).

With regard to the control of morphogenesis in cultured embryos by exogenously applied substances, both IBA and NAA increase the number of roots in the embryos of *Iris* (Stoltz 1977). The invariable response of cereal embryos to the presence of auxin in the medium is the regeneration of callus (Schaeffer et al. 1979; Bayliss and Dunn 1979; Granatek and Cockerline 1978, 1979). According to Granatek and Cockerline (1978), auxin-induced callus formation in cultured barley embryos can be reversed by the simultaneous addition of GA to the medium, resulting in plantlet regeneration. An interesting aspect of the hormonal interaction in this system is the changing osmolarity of the medium that accompanied callus growth and plantlet formation. Robichaud et al. (1980) have shown that cultured embryos of a viviparous mutant of corn is insensitive to added abscisic acid (ABA) to suggest that a defect in the biosynthetic pathway of action of this hormone might account for vivipary.

Shimada and Yamada (1979) observed regeneration of plantlets in cultures of young embryos of *Triticum aestivum* cv. Chinese Spring, and cv. Salmon. Fourteen-day-old embryos (1.0 mm long) proved most responsive. The cultures of both the cultivars were maintained, by regular subcultures (every 30 to 40 days), on Linsmaier and Skoog's medium (1965) containing 2,4-D (2.0 mg/l). When subcultured calluses were grown on 2,4-D-free medium, shoots differentiated, and maintained this ability for eight months. In *Sorghum bicolor* (Dunstan et al. 1979) there are two distinct methods of plantlet production from immature embryos (with coleoptile-coleorhiza). Dunstan et al. point out that shoot primordia and embryo-like structures arise de novo from the cells of scutellum, on MS-medium supplemented with adenine sulphate (100 mg/l). Over 300 plantlets were grown to maturity on transfer to pots containing compost. Fifteen plants were sterile which could be due to the segregation of restorer genes in the explants. The production of shoots continued for 18 months on transfer of "morphogenic explants" to MS-medium containing 6-benzylaminopurine (0.5 mg/l) and either 2,4-D (2.0 mg/l) or NAA (5.0 mg/l). As in other plant tissue cultures, activated charcoal promotes the growth of embryos of *Cocos nucifera*, on $MS + CM + IAA + N-(\Delta^2$-isopentyl) adenine (Fisher and Tsai 1978). These authors suggest that activated charcoal absorbs inhibitory substances present in traces in the medium, or toxic products released by the cultured tissues.

A study on the utilization of sugar by embryos of corn has confirmed the often-reported superiority of sucrose over other sugars, especially in enhancing elongation of root (Burghardtová and Tupý 1980). According to Landová and Ondřej (1979), irrespective of the type of medium used, best growth of cultured embryos of barley occurred in shake cultures providing for proper aeration of the medium. A temperature of 25 °C is optimal for the growth of isolated embryos of olive (Diamantoglou and Mitrakos 1979). Although dormancy of cultured embryos of tulip is broken by treatment at both high and low temperatures, initiation of a bulb-primordium occurred only at low temperature. A period at high temperature following chilling accelerated the subsequent development of bulb primordium (Niimi 1978).

Farnesol is a sesquiterpenoid which apparently functions as an antitranspirant through the release of endogenous abscisic acid (Mansfield et al. 1978). Addition of farnesol stimulates root growth in cultured embryos of *Phaseolus vulgaris* and *Hordeum vulgare* (Wardle and Simpkins 1980). Since the survival of seedlings obtained by embryo culture methods is often hampered by excessive loss of water, incorporation of farnesol in the medium may appear to be of survival value to the seedlings.

Bennici and Cionini (1979) report that the addition of zeatin, or zeatin riboside, to the medium favours growth of early-stage of embryos of *Phaseolus coccineus*, while later-stage embryos are insensitive to the hormones. As with embryos of several other species, the use of a medium of high osmolarity is essential for the growth of early-division-phase embryos of pea (Stafford and Davies 1979; see also Srivastava et al. 1980).

A further study of the suspensor has demonstrated the possibility of obtaining a callus exclusively from suspensor cells of embryos of *Phaseolus coccineus* cultured in a medium enriched with 12% sucrose and exposed to continuous light (Bennici et al. 1976). Since suspensors separated from the organogenic part of the embryo do not form a callus, integrity of the suspensor-embryo complex appears essential for this morphogenetic event. Callus growth occurred in embryo suspensors known to be rich in GA; by combined gas-chromatography-mass spectrometry, the specific gibberellin of the suspensor of *P. coccineus* has been identified as GA_1 (Alpi et al. 1979).

Crouch and Sussex (1981) compared the embryogeny in vitro and in vivo by monitoring the synthesis and accumulation of storage-proteins in developing embryos of *Brassica napus* Immature embryos, 15 (heart-shaped) to 40-day-old (cotyledon stage), were cultured on Monnier's (1976) embryo culture medium with, and without, different concentrations of an inhibitor of germination, ABA. Two major proteins (12S and 1.7S) could be first detected when the cotyledons were well developed. The accumulation of 1.7S protein stopped when water content of the embryo began to decline, whereas 12S proteins continued until the maturity of seed. Twentyseven-day-old embryos (just as storage proteins were beginning to accumulate) cultured on 2% sucrose, without hormones, germinated precociously within three days. The incorporation of amino acids into 12S storage-proteins dropped from 3% of total incorporation to less than 1%. The addition of $10^{-6}M$ ABA into the medium increased amino acid incorporation from 3% to 18%, and this rate corresponded to the maximal rate in seeds in vivo. Such studies are useful in a comparison of growth rate of embryos developing in vivo and in vitro, and the biochemical changes taking place in the developing seeds.

It is hoped that studies on embryo culture will be helpful in physiological and developmental studies.

References

Abou-Zeid A (1972) Embryoachsenkultur von Kirschen in flüssiger Nährlösung. Gartenbauwissenschaft 37:273–280

Alexander LJ (1956) Embryo culture of tomato interspecific hybrids. Abstr Phytopathology 46:6

Alpi A, Tognoni F, D'Amato F (1975) Growth regulator levels in embryo and suspensor of *Phaseolus coccineus* at two stages of development. Planta 127:153–162

Alpi A, Lorenzi R, Cionini PG, Bennici A, D'Amato F (1979) Identification of gibberellin A_1 in the embryo suspensor of *Phaseolus coccineus*. Planta 147:225–228

Amemiya A (1964) Effect of peptone on growth of rice embryo. IV. Studies on the embryo culture in rice plant. Bull Natl Inst Agric Sci, Tokyo Ser D, Plant Physiol 11:151–210

Andronescu DI (1919) Germination and further development of the embryo of *Zea mays* separated from the endosperm. Am J Bot 6:443–452

Bajaj YPS (1966) Behaviour of embryo segments of *Dendrophthoe falcata* (L.f.) Ettings. in vitro. Can J Bot 44:1127–1131

Bajaj YPS (1967) In vitro studies on the embryos of two mistletoes, *Ameyma pendula* and *A. miquelii*. New Zealand J Bot 5:49–56

Ball E (1946) Development in sterile culture of stem tips and subjacent regions of *Tropaeolum majus* L. and *Lupinus albus* L. Am J Bot 33:301–318

Bayliss MW, Dunn SDM (1979) Factors affecting callus formation from embryos of barley *(Hordeum vulgare)*. Plant Sci Lett 14:311–316

Bennici A, Cionini PG (1979) Cytokinins and in vitro development of *Phaseolus coccineus* embryos. Planta 147:27–29

Bennici A, Cionini PG, D'Amato F (1976) Callus formation from the suspensor of *Phaseolus coccineus* in hormone-free medium: A cytological and DNA cytophotometric study. Protoplasma 89:251–261

Bhojwani SS (1968) Morphogenetic studies on cultured mature endosperm and embryo of some angiosperms. PhD Thesis, Univ Delhi, Delhi

Blakeslee AF, Satina S (1944) New hybrids from incompatible crosses in *Datura* through culture of excised embryos on malt media. Science 99:331–334

Borriss H (1967) Untersuchungen über die Steuerung der Enzymaktivität in pflanzlichen Embryonen durch Cytokinine. Wiss Z Univ Rostock Math Naturwiss 16:629–639

Brink RA, Cooper DC (1947) The endosperm in seed development. Bot Rev 13:423–541

Brink RA, Cooper DC, Ausherman LE (1944) A hybrid between *Hordeum jubatum* and *Secale cereale*. J Heredity 35:67–75

Brown HT (1906) On the culture of the excised embryos of barley on nutrient solutions containing nitrogen in different forms. Trans Guinness Res Lab 1:288–299

Buckner GD, Kastle JH (1917) The growth of isolated plant embryos. J Biol Chem 29:209–213

Bulard C (1967) Un cas d'inhibition de croissance de l'épicotyle chez *Ginkgo biloba* L. obtenu sous l'influence de gibbérellines. CR Acad Sci Paris 265:1301–1304

Burghardtová K, Tupý J (1980) Utilization of exogenous sugars by excised maize embryos in culture. Biol Plant 22:57–64

Button J, Bornman CH, Carter M (1971) *Welwitschia mirabilis:* Embryo and free-cell culture. J Exp Bot 22:922–924

Chang CW (1963) Comparative growth of barley embryos in vitro and in vivo. Bull Torrey Bot Club 90:385–391

Choudhury B (1955) Embryo culture technique. II. "Embryo factors" and immature tomato embryos. Indian J Hortic 12:152–154

Chueca M-C, Cauderon Y, Tempe J (1977) Technique d'obtention d'hybrides Blé tendre x *Aegilops* par culture in vitro d'embryons immatures. Ann Amélior Plantes 27:537–547

Cionini PG, Bennici A, Alpi A, D'Amato F (1976) Suspensor, gibberellin and in vitro development of *Phaseolus coccineus* embryos. Planta 131:115–117

Cooper KV, Dale JE, Dyer AF, Lyne RL, Walker JT (1978) Hybrid plants from the barley × rye cross. Plant Sci Lett 12:293–298

Corsi G (1972) The suspensor of *Eruca sativa* Miller (Cruciferae) during embryogenesis in vitro. G Bot Ital 106:41–54

Crouch ML, Sussex IM (1981) Development and storage-protein synthesis in *Brassica napus* L. embryos in vivo and in vitro. Planta 153:64–74

Degivry MT (1966) Revue bibliographique sur les problèmes liés cultures in vitro d'embryons immatures. Bull Sci Bourgogne 24:57–87

De Lautour G, Jones WT, Ross MD (1978) Production of interspecific hybrids in *Lotus* aided by endosperm transplants. New Zealand J Bot 16:61–68

DeMaggio AE, Wetmore RH (1961a) Growth of fern embryos in sterile culture. Nature (London) 191:94–95

DeMaggio AE, Wetmore RH (1961b) Morphogenetic studies on the fern *Todea barbara*. III. Experimental embryology. Am J Bot 48:551–565

Devine V (1949) Notes on the culture of *Lychnis* embryos. Proc Iowa Acad Sci 55:95–97

Dieterich K (1924) Über Kultur von Embryonen außerhalb des Samens. Flora 117:379–417
Diamantoglou S, Mitrakos K (1979) Sur la culture in vitro de l'embryon d'olivier (Olea europaea L. var Oleaster). CR Acid Sci Paris 288:1537–1540
Doležel J, Novák FJ, Lužný J (1980) Embryo development and in vitro culture of Allium cepa and its interspecific hybrids. Z Pflanzenzutg 85:177–184
Dubard M, Urbain J-A (1913) De l'influence de l'albumen sur développement de l'embryon. CR Acad Sci Paris 156:1086–1089
Dunstan DI, Short KC, Dhaliwal H, Thomas E (1979) Further studies on plantlet production from cultured tissues of Sorghum bicolor. Protoplasma 101:355–361
Dure LS, Jensen WA (1957) The influence of gibberellic acid and indoleacetic acid on cotton embryos cultured in vitro. Bot Gaz 118:254–261
Emsweller SL, Asen S, Uhring J (1962) Lilium speciosum × L. auratum. Lily yearbook N Am Lily Soc 15:7–15
Farquharson LI (1957) Hybridization of Tripsacum and Zea. J Heredity 48:295–299
Fisher JB, Tsai JH (1978) In vitro growth of embryos and callus of coconut palm. In vitro 14:306–311
Fridriksson S, Bolton JL (1963) Preliminary report on the culture of alfalfa embryos. Can J Bot 41:439–440
Furuya M, Soma K (1957) The effects of auxins on the development of bean embryos cultivated in vitro. J Fac Sci Univ Tokyo Sect III Bot 7:163–198
Granatek GH, Cockerline AW (1978) Callus formation versus differentiation of cultured barley embryos: Hormonal and osmotic interactions. In vitro 14:212–217
Granatek GH, Cockerline AW (1979) Callus formation of cultured early differentiating barley embryos. Bull Torrey Bot Club 106:85–96
Guha S, Johri BM (1966) In vitro development of ovary and ovule of Allium cepa L. Phytomorphology 16:353–364
Guttenberg H, Wiedow HL (1952) Über den Wirkstoffbedarf isolierter Haferembryonen. Planta 41:145–166
Guzowska I, Zenkteler M (1969) In vitro development of immature embryos of Cuscuta lupuliformis Krock. Bull Soc Sci Lett Poznan Ser D 9:37–48
Györffy B, Rédei G, Rédei G (1955) La substance de croissance du maïs laiteux. Acta Bot Acad Sci Hung 2:57–76
Haagen-Smit AJ, Siu R, Wilson G (1945) A method for the culturing of excised, immature corn embryos in vitro. Science 101:234
Haccius B (1955) Experimentally-induced twinning in plants. Nature (London) 176:355–357
Haccius B, Trompeter G (1960) Experimentelle induzierte Einkeimblättrigkeit bei Eranthis hiemalis. I. Synkotilie durch 2,4-D. Planta 54:466–481
Hanawa J, Ishizaki M (1953) Malformation in Sesamum indicum L. caused by the operation on the embryo. Sci Rep Fac Lib Art Edu Gfu Univ 1:55–61
Hannig E (1904) Zur Physiologie pflanzlicher Embryonen. I. Über die Kultur von Cruciferen-Embryonen außerhalb des Embryosacks. Bot Z 62:45–80
Hirschberg K, Hübner G, Borriss H (1972) Cytokinin-induzierte de novo-Synthese des Nitratreductase in Embryonen von Agrostemma githago. Planta 108:333–337
Honma S (1955) A technique for artificial culturing of bean embryos. Proc Am Soc Hort Sci 65:405–408
Hotta Y (1957) Roles of the cotyledon in the morphological differentiation of bean seedlings (Morphogenetical studies in Vigna sesquipedalis, II). Bot Mag Tokyo 70:383–390
Inomata N (1978) Production of interspecific hybrids in Brassica campestris × B. oleracea by culture in vitro of excised ovaries I. Development of excised ovaries in the crosses of various cultivars. Jap J Genet 53:161–173
Islam AS (1964) A rare hybrid combination through application of hormone and embryo culture. Nature (London) 201:320
Ivanička J, Pretová A (1980) Embryo culture and micropropagation of cherries in vitro. Scient Hort 12:77–82
Ivanovskaya EV (1962) The method of raising embryos on an artificial nutrient medium and its application to wide hybridization. In: Tsitsin NV (ed) Wide hybridization in plants. Israel Programm for Scientific Translations, Jerusalem, pp 134–142

Iyer RD, Sulbha K, Ramanujam S (1959) Embryo culture studies in jute and tomato. Mem Indian Bot Soc No 2:30–35

Johri BM, Bajaj YPS (1963) In vitro response of the embryo of *Dendrophthoe falcata* (L.f.) Ettings. In: Maheshwari P, Rangaswamy NS (eds) Plant tissue and organ culture – A symposium. Int Soc Plant Morphologists, Univ Delhi, Delhi, pp 292–301

Johri BM, Bajaj YPS (1964) Growth of embryos of *Amyema*, *Amylotheca*, and *Scurrula* on synthetic media. Nature (London) 204:1220–1221

Johri BM, Bajaj YPS (1965) Growth responses of globular proembryos of *Dendrophthoe falcata* (L.f.) Ettings. in culture. Phytomorphology 15:292–300

Joshi PC, Pundir NS (1966) Growth of ovules in the cross *Gossypium arboreum* × *G. hirsutum* in vivo and in vitro. Indian Cotton J 20:23–29

Kaneko K (1957) Studies of the embryo culture on the interspecific hybridization of *Chrysanthemum*. Jap J Genet 32:300–305

Kano K (1965) Studies on the media for orchid seed germination. Mem Fac Agric Kagawa Univ No 20:1–68

Kanta K, Padmanabhan D (1964) In vitro culture of embryo segments of *Cajanus cajan* (L.) Millsp. Curr Sci 33:704–706

Kende H, Hahn H, Kays SE (1971) Enhancement of nitrate reductase activity by benzyladenine in *Agrostemma githago*. Plant Physiol 48:702–706

Kent NF, Brink RA (1947a) Embryonic growth in vitro of immature *Hordeum* embryos. Abstr Am J Bot 34:600–601

Kent NF, Brink RA (1947b) Growth in vitro of immature *Hordeum* embryos. Science 106:547–548

Kerr T, Anderson DB (1944) Osmotic quantities in growing cotton bolls. Plant Physiol 19:338–349

Kester DE (1953) Factors affecting the aseptic culture of lovell peach seedlings. Hilgardia 22:335–365

Khan AA, Gasper T, Roe CH, Bouchet M, Dubucq M (1972) Synthesis of isoperoxidases in lentil embryonic axis. Phytochemistry 11:2963–2969

Knudson L (1922) Nonsymbiotic germination of orchid seeds. Bot Gaz 73:1–25

Kochba J, Button J, Spiegel-Roy P, Bornman CH, Kochba M (1974) Stimulation of rooting of *Citrus* embryoids by gibberellic acid and adenine sulphate. Ann Bot 38:795–802

Konzak CF, Randolph LF, Jensen LF (1951) Embryo culture of barley species hybrids. Cytological studies of *Hordeum sativum* × *Hordeum bulbosum*. J Heredity 42:125–134

Kravtsov PV, Kas'yanova VG (1968) Culture of isolated embryos as a method for prevention of sterility in distant hybrids of fruit plants. Fiziol Rast 15:786

Kruyt W (1952) Effects of some plant growth substances on early growth of pea in sterile culture; a study in connection with the problem of hormonisation of seeds. K Ned Akad Wet Proc C 55:503–513

Kruyt W (1954) A study in connection with the problem of hormonisation of seeds. Acta Bot Neerl 3:1–82

Laibach F (1925) Das Taubwerden von Bastardsamen und die künstliche Aufzucht früh absterbender Bastardembryonen. Z Bot 17:417–459

Lammerts WE (1942) Embryo culture an effective technique for shortening the breeding cycle of deciduous trees and increasing germination of hybrid seeds. Am J Bot 29:166–171

Landová B, Ondřej M (1979) The growth of isolated barley embryos cultivated under different conditions. Biol Plant 21:27–34

LaRue CD (1936) The growth of plant embryos in culture. Bull Torrey Bot Club 63:365–382

LaRue CD (1952) Growth of the scutellum of maize in culture. Science 115:315–316

LaRue CD, Avery Jr GS (1938) The development of the embryo of *Zizania aquatica* in the seed and in artificial culture. Bull Torrey Bot Club 65:11–21

Lee AE (1955) Growth in culture of excised portions of lupine embryos. Bot Gaz 116:359–364

Li TT (1934) The development of embryo of *Ginkgo biloba* in vitro. Sci Rep Natl Tsing Hua Univ Ser B 2:29–35

Li TT, Shen T (1934) The effect of "pantothenic acid" on the growth of the yeast and on the growth of the radicle of *Ginkgo* embryos in artificial media. Sci Rep Natl Tsing Hua Univ Ser B 2:53–60

Linsmaier EM, Skoog F (1965) Organic growth factor requirements of tobacco tissue cultures. Physiol Plant 18:100–127

Lofland Jr HB (1950) In vitro culture of the cotton embryo. Bot Gaz 111:307–311

Loo SW, Wang FH (1943) The culture of young conifer embryos in vitro. Science 98:544

Maheshwari P (1966) The embryology of angiosperms – a retrospect and prospect. In: Cutter EG (ed) Trends in plant morphogenesis. Longman, London, pp 97–112

Maheshwari P (1966) The embryology of angiosperms – A retrospect and prospect. In: Cut- *Cuscuta reflexa*. In: Plant embryology – A symposium. Council of Scientific and Industrial Research, New Delhi, pp 129–138

Maheshwari P, Rangaswamy NS (1965) Embryology in relation to physiology and genetics. In: Preston RD (ed) Advances in botanical research, vol II. Academic Press, London New York, pp 219–321

Mansfield TA, Wellburn AR, Moreira TJS (1978) The role of abscisic acid and farnesol in the alleviation of water stress. Phil Trans R Soc London B284:471–482

Matsubara S (1962) Studies on a growth-promoting substance, "embryo factor," necessary for the culture of young embryos of *Datura tatula* in vitro. Bot Mag Tokyo 75:10–18

Matsubara S (1964) Effect of *Lupinus* growth factor on the in vitro growth of embryos of various plants and carrot root tissue. Bot Mag Tokyo 77:403–411

Matsubara S, Nakahira R (1965) Some factors affecting the growth of young embryo in vitro. Sci Rep Kyoto Univ Ser A 16:1–6

Matsuzawa Y (1978) Studies on the interspecific hybridization in genus *Brassica*. I. Effects of temperature on the development of hybrid embryos and the improvement of crossability by ovary culture in interspecific cross, *B. campestris* × *B. oleracea*. Jap J Breed 28:186–196

Mauney JR (1961) The culture in vitro of immature cotton embryos. Bot Gaz 122:205–209

McLane SR, Murneek AE (1952) The detection of syngamin, an indigenous plant hormone by culture of immature corn embryos. Univ Mo Agric Sta Res Bull 496

Merry J (1942) Studies in the embryo of *Hordeum sativum*. II. The growth of the embryo in culture. Bull Torrey Bot Club 69:360–372

Modrzejewski R, Guzowska I, Zenkteler M (1970) Regeneracja fragmentów dojrzalego zarodka *Cuscuta lupuliformis* Krock. w hodowli in vitro. Poznan Towar Przy Nauk Biol 33:39–53

Monnier M (1968) Comparison du développement des embryons immatures de *Capsella bursa-pastoris* in vitro et in situ. Bull Soc Bot Fr 115:15–29

Monnier M (1976) Culture in vitro d l'embryon immature de *Capsella bursa-pastoris* Moench (L). Rev Cytol Biol Veg 39:1–120

Muzik TJ (1956) Studies on the development of the embryo and seed of *Hevea brasiliensis* in culture. Lloydia 19:86–91

Nag KK (1970) Morphogenic studies on endosperm, embryo, and other sporophytic tissues of some parasitic angiosperms. PhD Thesis, Univ Delhi, Delhi

Nagl W (1974) The *Phaseolus* suspensor and its polytene chromosomes. Z Pflanzenphysiol 73:1–44

Nakajima T (1962) Physiological studies of seed development, especially embryonic growth and endosperm development. Bull Univ Osaka Prefect Ser B 13:13–48

Narayanaswami S (1959) Experimental studies on growth of excised grass embryo in vitro. II. Effect of maleic hydrazide on embryo growth. Bull Torrey Bot Club 86:248–258

Narayanaswami S (1963) Studies on growth of excised grass embryos in culture. In Maheshwari P, Rangaswamy NS (eds) Plant tissue and organ culture – A symposium. Int Soc Plant Morphologists, Univ Delhi, Delhi, pp 302–313

Narayanaswami S, Norstog K (1964) Plant embryo culture. Bot Rev 30:587–628

Niimi Y (1978) Influence of low and high temperature on the initiation and development of a bulb primordium in isolated tulip embryos. Scien Hort 9:61–69

Norstog KJ (1956a) The growth of barley embryos on coconut milk media. Bull Torrey Bot Club 83:27–29

Norstog KJ (1956b) Growth of rye-grass endosperm in vitro. Bot Gaz 117:253–259

Norstog KJ (1961) The growth and differentiation of cultured barley embryos. Am J Bot 48:876–884

Norstog K (1965) Development of cultured barley embryos. I. Growth of 0.1–0.4 mm embryos. Am J Bot 52:538–546

Norstog K (1967) Studies on the survival of very small barley embryos in culture. Bull Torrey Bot Club 94:223–229

Norstog K, Smith J (1963) Culture of small barley embryos on defined media. Science 142:1655–1656

North C (1975) Embryo culture as an aid to breeding in *Lilium*. Acta Hort 47:187–191

Nuchowicz A (1965) Recherches sur la culture d'embryons et de fragments d'embryons d'*Arachis hypogaea* L. I. Essais orientatifs. Agricultura 3:3–37

van Overbeek J (1942) Hormonal control of embryo and seedling. Cold Spring Harbor Symp Quant Biol 10:126–133

van Overbeek J, Conklin ME, Blakeslee AF (1941) Factors in coconut milk essential for growth and development of very young *Datura* embryos. Science 94:350–351

van Overbeek J, Conklin ME, Blakeslee AF (1942) Cultivation in vitro of small *Datura* embryos. Am J Bot 29:472–477

van Overbeek J, Siu R, Haagen-Smit AJ (1944) Factors affecting the growth of *Datura* embryos in vitro. Am J Bot 31:219–224

Pecket RC, Selim ARAA (1965) Embryo culture in *Lathyrus*. J Exp Bot 16:325–328

Pickering RA, Thomas HM (1979) Crosses between tetraploid barley and diploid rye. Plant Sci Lett 16:291–296

Pieczur EA (1952) Effect of tissue cultures of maize endosperm on the growth of excised maize embryos. Nature (London) 170:241–242

Pinfield NJ, Stobart AK (1972) Hormonal regulation of germination and early seedling development in *Acer pseudoplatanus* (L.). Planta 104:134–145

Poddubnaya-Arnoldi VA (1959) Study of fertilization and embryogenesis in certain angiosperms using living material. Am Nat 93:161–169

Pollard JF, Shantz EM, Steward FC (1961) Hexitols in coconut milk: Their role in nurture of dividing cells. Plant Physiol 36:492–501

Rabéchault H (1962) Recherches sur la culture in vitro des embryons de palmier à huile (*Elaeis guineensis* Jacq.). I. Effets de l'acide β-indolylacetique. Oléaginéux 17:757–764

Radforth NW (1936) The development in vitro of the proembryo of *Ginkgo*. Trans Can Inst 21:87–94

Radforth NW, Pegoraro LC (1955) Assessment of early differentiation in *Pinus* proembryos transplanted to in vitro conditions. Trans R Soc Can 49:69–82

Raghavan V (1966) Nutrition, growth and morphogenesis of plant embryos. Biol Rev 41:1–58

Raghavan V (1967) Plant embryo culture. In: Wilt FH, Wessels NK (eds) Methods in developmental biology. Thomas Y Crowell, New York, pp 413–424

Raghavan V (1976) Experimental embryogenesis in vascular plants. Academic Press, London New York

Raghavan V (1977) Applied aspects of embryo culture. In: Reinert J, Bajaj YPS (eds) Applied and fundamental aspects of plant cell, tissue and organ culture. Springer, Berlin Heidelberg New York, pp 375–397

Raghavan V (1980) Embryo culture. Int Rev Cytol Suppl 11B:209–240

Raghavan V, Torrey JG (1963) Growth and morphogenesis of globular and older embryos of *Capsella* in culture. Am J Bot 50:540–551

Raghavan V, Torrey JG (1964) Effects of certain growth substances on the growth and morphogenesis of immature embryos of *Capsella* in culture. Plant Physiol 39:691–699

Rangan TS, Rangaswamy NS (1968) Morphogenic investigations on parasitic angiosperms. I. *Cistanche tubulosa* Wight (family Orobanchaceae). Can J Bot 46:263–266

Rangaswamy NS (1961) Experimental studies on female reproductive structures of *Citrus microcarpa* Bunge. Phytomorphology 11:109–127

Rangaswamy NS (1963) Studies on culturing seeds of *Orobanche aegyptiaca* Pers. In: Maheshwari P, Rangaswamy NS (eds) Plant tissue and organ culture – A symposium. Int Soc Plant Morphologists, Univ Delhi, Delhi, pp 345–354

Rangaswamy NS (1967) Morphogenesis of seed germination in angiosperms. Phytomorphology 17:477–487

Rangaswamy NS, Promila (1972) Morphogenesis of the adult embryo of *Azadirachta indica* A. Juss. Z Pflanzenphysiol 67:377–379

Rangaswamy NS, Rangan TS (1963) In vitro culture of embryos of *Cassytha filiformis* L. Phytomorphology 13:445–449

Rangaswamy NS, Rangan TS (1969) Morphogenic investigations on parasitic angiosperms. II. *Striga angustifolia* (Don) Saldhana (Scrophulariaceae). Flora 160:448–456

Rangaswamy NS, Rangan TS (1971) Morphogenic investigations on parasitic angiosperms. IV. Morphogenesis in decotylated embryos of *Cassytha filiformis* L. (Lauraceae). Bot Gaz 132:113–119

Rappaport J (1954) In vitro culture of plant embryos and factors controlling their growth. Bot Rev 20:210–225

Rappaport J, Satina S, Blakeslee AF (1950) Extracts of ovular tumors and their inhibition of embryo growth in *Datura*. Am J Bot 37:586–595

Rédei G (1955) *Triticum durum abyssinicum × Secale cereale* hybridek elöallitasa mesterséges embryo nevelés segitségerel. Novenytermeles 4:365–367

Rietsema J, Satina S, Blakeslee AF (1953) The effect of sucrose on the growth of *Datura stramonium* embryos in vitro. Am J Bot 40:538–545

Rijven AHGC (1952) In vitro studies on the embryo *Capsella bursa-pastoris*. Acta Bot Neerl 1:157–200

Robichaud CS, Wong J, Sussex IM (1980) Control of in vitro growth of viviparous embryo mutants of maize by abscisic acid. Develop Genet 1:325–330

Ryczkowski M (1960) Changes of the osmotic value during development of the ovule. Planta 55:343–356

Ryczkowski M (1961) Changes in the specific gravity of the central vacuolar sap in developing ovules. Bull Acad Polon Sci Ser Sci Biol 9:261–266

Ryczkowski M (1965) Changes in osmotic value of the endosperm sap and differentiation of the egg cell in developing ovules of *Cycas revoluta* (Gymnospermae). Bull Acad Polon Sci Ser Sci Biol 13:557–559

Ryczkowski M (1969) Changes in osmotic value of the central vacuole and endosperm sap during growth of the embryo and ovule. Z Pflanzenphysiol 61:422–429

Sanders ME (1950) Development of self and hybrid *Datura* embryos in artificial culture. Am J Bot 37:6–15

Sanders ME, Ziebur NK (1963) Artificial culture of embryos. In: Maheshwari P (ed) Recent advances in the embryology of angiosperms. Int Soc Plant Morphologists, Univ Delhi, Delhi, pp 297–325

Schaeffer GW, Baenziger PS, Worley J (1979) Haploid plant development from anthers and in vitro embryo culture of wheat. Crop Sci 19:697–702

Schlosser-Szigat G (1962) Artbastardierung mit Hilfe der Embryokultur bei Steinklee (*Melilotus*). Naturwissenschaften 49:452–453

Sen B, Verma G (1959) Cultivation of mustard embryo and seedling fragments. Mem Indian Bot Soc No 2:36–39

Sen B, Verma G (1963) Studies on embryo fragments. In: Maheshwari P, Rangaswamy NS (eds) Plant tissue and organ culture – A symposium. Int Soc Plant Morphologists, Univ Delhi, Delhi, pp 326–331

Sharma DR, Chawdhury JB, Ahuja U, Dankhar BS (1980) Interspecific hybridization in genus *Solanum* – A cross between *S. melongena* and *S. khasiana* through embryo culture. Z Pflanzenzutg 85:248–253

Shimada T, Yamada Y (1979) Wheat plants regenerated from embryo cell cultures. Jap J Genet 54:379–385

Skene KGM (1969) Stimulation of germination of immature bean embryos by gibberellic acid. Planta 87:188–192

Smirnov AM, Pavlov AM (1964) Cultivation of corn embryos without scutella from immature seeds under sterile conditions. Fiziol Rast 11:347–351

Smith JG (1973) Embryo development in *Phaseolus vulgaris*. II. Analysis of selected inorganic ions, ammonia, organic acids, amino acids, and sugars in the endosperm liquid. Plant Physiol 51:454–458

Solomon B (1950) Inhibiting effect of autoclaved malt preventing the in vitro growth of *Datura* embryos. Am J Bot 37:1–5

Srivastava PS, Varga A, Bruinsma J (1980) Growth in vitro of fertilized ovules of pea, *Pisum sativum* L., with and without pods. Z Pflanzenphysiol 98:347–354

Staden J van, Brown NAC, Button J (1972) The effects of applied hormones on germination of excised embryos of *Protea compacta* R.Br. in vitro. J S Afr Bot 38:211–214

Stafford A, Davies DR (1979) The culture of immature pea embryos. Ann Bot 44:315–321

Sterling C (1949) Preliminary attempts in larch embryo culture. Bot Gaz 111:90–94

Stewart JM, Hsu CL (1978) Hybridization of diploid and tetraploid cottons through in-ovulo embryo culture. J Heredity 69:404–408

Stingl G (1907) Experimentelle Studie über die Ernährung von pflanzlichen Embryonen. Flora 97:308–331

Stoltz LP (1977) Growth regulator effects on growth and development of excised mature *Iris* embryo in vitro. Hort Sci 12:495–496

Taira T, Larter EN (1977) Effects of ε-amino-*n*-caproic acid and L-lysine on the development of hybrid embryos of *Triticale* (× *Triticosecale*). Can J Bot 55:2330–2334

Taira T, Larter EN (1978) Factors influencing development of wheat-rye hybrid embryos in vitro. Crop Sci 18:348–350

Takeshita M, Kato M, Tokumasu S (1980) Application of ovule culture to the production of intergeneric or interspecific hybrids in *Brassica* and *Raphanus*. Jap J Genet 55:373–387

Thomas MJ (1972) Comportement des embryons de trois espèces de Pins (*Pinus mugo*, Turra, *Pinus silvestris* L. et *Pinus nigra* Arn.) isolés au moment de leur clivage et cultivés in vitro, en présence de cultures–nourices. CR Acad Sci Paris 274:2655–2658

Torrey JG (1973) Plant embryos. In: Kruse Jr PF, Patterson Jr MK (eds) Tissue culture: methods and applications. Academic Press, London New York, pp 166–170

Truscott FH (1966) Some aspects of morphogenesis in *Cuscuta gronovii*. Am J Bot 53:739–750

Tukey HB (1934) Artificial culture methods for isolated embryos of deciduous fruits. Proc Am Soc Hort Sci 32:313–322

Tukey HB (1938) Growth patterns of plants developed from immature embryos in artificial culture. Bot Gaz 99:630–665

Uttaman P (1949 a) The effect of coconut water on the growth of immature embryos of corn (maize). Curr Sci 18:251–252

Uttaman P (1949 b) A study in contrast of the effects of coconut water on the growth of immature embryos of corn (maize) when applied before and after germination of the embryo. Curr Sci 18:343–344

Vasil V (1963) In vitro culture of embryos of *Gnetum ula* Brongn. In: Maheshwari P, Rangaswamy NS (eds) Plant tissue and organ culture – A symposium. Int Soc Plant Morphologists, Univ Delhi, Delhi, pp 278–280

Veen H (1961) The effect of gibberellic acid on the embryo growth of *Capsella bursa-pastoris*. K Ned Akad Wet Proc 64:79–85

Veen H (1962) Preliminary report on effects of kinetin on embryonic growth in vitro of *Capsella* embryos. Acta Bot Neerl 11:228–229

Veen H (1963) The effect of various growth regulators on embryos of *Capsella bursa-pastoris* growing in vitro. Acta Bot Neerl 12:129–171

Voss H (1939) Nachweis des inaktiven Wuchsstoffes, eines Wuchsstoffantagonisten und dessen wachstums-regulatorische Bedeutung. Planta 30:252–285

Wang T-Y, Chang C-J (1975) Artificial cultures of embryo in *Citrus*. Acta Bot Sinica 17:149–152

Wang T-Y, Chang C-J (1978) Propagation in vitro in *Citrus*. Acta Genet Sinica 5:135–137

Wardlaw CW (1965) General physiological problems of embryogenesis in plants. In: Ruhland W (ed) Encyclopedia of plant physiology, vol. XV. Springer, Berlin Heidelberg New York, pp 424–442

Wardle K, Simpkins I (1980) Response of cultured embryos of *Phaseolus vulgaris* and *Hordeum vulgare* to farnesol. Ann Bot 46:505–510

Warmke H, Rivera-Perez E, Ferrer-Monge JA (1946) The culture of sugarcane embryos in vitro. Inst Trop Agric Univ Puerto Rico 4th Ann Rep 22–23

White DWR, Williams E (1976) Early seed development after crossing of *Trifolium semipilosum* and *T. repens*. New Zealand J Bot 14:161–168

White PR (1932) Plant tissue cultures. A preliminary report of results obtained in the culturing of certain plant meristems. Arch Exp Zellforsch 12:602–620

Williams E (1978) A hybrid between *Trifolium repens* and *T. ambiguum* obtained with the aid of embryo culture. New Zealand J Bot 16:499–506

Williams EG (1980) Hybrids between *Trifolium ambiguum* and *T. hybridum* obtained with the aid of embryo culture. New Zealand J Bot 18:215–220

Williams EG, de Lautour G (1980) The use of embryo culture with transplanted nurse endosperm for the production of interspecific hybrids in pasture legumes. Bot Gaz 141:252–257

Williams E, White DWR (1976) Early seed development after crossing of *Trifolium ambiguum* and *T. repens*. New Zealand J Bot 14:307–314

Yoshii Y (1925) Über die Reifungsvorgänge des Pharbitissamens mit besonderer Rücksicht auf die Keimungsfähigkeit des unreifen Samens. J Fac Sci Tokyo Univ Sect III Bot 1:1–139

Zenkteler M, Hildebrandt AC, Cooper DC (1961) Growth in vitro of mature and immature carrot embryos. Phyton 17:125–128

Ziebur NK (1951) Factors influencing the growth of plant embryos. In: Skoog F (ed) Plant growth substances. Univ Wisconsin Press, Madison, pp 253–261

Ziebur NK, Brink RA (1951) The stimulative effect of *Hordeum* endosperms on the growth of immature plant embryos in vitro. Am J Bot 38:253–256

Ziebur NK, Brink RA, Graf LH, Stahmann MA (1950) The effect of casein hydrolysate on the growth in vitro of immature *Hordeum* embryos. Am J Bot 37:144–148

10. Protoplast Culture

P.S. RAO

A chief distinguishing feature of the higher plant cell is that it possesses a rigid cell wall. Plasmolysis, a phenomenon in which the protoplast of the cell draws away from the cell wall, is in fact dependent on the presence of this cell wall. Hanstein (1880; cited in Cocking 1972) first used the term protoplast (protos meaning first, plastos meaning being formed). Klercker (1892; cited in Cocking 1972) described "Eine Methode zur Isolierung lebender Protoplasten" in which he referred to them as isolated protoplasts. Evidently, isolated protoplasts are obtained from cells from which the cell wall has been removed, and most of the investigators refer to such isolated protoplasts as naked protoplasts.

Isolation of Protoplasts

Investigations on the methods to isolate protoplasts have been extensive (see Cocking and Evans 1973). The techniques used for the isolation have a bearing on their subsequent behaviour and development and, therefore, it is important that considerable attention should be paid to the procedural details. There are principally two methods: mechanical and enzymatic.

Mechanical Method

The mechanical method was described by Klercker in 1892 (cited in Cocking 1972). He plasmolysed leaf tissues of water warrior *(Stratiotes aloides)*, cut the cell walls, and observed the extrusion of protoplasts. Subsequently, attempts were made by many workers to improve and modify the techniques (see Cocking 1972). Mechanical methods for the isolation of protoplasts have certain limitations: (1) only a small number of protoplasts can be obtained, (2) they are limited to tissues in which extensive plasmolysis takes place, and (3) are not amenable to isolation of protoplasts from mature and meristematic cells. Besides, the methods are time-consuming and tedious.

Enzymatic Method

Enzymes have been used to remove cell walls and, principally, two types of enzyme preparations have been used: cellulases and pectinases. Cocking (1960), for the first time, demonstrated the large scale isolation of protoplasts from root tips of tomato *(Lycopersicon esculentum)* by using concentrated solutions of *Myrothecium verru-*

casa cellulase. Enzymatic preparation of protoplasts has certain advantages: (1) large quantities of protoplasts can be isolated easily, (2) there is less osmotic shrinkage, and (3) cells are relatively intact and are not injured as in the case of mechanical methods (Ruesink 1971).

Pectinase enzyme preparations are especially suitable for obtaining the protoplasts from placental tissues of solanaceous fruits which generally have a high pectin content. By using a three-step procedure, viz. (1) enzyme treatment, (2) release of protoplasts from the weakened cell walls, and (3) separation of intact protoplasts from cell debris, Gregory and Cocking (1965) obtained a very large yield of protoplasts from the locule tissue of the mature green tomato fruits. Raj and Herr (1970) also successfully isolated protoplasts from the interplacental regions of berries of *Solanum*, by digesting the walls with 12% of Sigma pectinase in a sucrose solution. Although it was possible to isolate protoplasts readily and in large quantities from tomato fruit tissues, isolation of protoplasts from the leaves had proved difficult in initial stages. Takebe et al. (1968) developed a technique for the isolation of protoplasts from the mesophyll cells of tobacco, *Nicotiana tabacum*. Essentially, the technique involves the following steps: (1) surface sterilisation of the leaves, (2) peeling off the lower epidermis, (3) incubation of stripped leaf tissue in an enzyme solution, and (4) isolation of protoplasts.

Fully-expanded leaves are excised from healthy plants (60–80 days old), given a quick dip in 70% ethanol, and placed in a 10% solution of calcium hypochlorite for 15–20 min. The leaves are then repeatedly washed in sterile distilled water. The lower epidermis of the sterilised leaves is carefully peeled with the help of a forcep, and the stripped leaves cut into small pieces (4 cm^2). Peeling of the epidermis would be relatively easy if the sterilised leaves were allowed to become flaccid, or if the water supply to the plants is limited before the leaves are excised. From the peeled leaf tissues protoplasts are isolated by a two-step enzymatic treatment. First 2 g of stripped leaf pieces are placed in a 20 ml enzyme medium (0.5% of fungal pectinase macroenzyme, 0.8 M D-mannitol and 0.3% of potassium dextran sulphate at pH 5.8) in 100 ml flasks, and gently shaken in a water bath at 25 °C by a reciprocal shaker (4.5 cm/stroke). Fresh medium is replenished at 30 min intervals until maceration of mesophyll cells is complete (2 h). The mesophyll cells thus isolated by maceration are washed repeatedly with 0.8 M mannitol containing 0.1 mM CaCl$_2$, and suspended in a 2% solution of cellulase Onozuka P 1,500 (a crude cellulase preparation from *Trichoderma viride*) in 0.8 M mannitol at pH 5.4. The suspension is incubated at 36 °C for 2–3 h with occasional shaking. The cellulase digests the cell walls resulting in the release of protoplasts.

Power and Cocking (1969) slightly modified the above method. The peeled leaf tissues were placed directly in an enzyme mixture consisting of 0.5% pectinase enzyme (macroenzyme) and 2% cellulase (Onozuka P 1,500 obtained from *Trichoderma viride* in 0.7 M sorbitol or mannitol solution). The cell separation was considerably enhanced if potassium dextran sulphate (0.3%) was added to the enzyme mixture. About 1 g leaf material was added to 10 ml of the enzyme mixture (pH 5.4), incubated at 25 °C for 4–6 h, and gently teased to facilitate liberation of more protoplasts. The leaf debris is removed by filtration, and the filtrate is transferred to tubes and centrifuged at least thrice at 100g for 1 min. Each time the protoplasts sink to the bottom, the supernatant enzyme mixture and debris is de-

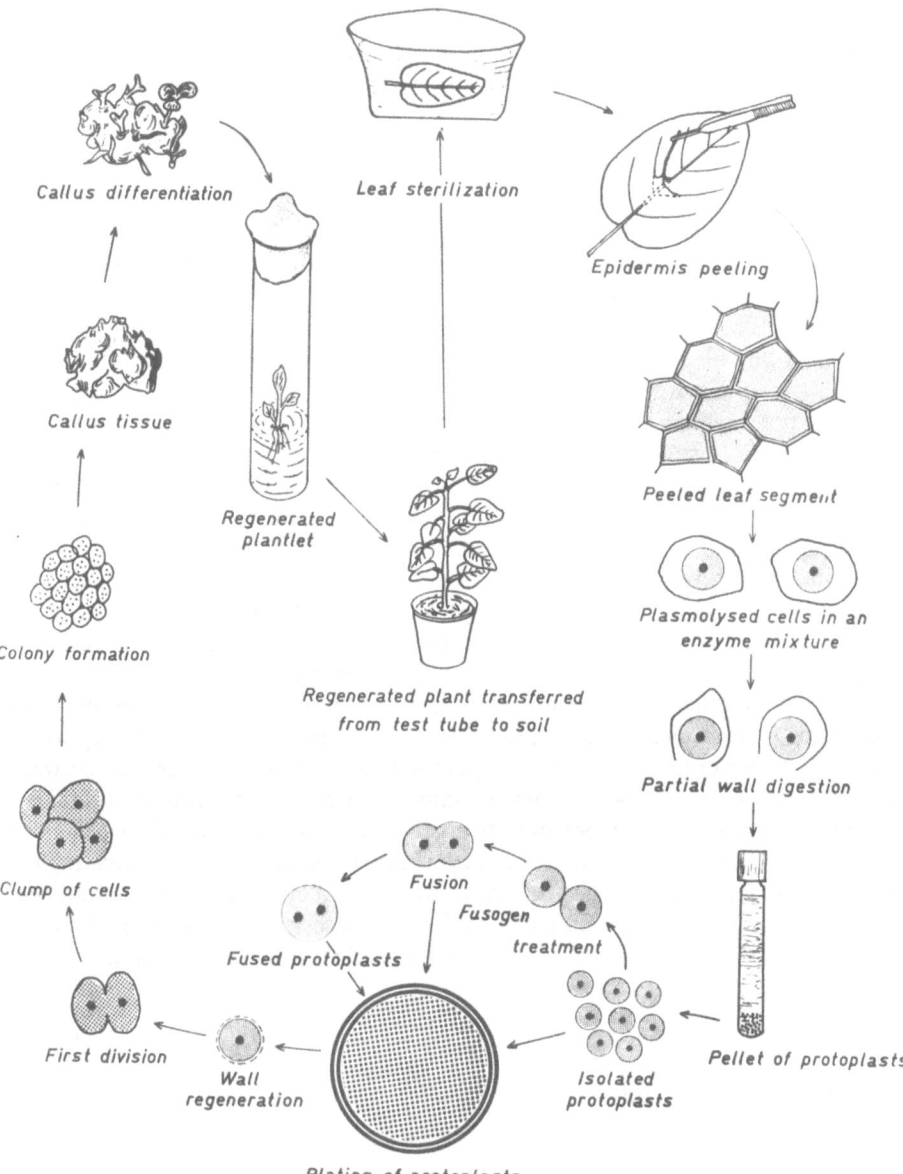

Fig. 1. Isolation, culture, and fusion of leaf-cell protoplasts. (After Bajaj 1974b)

canted. The protoplasts are washed with 0.7 M sorbitol and, finally, sorbitol is replaced by a 20% sucrose solution and centrifuged at 200g for 1 min. The cleaned protoplasts float, they are gently pipetted out with Pasteur pipettes, and resuspended in a culture medium. A diagrammatic representation of the isolation, culture, and fusion of leaf cell protoplasts is given in Fig. 1.

Mixtures of enzyme preparations are not, however, effective in all situations, for example pectinases inhibit the release of protoplasts from *Avena* coleoptiles by cellulase (Ruesink and Thimann 1966). Chupeau and Morel (1970) obtained protoplasts from carrot callus using a mixture of crude enzymes, 5% cellulase Onozuka, 5% cellulase 5 (Industrie Biologique Francaise), 1% Sigma pectinase, and 0.1% helicase.

Glusulase, a commercially available snail enzyme preparation, has also been used for the isolation of protoplasts from a number of leaf and root tissues (Pinto da Silva 1969). A combination of mechanical and enzymatic techniques was used by Harada (1973) in the isolation of mesophyll protoplasts of *Ipomea* and *Calystegia*. Harada's technique involved two steps: (1) mechanical isolation of free cells, and (2) enzymatic conversion of the free cells into viable protoplasts.

The use of enzymes in the degradation of cell walls for obtaining protoplasts has now become an acceptable laboratory procedure. However, each tissue must be investigated in relation to the optimum conditions for protoplast release. Since the isolated protoplasts are to be used for various critical experiments, it is important that protoplasts should be exposed for the minimum period of enzyme treatment and also washed thoroughly thereafter.

Although the isolation of protoplasts can most readily be accomplished, the yield of protoplasts depends upon several factors, like the physiological state of the experimental material, the purity and composition of enzymes, the choice of the osmotic solutions, and the pH (Schenk and Hildebrandt 1969).

Osmotic stabilization is an important factor in the culture of protoplasts. Generally, sorbitol or mannitol are used as osmotic stabilizers at an osmolality which is slightly hypertonic (0.5–0.6 osm). A pH range of 5–6 and a temperature between 25°–30 °C have been found to provide optimum conditions for protoplast stability. During the preparation of protoplasts a gentle agitation at intervals is beneficial for enhancing the penetration of the enzyme into cells and, thereby, liberating the protoplasts from the weakened cell walls. Removal of enzymes, once the wall degradation is completed, is a prerequisite for the successful culture of protoplasts. This is generally accomplished by washing by gentle filtration on 8 μm filter, or by low-speed centrifugation.

Culture of Protoplasts

Procedures

There are two techniques which are generally used for culturing protoplasts: liquid drop method and plating method.

The liquid drop method consists in placing the droplets of protoplast suspension in liquid medium in plastic dishes. This method provides for exchange of gases and diffusion of excreted products, besides facilitating addition of fresh medium. It has successfully been used for growing protoplasts of soybean and *Haplopappus gracilis* (Kao et al. 1971). V. Vasil and I.K. Vasil (1974) also employed the liquid droplet method for growing mesophyll protoplasts of tobacco and *Petunia* in microchambers.

Plating method involves the same procedure as that of the Bergmann cell plating technique (Bergmann 1960) with a slight modification devised by Nagata and Takebe (1971) for the culture of isolated mesophyll protoplasts of tobacco. A 2-ml aliquot of protoplasts is suspended in liquid culture medium, poured into Falcon plastic petri dishes, and mixed gently with an equal volume of the same medium containing 1% agar (temperature of the medium to be less than 45 °C). The petri dishes are sealed with Parafilm. The dishes are kept in an inverted position in a growth room at 28 °C at a continuous light intensity of 2,300 lx, and further development is followed. There are two advantages in plating protoplasts: (1) the entire sequence of division, growth, and development of a plant from a single protoplast can be easily observed, and (2) a higher plating efficiency in protoplast cultures can be achieved than with single cells. Plating method has been used also for growing haploid tobacco protoplasts (Nitsch and Ohyama 1971), *Petunia* (Potrykus and Durand 1972), and cell protoplasts of *Daucus carota* (Wallin and Eriksson 1972).

Protoplast concentration is critical for division. In liquid droplets the minimum concentration necessary is between 10^4 and 10^5 protoplasts per ml, whereas for agar plating it is between 10^3 and 10^4 per ml.

Nutrient Media

The basic nutritional requirements for culture of protoplasts are more or less similar to those of cell cultures. Murashige and Skoog (1962) or Nagata and Takebe's (1971) medium has generally been used with some modification by most of the investigators. This medium contains major and minor elements, vitamins, myo-inositol, FeEDTA, and sucrose. Since plant hormones have been demonstrated to be essential for division in cell cultures, protoplasts also require the presence of growth substances in the medium for division. The synthetic auxin 2,4-dichlorophenoxyacetic acid (2,4-D), and cytokinins such as 6-benzylaminopurine (BAP) or kinetin (KN) have been very effective for inducing division in protoplast cultures of many species (Nitsch and Ohyama 1971).

Cell Wall Formation and Division

Cell wall formation in cultured protoplasts usually takes place once the enzymes are removed. The newly synthesized cell wall can be observed under fluorescence microscope using calcofluor white [0.1% as a reagent (Nagata and Takebe 1970)]. Although Cocking first reported a technique for the isolation of protoplasts, it was Eriksson and Jonasson (1969) who, for the first time, reported nuclear division in the isolated protoplasts of *Haplopappus gracilis*. Subsequently, Reinert and Hellman (1971) made similar observations in isolated carrot protoplasts. However, these nuclear divisions in protoplasts were not followed by cytokinesis. Sustained division followed by cytokinesis was first reported in soybean protoplasts by Kao et al. (1970) and, later, by Takebe et al. (1971) in tobacco mesophyll protoplasts. Since then many reports have followed on division in cultured protoplasts of several plant genera (see Cocking 1972).

Table 1. Examples of regeneration obtained in protoplast cultures

Plant	Source	Type of regeneration	References
Antirrhinum majus	Mesophyll	Embryoids	Poirier et al. (1973, 1974)
Asparagus officinalis	Cladodes	Roots, shoots, embryoids and plantlets	Bui Dang Ha and Mackenzie (1973), Bui Dang Ha et al. (1975)
Atropa belladonna	Cell suspension	Embryos and plantlets	Gosch et al. (1975b)
Brassica napus	Mesophyll	Callus plantlets	Kartha et al. (1974a)
Bromus inermis	Cell suspension	Embryos	Kao et al. (1973)
Citrus sinensis	Ovular callus	Embryos	Vardi et al. (1975)
Datura innoxia	Mesophyll	Callus and plantlets	Schieder (1975a)
Datura innoxia	Suspension	Callus and plantlets	Furner et al. (1978)
Datura meteloides	Mesophyll	Callus and plantlets ⎱	Schieder (1977)
Datura metel	Mesophyll	Callus and plantlets ⎰	
Daucus carota	Root tissues	Embryos and plantlets	Kameya and Uchimiya (1972)
Daucus carota	Root and petiole callus	Embryos and plantlets	Grambow et al. (1972)
Daucus carota	Cell suspension	Embryos	Gosch et al. (1975a)
Hyoscyamus albus	Mesophyll	Callus and roots ⎱	Lorz et al. (1979)
Hyoscyamus muticus	Mesophyll	Callus and plantlets ⎰	
Nicotiana tabacum	Mesophyll	Callus and plantlets	Nagata and Takebe (1971)
Nicotiana tabacum	Mesophyll	Callus and plantlets	Ohyama and Nitsch (1972)
Nicotiana tabacum	Mesophyll	Callus and plantlets	Chupeau et al. (1973)
Nicotiana tabacum	Mesophyll	Callus and plantlets	Vasil and Vasil (1974)
Nicotiana tabacum	Callus	Callus and plantlets	Koblitz (1978)
Nicotiana tabacum	Mesophyll	Plantlets	Bajaj (1972)
Nicotiana tabacum	Mesophyll	Plantlets	Raveh and Galun (1975)
Nicotiana tabacum	Epiderm. cells	Plantlets	Davey et al. (1974)
Nicotiana tabacum	Mesophyll	Plantlets	Gleba et al. (1974)
Nicotiana sylvestris	Mesophyll	Callus and plantlets	Nagy and Maliga (1976)
Petunia hybrida	Mesophyll	Callus and plantlets	Frearson et al. (1973)
Petunia hybrida	Mesophyll	Callus and plantlets	Durand et al. (1973)
Petunia hybrida	Mesophyll	Callus and plantlets	Doon et al. (1973)
Petunia hybrida	Shoot tip	Callus and plantlets	Binding (1974a, b)
Petunia hybrida	Stem callus	Callus and plantlets	Vasil, V. and Vasil, I. K. (1974)
Petunia parodii	Mesophyll	Callus and plantlets	Hayward and Power (1975)
Petunia parviflora	Mesophyll	Callus and plantlets	Sink and Power (1977)
Pharbitis nil	Cotyledon	Callus and roots	Messerschmidt (1974)
Ranunculus sceleratus	Mesophyll	Embryos and plantlets	Dorion et al. (1975)
Solanum	Mesophyll	Callus and plantlets	Grun and Chu (1978)
Solanum tuberosum	Mesophyll	Plantlets	Shepard and Totten (1977)
Solanum tuberosum	Mesophyll	Plantlets	Binding et al. (1978)

Regeneration in Protoplast Cultures

Although protoplasts seem easy to isolate with several enzyme preparations described above, the major problem is one of inducing protoplasts to grow a new cell wall, to divide and, eventually, regenerate into a new plant. Research in the area of protoplast would become meaningful and useful if the protoplasts remained totipotent, as has been demonstrated for plant cells (Steward 1970; I.K Vasil, and V. Vasil 1972). Thus far regeneration of plants either through embryogenesis, or via callus development and organogenesis, has been obtained in protoplast cultures of only a few species of higher plants (Table 1).

Embryogenesis

The formation of embryos in protoplast cultures has been reported only in a few species. Also, no detailed studies have been made regarding the ontogenetic stages leading to embryogenesis.

Several investigators have demonstrated the totipotency of the tissue and cell cultures of carrot, *Daucus carota*. Grambow et al. (1972) enzymatically isolated carrot protoplasts from carrot cell suspension cultures, and successfully grew the protoplasts on a medium containing 0.2 mg/l 2,4-D. After 8–10 days, 5%–30% of the dividing protoplasts had formed cell clusters which produced embryoids. The embryoids developed into tiny plantlets which were transferred to a 2,4-D-free medium and, thus, adult carrot plants were obtained (Fig. 2 A–D). Kameya and Uchimiya (1972), also working with carrot *(Daucus carota)*, adopted a slightly different approach. They isolated protoplasts from carrot root slices which were placed in an enzyme mixture, and the liberated protoplasts were cultured on a nutrient medium containing 1% of coconut milk (CM), 0.1 mg/l naphthaleneacetic acid (NAA) or 2,4-D. The dividing protoplasts developed into clusters which were plated on a 0.6% nutrient agar medium fortified with 500 ppm casein hydrolysate (CH) and 5% CM, or 1 ppm KN. Cell clusters developed 4–8 weeks after plating, into small callus masses in which embryoid differentiation occurred. The embryoids, eventually, developed into plants. Gosch et al. (1975 a) have also reported the induction of embryos in protoplasts isolated from cell suspensions of an anthocyanin-containing cultivar of *Daucus carota*.

Differentiation of roots, shoots, and embryoids culminating in whole plants in tissues derived from the protoplasts of a monocotyledon, *Asparagus officinalis*, has been described by Bui Dang Ha and Mackenzie (1973) and Bui Dang Ha et al. (1975). Protoplasts were isolated from asparagus cladodes and grown in a medium supplemented with NAA (1 mg/l), zeatin (ZN-0.3 mg/l), and L-glutamine (1,000 mg/l). It was determined that glutamine was effective in inducing repeated divisions in isolated protoplasts. Once the protoplasts embarked upon division, small colonies were formed. Root differentiation was achieved when these colonies were picked up and transferred to a solid medium in which glutamine level was re-

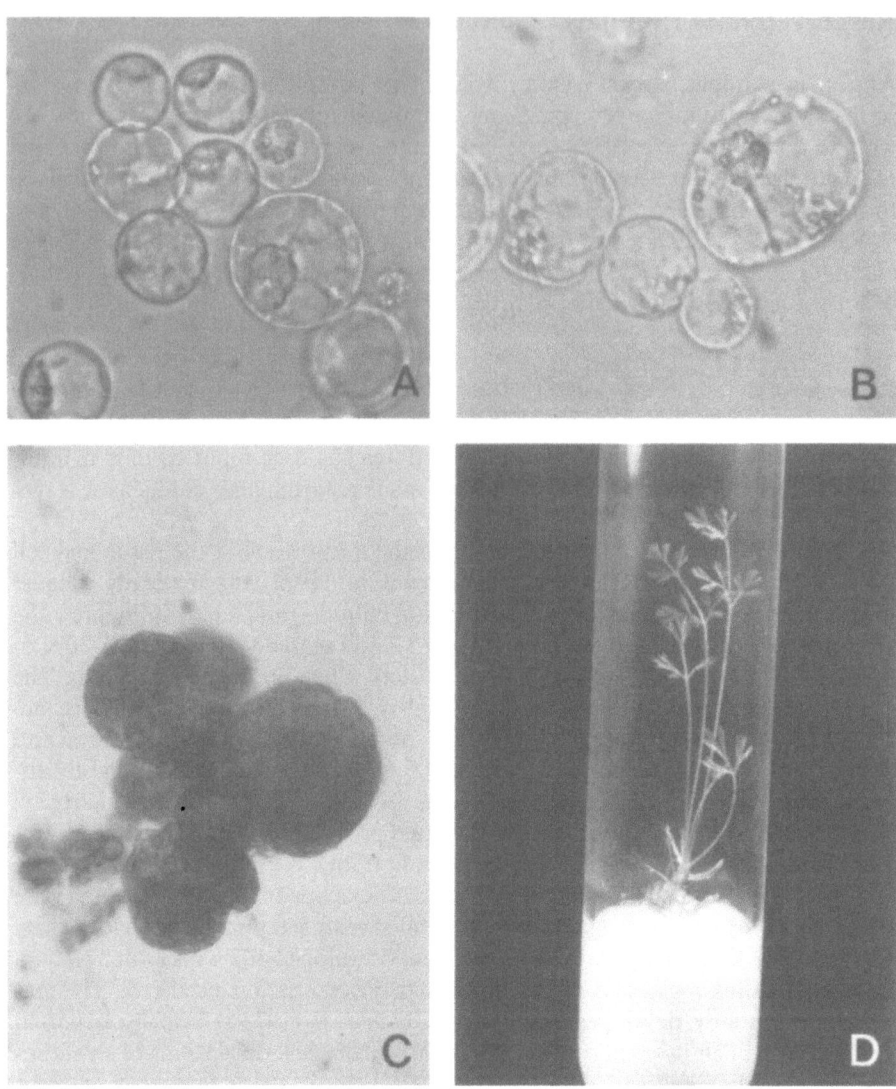

Fig. 2 A–D. Embryogenesis and formation of plants in carrot protoplast cultures. **A** Protoplasts from carrot cell cultures. **B** Protoplasts after wall formation. **C** Embryoids from carrot protoplasts, 3–4 weeks old. **D** Diploid carrot plant with 18 chromosomes. (**A–C** after Grambow et al. 1972; **D** after Dudits et al. 1976)

duced to 200 mg/l. Shoot production was initiated in the same medium when BAP (4 mg/l) was substituted for ZN and adenine (AD-40 mg/l) was incorporated in the medium. Regeneration of embryoids in protoplast-derived tissues was achieved by slightly varying the hormonal levels. Tissues were first grown for 6–8 weeks on a medium containing either 8×10^{-6} M ZN and 3×10^{-6} M NAA, or on 5×10^{-6} M BAP and 5×10^{-6} M NAA with 40 mg/l AD. When the tissues growing on the

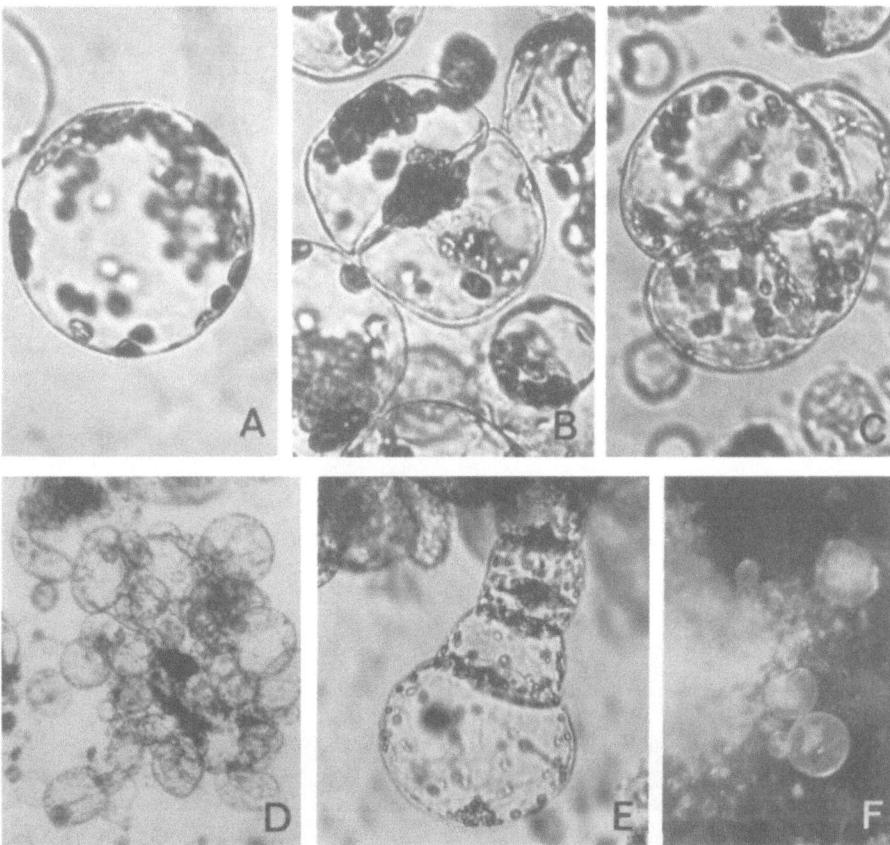

Fig. 3 A–F. Culture of mesophyll protoplasts of *Antirrhinum majus*. **A** Protoplast immediately after isolation showing the peripheral distribution of chloroplast. **B** 5-day-old protoplast after cell wall regeneration and first division. **C** Repeated divisions in a protoplast. **D** Formation of a small colony from protoplasts. **E** A group of cells showing a proembryolike structure. **F** Organization of globular embryoids in a 6-week-old culture. (After Poirier-Hamon et al. 1974)

above media were transferred to a medium lacking the growth substances, differentiation of a large number of embryoids occurred. The embryoids were of various developmental stages, of globular and elongated forms, resembled the normal zygotic embryos of monocotyledons, and possessed embryonic shoot and root meristems. By further transferring such embryoids to semiliquid media containing IAA and ZN, subsequent development of the embryoids into whole plants was achieved. A point of considerable interest in this work is that adenine in conjunction with BAP and indoleacetic acid (IAA) was effective in inducing embryoid formation.

Regeneration of embryoids in protoplast cultures of an ornamental snapdragon *(Antirrhinum majus)* has been reported by Poirier-Hamon et al. (1973, 1974). These investigators isolated protoplasts from the leaf mesophyll cells and cultured

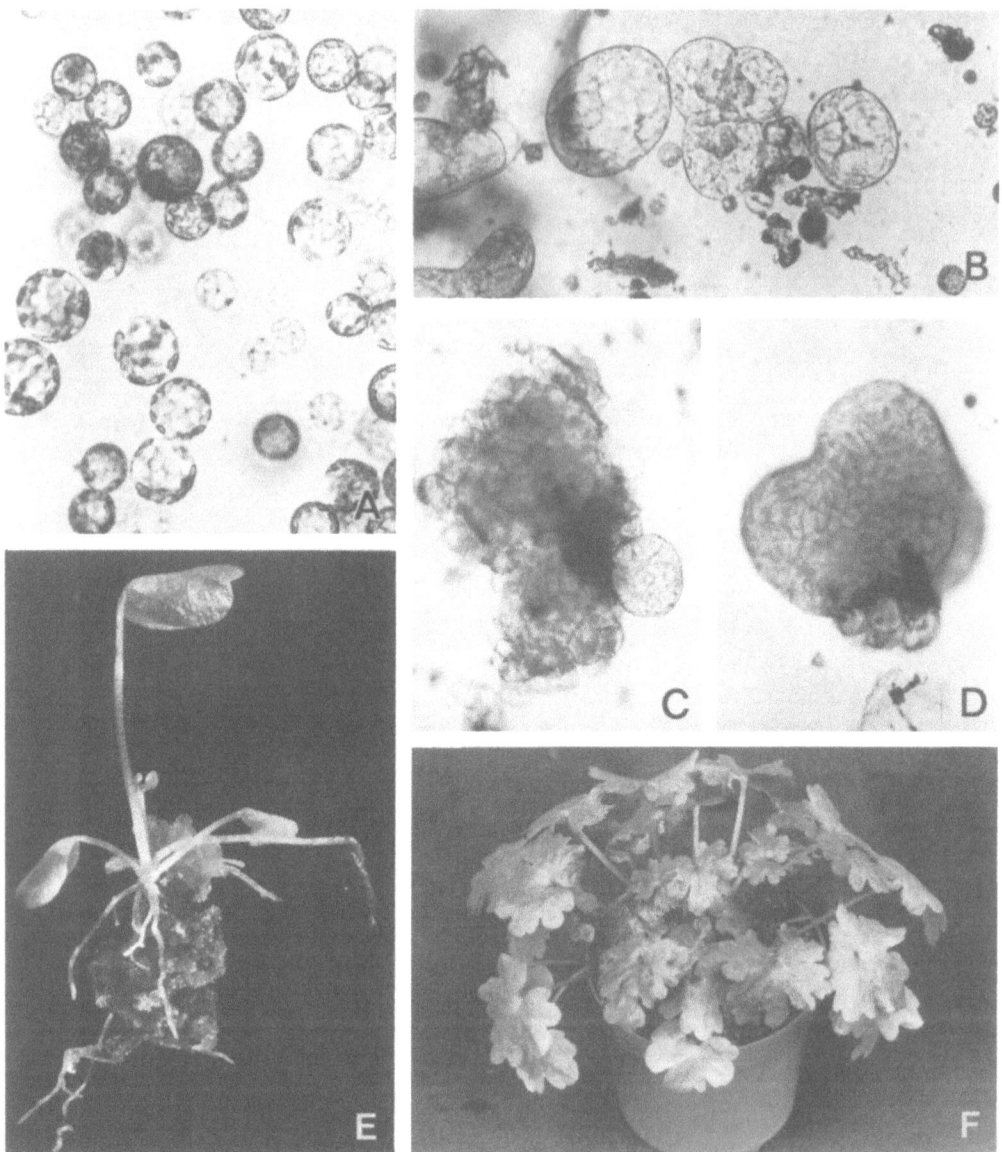

Fig. 4 A–F. Regeneration of plants in leaf protoplast cultures of *Ranunculus sceleratus*. **A** Freshly isolated leaf protoplasts. **B** First cell divisions observed within 10 days of culturing. **C** and **D** Cell clusters obtained within 2 months of culturing, and displaying presumed somatic embryos. **E** Plantlet differentiated from callus originating from leaf protoplasts. **F** Adult plant regenerated from leaf protoplasts. (After Dorion et al. 1975)

them on a medium supplemented with 2,4-D (1 mg/l) and BAP (0.5 mg/l). The protoplasts reformed a cell wall and divided rapidly giving rise to a cluster of cells. In 6 weeks macroscopically visible small, globular and heart-shaped embryoids differentiated in the small calluses derived from the protoplasts (Fig. 3 A–F).

Excised tissues and organs of *Ranunculus sceleratus* when grown in vitro readily yield embryos and plants (Konar and Nataraja 1969). Dorion et al. (1975) investigated the potential of protoplasts of *Ranunculus*. Mesophyll protoplasts were obtained by enzymatic treatment of leaf tissues, and cultured on a nutrient medium supplemented with 6-BAP (1 mg/l), NAA (3 mg/l), and glutamine (50 mg/l). The cultured protoplasts resynthesized a wall, and in 4 weeks small cell clusters were obtained which eventually developed into a callus. At the end of 8 weeks the microcalluses were transferred to a modified Murashige and Skoog's medium containing half-strength macronutrients, and no growth substances and glutamine. Plantlets developed on this medium. On subsequent transfer to soil, they gave rise to normal plants (Fig. 4A–F).

Embryogenesis and development of plantlets has also been reported in protoplasts isolated from actively growing cell suspension cultures of *Atropa belladonna* (Gosch et al. 1975b). Soon after reconstituting the cell wall the isolated protoplasts first divided in 2–3 days, and exhibited both unequal and equal divisions. In 1 week they had divided two to three times, resulting in small colonies of four to eight cells. Three weeks after culture the cell colonies were transferred to a medium with relatively low osmotic value (0.2 *M* sorbitol) which stimulated the growth of the colonies into callus masses. In small pieces of such calluses, when grown on NAA-free liquid medium, embryogenesis was observed and, within 2 weeks, globular proembryos, heart and torpedo-shaped embryos were observed. The embryos were grown on an auxin-free medium and, eventually, whole plants were recovered. Kao et al. (1973) have demonstrated the organization of embryos in protoplasts isolated from cell cultures of brome grass *(Bromus inermis)*. They observed that a combination of high sucrose and high $CaCl_2$ was especially beneficial to sustained cell division, and embryo formation in brome grass. By the 10th day the regenerated cells formed many proembryoids. The proembryoids developed into plantlets on a 2,4-D-free medium.

Most of the attempts on protoplast regeneration have met with success with herbaceous plants, and examples of protoplast regeneration in woody species are very rare and studied only in *Citrus*. Ovular callus from shamouti orange *(Citrus sinensis)* has been reported to possess regenerative capacity (Kochba et al. 1972). Vardi et al. (1975) utilised the callus tissues as the source material, and developed a method for protoplast isolation and regeneration. Undifferentiated callus tissues were incubated in an enzyme mixture, and the released protoplasts were plated on a solidified Nagata and Takebe's medium. Colonies were observed 14 days after plating which, eventually, developed into individual calluses. Irradiation of these calluses with X-rays (4,500 R to 7,500 R) promoted regeneration of embryoids, while embryoids did not develop in non-irradiated protoplast colonies (Fig. 5A–E).

Organogenesis

Nagata and Takebe (1971) described, for the first time, the regeneration of whole plants from mesophyll protoplasts of *Nicotiana tabacum*. Freshly isolated protoplasts were suspended in a modified MS medium with 3 mg/l NAA and 1 mg/l

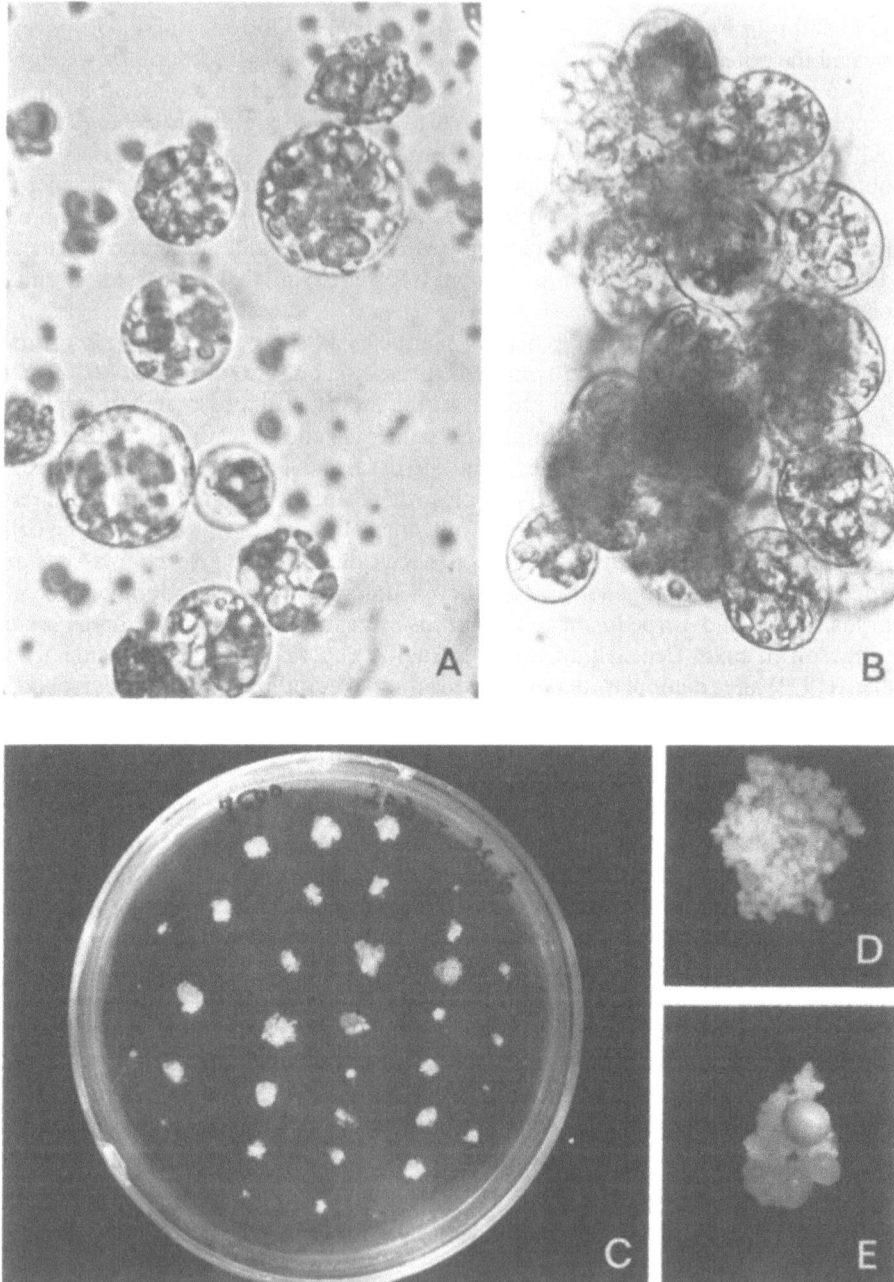

Fig. 5A–E. In vitro cultures of protoplasts isolated from "Shamouti" orange ovular callus. **A** Isolated protoplasts at the beginning of culture. **B** Young colony after 4 weeks. **C** Calluses after 4 months obtained from X-ray irradiated (7,500 R) protoplasts. **D** One of the enlarged undifferentiated calluses. **E** Note embryogenesis in another callus. (After Vardi et al. 1975)

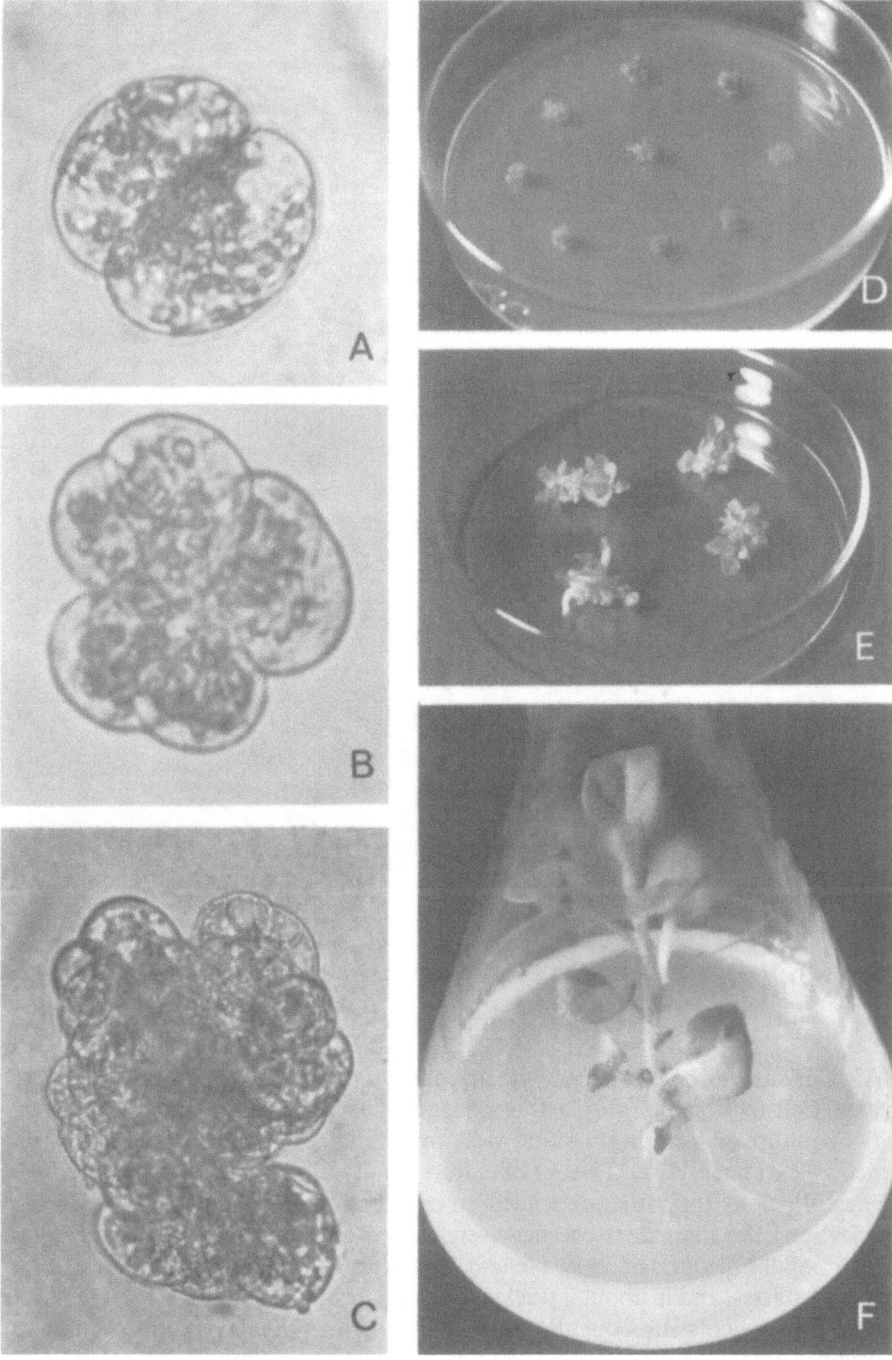

Fig. 6 A–F. Regeneration of plants in isolated tobacco mesophyll protoplasts. **A–C** Successive stages of colony formation by protoplasts embedded in agar medium. **D** Colonies transferred on to a plate of B_3 medium, and growing as callus masses; note indication of shoot differentiation in some calluses. **E** Shoots differentiated from callus masses on B_3 medium. **F** Plantlet of tobacco regenerated from single mesophyll protoplast. (After Nagata and Takebe 1971)

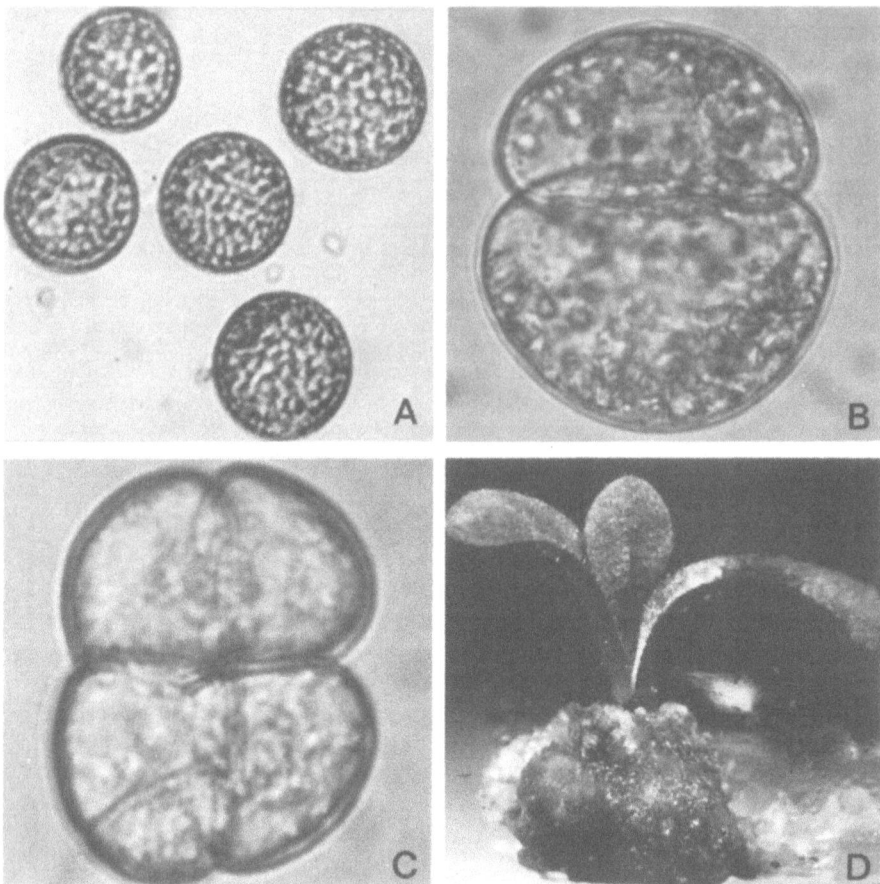

Fig. 7 A–D

BAP. The suspension was mixed with an equal volume of melted agar medium and plated. Cell division of protoplasts embedded in the agar medium began on the 3rd day of culture and, during the first week, more than 70% of protoplasts had divided at least once. After 3 weeks of culture small visible colonies appeared. Efficiency of colony formation depended on cell density, and light condition during incubation. The individual colonies were picked out, and placed on agar plates containing the medium of Sacristan and Melchers. The colonies grew actively first as callus masses, and in about 3 weeks, differentiated many shoots, mostly without roots. These shoots subsequently produced roots on White's basal medium, and developed into plantlets (Fig. 6 A–F) which could be transplanted into soil and grown in the greenhouse.

V. Vasil and I.K. Vasil (1974) have also successfully grown the mesophyll protoplasts of *Nicotiana tabacum* in liquid droplets, and obtained colony development. Plating of cell colonies resulted in callus tissues which, upon transfer to MS medium supplemented with IAA (4 mg/l) and KN (2.56 mg/l), regenerated shoots

Fig. 7 E–F

Fig. 7 A–F. Culture of mesophyll protoplasts of *Petunia parodii*. A Freshly isolated mesophyll protoplasts in liquid medium. B First division of protoplast after 7 days. C Second
division to produce four daughter cells after 10 days. D Shoot differentiation on callus
derived from protoplasts; MS medium; 3 months. E Adventitious root formation giving rise
to plantlet. F Normal, fertile plant regenerated from mesophyll protoplast; 6 months. (After
Hayward and Power 1975)

and, eventually, plantlets. Chupeau et al. (1973) also reported the regeneration of
entire plants in mesophyll protoplast cultures of *Nicotiana tabacum*.

Protoplasts derived from haploid tobacco *(Nicotiana tabacum)* have also
shown regenerative potential (Ohyama and Nitsch 1972). Like the diploid species,
haploid protoplasts also required the presence of an auxin (2,4-D, 0.1 mg/l) and
a cytokinin BAP (1 mg/l) for the sustained division and colony formation. Cell colonies derived from protoplasts were induced to form shoots on a medium containing IAA (1 mg/l), ZN (0.1 mg/l), and AD (40 mg/l). Roots were induced by transferring the shoots on the same medium, but devoid of ZN and AD. The plants obtained in vitro were transplanted to pots in the greenhouse, where they flowered
normally.

Next to tobacco, *Petunia*, yet another solanaceous member, seems to be an
ideal choice material for experimental investigations. The regenerative potential of
cells and tissues of *Petunia* has been extensively investigated (Handro et al. 1972,
1973; Rao et al. 1973 a, b; Rao and Harada 1974), and considerable work has followed on the protoplasts. Durand et al. (1973) isolated (enzymatically) protoplasts

from the mesophyll cells of the leaves of *Petunia hybrida*, and obtained continued divisions on a medium supplemented with 2,4-D (1.4 mg/l) and 6-BAP (0.4 mg/l) resulting in small calluses. Differentiation of shoots were induced in such calluses on Nagata and Takebe's medium (Nagata and Takebe 1971) with 1 mg/l NAA and 0.2 mg/l BAP, and roots on the same medium with 0.1 mg/l NAA and without cytokinin. Young plants developed in vitro were successfully established in greenhouse.

Mesophyll protoplasts of *Petunia hybrida* were also successfully grown by Frearson et al. (1973) on a basal medium supplemented with 2,4-D (0.5 mg/l), BAP (1 mg/l), and NAA (1.5 mg/l). The protoplasts regenerated cell walls, actively divided and developed into colonies which at 28 days were placed on MS medium containing 2.56 mg/l KN and 4 mg/l IAA. The cell colonies grew rapidly and, upon frequent subculturing, produced many shoots. The shoots were induced to root on MS medium supplemented with NAA (0.1 mg/l) and by this sequence entire plants were derived. Using similar conditions Hayward and Power (1975) demonstrated regeneration of plants in another species of *Petunia, P. parodii* (Fig. 7 A–F). It is interesting to note that protoplasts of *P. parodii* and *Nicotiana tabacum* have shown regenerative abilities under identical conditions. This indicates that there is a certain amount of biological compatibility between *P. parodii* and *N. tabacum* which would form an excellent system for somatic cell hybridisation between these two species.

Besides mesophyll cells, callus tissues of *Petunia hybrida* have also been used as a source material for the isolated protoplasts and its successful regeneration into whole plants (V. Vasil and I.K. Vasil 1974).

Binding (1974a, b) used aseptic shoot tip cultures of diploid and haploid *Petunia hybrida* as source material. A pH of 5.4–5.6 at an osmolality of 0.5 *M* mannitol and 4 h incubation period at a shaking rate of 80 strokes/min and a temperature of 28 °C and light regime of 1,500 lx helped in obtaining a consistently high yield of viable protoplasts. The yield was dependent on the composition of the culture medium, light intensity, and temperature regime. The isolated protoplasts were plated in liquid medium with enhanced levels of calcium and magnesium ions and supplemented with 2,4-D and 6-BAP. Optimal plating efficiencies of 50%–60% and 80%–90% for haploid and diploid protoplasts, respectively, were obtained. Differentiation of shoots occurred when cell colonies were transferred to Nagata and Takebe's medium with 0.5 mg/l NAA and 1 mg/l BAP. Roots differentiated in cultures on the same medium but with 0.1 mg/l NAA and without cytokinin. The haploid and diploid plants were grown in greenhouse.

Callus formation and plant regeneration has also been reported from mesophyll protoplasts of rape plants, *Brassica napus* (Kartha et al. 1974a). Protoplasts from mesophyll cells were isolated by enzymatic removal of cell walls, and cultured on a medium supplemented with BAP and NAA (both at 10^{-6} *M*). Within 24 h more than 80% of protoplasts assumed an oval shape, and soon wall formation followed. By repeated divisions clusters of cells were produced within 15 days which later developed into callus. Regeneration of plants from the callus was influenced by several factors. BAP alone did not induce shoot formation, although it did in cultures of the stem segments of the same species. This would imply that organogenetic reactions are considerably influenced by the source of explant.

BAP 5×10^{-6} M in conjunction with gibberellic acid (GA_3) 10^{-7} M triggered bud initiation. The buds remained dormant as long as growth regulators were present in the medium. Withdrawal of growth hormones allowed shoot elongation, and full plantlet development.

Other recent examples of plantlet regeneration in protoplast cultures are mentioned in Table 1 (see p. 236).

Applications

Since isolated protoplasts are generally freely separated from one another, they constitute the nearest possible approach to isolated cell suspension of higher plants, and, as such, are finding increasing use in several areas of plant cell biology. Some of these are briefly outlined here.

Somatic Cell Fusion and Hybridization

The most promising potential application of the culture of plant protoplasts is in inducing protoplasts from widely different genetic backgrounds to fuse and regenerate an intact hybrid plant which exhibits the characteristics of both parents. Such a process, known as "parasexual hybridization" holds immense possibilities in the area of genetic improvement of plants.

The phenomenon of induced fusion between the protoplasts from different taxa to produce heterokaryons was first reported by Power et al. in 1970. They demonstrated intra- and interspecific fusion between protoplasts obtained from root tips of maize and oat seedlings, and described the events leading to the fusion of protoplasts. Root tips 2–3 cm long were incubated for 24 h in an enzyme mixture (10% of Onozuka P 1,500 cellulase and 5% of maceroenzyme in 0.56 M sucrose), and washed twice with 0.25 M sodium nitrate. The root tips were then placed on a slide with a drop of 0.25 M sodium nitrate and a cover slip. The pressure facilitated liberation of protoplasts from the meristematic as well as the vacuolated regions of root tips. The protoplasts kept floating freely in the sodium nitrate medium, adhered to one another and, subsequently, fused. Sodium nitrate does not affect the viability of the protoplasts, and the overall effect is somewhat akin to induced fusion of animal cells by Sendai virus (Harris 1970). However, the rates of fusion obtained by sodium nitrate were very low and the resultant products failed to divide and, therefore, hybrids were not obtained. Induced fusion is also influenced by alkaline pH, and high calcium concentrations (Keller and Melchers 1973).

Polyethylene glycol (PEG) is yet another effective inducing agent for fusion of protoplasts of different genera (Gamborg et al. 1974). The process of fusion is considerably enhanced if the protoplast suspension is pretreated with a membrane activating agent such as lysozyme (Kao and Michayluk 1974; Constabel and Kao 1974). The rate of fusion depends upon the species of protoplasts under investigation, the concentration and mode of application of the inducing agent. Induced fusion between isolated protoplasts of the same species results in homokaryons, and with different species heterokaryons are formed. To date, intraspecific, interspe-

Table 2. Examples of intraspecific, interspecific and intergeneric induced-fusion in protoplasts

Plant species	Reference
Avena sp. + *Zea mays*	Power et al. (1970)
Torenia bailloni + *T. fournieri*	Potrykus (1971)
Petunia hybrida + *Nicotiana tabacum*	Power and Frearson (1973)
Pisum sativum + *Centaurea cyanus*	Davey and Short (1973)
Brassica napus + *Glycine max*	Kartha et al. (1974b)
Nicotiana langsdorffii + *N. glauca*[a]	Carlson et al. (1972)
Vicia hajastame + *Glycine max* ⎫	Kao et al. (1974)
Hordeum vulgare + *Glycine max* ⎪	
Pisum sativum + *Glycine max* ⎬	
Pisum sativum + *Vicia hajastane* ⎪	
Zea mays + *Glycine max* ⎭	
Vicia villosa + *V. hajastane* ⎫	Gamborg et al. (1974)
Medicago sativa + *Glycine max* ⎪	
Melilotus alba + *Glycine max* ⎬	
Cicer arietinum + *Glycine max* ⎪	
Angelica archangelica + *Glycine max* ⎭	
Nicotiana tabacum + *Daucus carota*	Gosch et al. (1975a)
Petunia hybrida + *Parthenocissus tricuspidata*	Power et al. (1975)
Nicotiana sylvestris + *N. knightiana*	Maliga et al. (1977)
Daucus carota + *D. capillifolius*	Dudits et al. (1977)
Petunia hybrida + *P. parodii*	Power et al. (1977)
Datura innoxia + *D. stramonium*[a] ⎫	Schieder (1978)
Datura innoxia + *D. discolor*[a] ⎭	
Nicotiana tabacum + *N. knightiana*[a]	Maliga et al. (1978)
Nicotiana glauca + *N. langsdorffii*[a]	Chupeau et al. (1978)
Nicotiana tabacum + *N. rustica*[a]	Nagao (1978)
Lycopersicon esculentum + *Solanum tuberosum*[a]	Melchers et al. (1978)
Petunia hybrida + *P. axillaris*[a]	Izhar and Power (1979)
Daucus carota + *Aegopodium podagraria*[a]	Dudits et al. (1979)

[a] Regeneration of hybrid plants

cific, and intergeneric fusion has been achieved in many plant species (see Table 2) and, in many instances, the heterokaryons have been cultured and induced to divide a few times. The nuclei apparently fuse in those in which division takes place, but it has not been established if nuclear division occurs prior to mitosis.

A critical factor in studies on induced-fusion between two unrelated protoplasts relates to the "selective markers" to identify the heteroplasmic fusion products. A couple of examples are cited here. In studies on PEG-induced fusion of rape-seed and soybean protoplasts (Kartha et al. 1974b) plastids served as markers to identify the heteroplasmic fusion products. The heterokaryons shared the chloroplasts from rape-seed, and dense cytoplasm and colourless plastids from the soybean protoplasts. Recently, Gosch et al. (1975a) reported intergeneric fusion from protoplasts of tobacco and carrot. The protoplasts were induced to agglutinate and fuse by PEG treatment. Agglutination as well as fusion could be enhanced by addition of $CaCl_2$ in glycine NaOH leading to a high pH value. Green chloroplasts of tobacco and the red anthocyanin of carrot protoplasts were used as selective markers.

Similarly, in experiments on interspecific hybridization between protoplasts of two species of *Torenia* the fusion products could be clearly identified; pure anthocyanin protoplasts of *Torenia fournieri* were combined with pure chromoplast or pure chloroplast protoplasts of *Torenia baillonii* (Potrykus 1971). Besides plastids, other biochemical and genetical markers such as isoenzyme and nucleic acid patterns, drug resistance, chromosome number, and karyotypes could also be used as aids in the identification of the somatic hybrid.

The first report of the successful production of a mature interspecific hybrid plant by fusion of leaf protoplasts is by Carlson et al. (1972). The species used were *Nicotiana glauca* (2 $N=24$) and *Nicotiana langsdorffii* (2 $N=18$). The amphidiploid (2 $N=42$) hybrid between these two species has been produced by sexual means, and the characteristics of the hybrid plant have been thoroughly studied (Schaeffar and Smith 1963). In Nagata and Tabeke's (1971) medium protoplasts of the two species, *N. glauca* and *N. langsdorffii*, regenerated a cell wall but failed to undergo more than one division, whereas protoplasts from the amphidiploid species divided repeatedly and gave rise to a callus mass. This difference in growth requirements constitutes a selection method for recovering preferentially hybrid individuals from a mixed population of protoplasts since the only ones capable of growing into viable callus are those with hybrid genetic constitution. Another characteristic feature of this hybrid callus is its capacity to grow vigorously on a medium without added hormones. This constitutes a second selective step. Against this background populations of isolated protoplasts of *Nicotiana glauca* and *N. langsdorffii* were stimulated to fuse by their suspension in 0.25 M of $NaNO_3$. After the fusion process the population which consisted of protoplasts of both parental types was plated on Nagata and Takebe's medium that caused growth of only hybrid cells. 33 calluses were obtained after 6 weeks and were placed on Linsmaier and Skoog's medium (1965) without added hormones. The calluses differentiated shoots and, eventually, complete hybrid plants which flowered and set seeds. The somatic hybrid plants were identical with the amphidiploid species (obtained through sexual means) in their morphology, chromosome number, and also peroxidase isoenzyme patterns.

Following the success reported by Carlson et al. (1972) parasexual hybridisation has been attempted by many investigators, and hybrid plants have been obtained in a few instances. A major problem has been one of developing suitable selection procedures. Some of the demonstrated selection techniques for plant somatic hybrids are given in Table 3.

Power et al. (1976) demonstrated interspecific somatic hybridisation of *Petunia hybrida* and *P. parodii* which resulted in mature plants. These investigators used two selection schemes for recovering the somatic hybrids. In the first scheme of Power et al. (1977), selection was based on differences in growth requirements of the protoplasts to nutrient media and actinomycin D. In the second scheme (Cocking et al. 1977) an albino complementation selection procedure was developed for the selection of somatic hybrids. This method involved fusing protoplasts of albino *P. hybrida* with leaf mesophyll protoplasts of *P. parodii*, and selecting green colonies formed as a result of complementation and selective growth. Plants produced by both the methods of selection were classed as somatic hybrids based on their isoperoxidase patterns, chromosome numbers, corolla colour, and length and gen-

Table 3. Selection systems for plant somatic hybrids

Selection system	Principle	Reference
Auxin Autotrophy	Parents require exogenous hormone while fusion product does not	Carlson et al. (1972)
Visual	Integration of parental characteristics in the hybrid observed visually at the light microscopic level	Kao et al. (1974)
Recessive gene complementation	Chlorophyll deficiency and light sensitivity in parents complemented in hybrid with normal leaf colour and light resistance	Melchers and Labib (1974)
Cytoplasmic factors	Transfer of male sterility to hybrid	Izhar and Power (1979)
Enzyme pattern	Ribulose 1,5-biphosphate carboxylase	Melchers et al. (1978)
Enzyme deficiency complementation	Nitrate reductase-deficient complementation	Glimelius et al. (1978)
Auxotrophic mutant complementation	Restoration of normal nutrient requirements	Schieder (1975b)
Biochemical and phenotypic characters	Restoration of photosynthetic apparatus, synthesis of specific RNA in hybrid plants	Dudits et al. (1979)
Kanamycin-resistance pigmentation	Ability to grow on an auxin-free high cytokinin medium	Maliga et al. (1977)

eral plant morphology. In a further study Power et al. (1978) compared the somatic and sexual hybrids of *P. hybrida* and *P. parodii* with respect to flower colour and segregation. Although minor variations were detected between individuals both for somatic and sexual hybrids, in the overall pattern both types of hybrids segregated for flower colour in a similar manner.

Somatic hybrids of *Daucus carota* and *D. capillifolius* have been reported by Dudits et al. (1977). Protoplasts isolated from cultured cells of albino carrot *(D. carota)* and normal green *D. capillifolius* were fused with polyethylene glycol. A chief index for the selection of somatic hybrid plants was the restoration of photosynthetic function in hybrids. The resultant green plants bore leaves which were intermediate between those of the parents.

Complementing chlorophyll-deficient mutants provide biological markers of a more general application. Melchers and Labib (1974) used this parameter in inducing fusion of protoplasts from two chlorophyll-deficient light-sensitive varieties of *Nicotiana tabacum* (haploid), and to selectively isolate somatic hybrids resistant to high light intensities and with normal green leaves. Similar results have been reported by Gleba et al. (1975).

More recently, Schieder (1978) succeeded in selecting somatic hybrids from *Datura innoxia*. He fused the protoplasts obtained from a chlorophyll-deficient diploid mutant of *D. innoxia* which had the ability to regenerate shoots with the green wild-type protoplasts of *D. stramonium* which did not have the regenerative ability, and obtained 49 green hybrid calluses on agar medium. Most of these somatic hybrid calluses gave rise to shoots. In an almost identical experiment, protoplasts of *D. innoxia* mutant were fused with the green wild-type protoplasts of *D. discolor* Bernh. which did not possess the ability to regenerate. The hybrid callus tissue regenerated shoots. The shoots obtained in both the combinations were rooted and successfully established in soil, and flowering plants were obtained.

Auxotrophic mutant complementation has been employed to select somatic hybrids in a liverwort *Sphaerocarpus donnellii* (Schieder 1975 b). The complemented fusion-hybrid was obtained by fusing protoplasts from a normal green nicotinic acid-deficient female plant, and a pale-green glucose-deficient male plant. This is somewhat analogous to the complementation in some animal systems, and is based upon the restoration of normal nutrient requirements in the somatic hybrid following complementation of the two parental auxotrophic strains.

Maliga et al. (1978) established somatic hybrids of *Nicotiana knightiana* ($2 n = 2 \times = 24$) and an albino mutant of *N. tabacum* ($2 n = 4 \times = 48$) through protoplast fusion by polyethylene glycol. The hybrid nature of the lines was confirmed by their characteristic isoenzyme patterns, the morphology of the regenerated plants, and by the appearance of heterochromatic blocks in the interphase nuclei. In another study Maliga et al. (1977) reported that somatic hybrid plants could be isolated from fused protoplasts of a kanamycin-resistant cell line of *N. sylvestris* and of a cell line of *N. knightiana*, both deficient in the potential to produce shoots. In the somatic hybrid callus the potential for shoot production was restored.

An outstanding example of intergeneric fusion of protoplasts is the work of Melchers et al. (1978) who have reported regeneration of somatic hybrid plants of potato and tomato. Mesophyll protoplasts of *Lycopersicon esculentum* Mill. var. Cerasiforme (Dunal) Alef, mutant yellow-green 6 rick, and protoplasts obtained from liquid callus culture of the dihaploid strain H H 258 of *Solanum tuberosum* L. were fused first in the presence of polyethylene glycol and, eventually, with a high calcium ion concentration. The fused protoplasts divided and developed into callus masses in some of which regeneration of normal green shoots occurred. The shoots were either transferred to soil, or grafted on to a tomato stock. In this case the analysis of ribulose triphosphate carboxylase proved to be a convenient marker to demonstrate the hybrid nature of the plants.

It is, therefore, apparent from the foregoing account that availability of sensitive and powerful selective techniques which permit preferential growth of hybrid cells would be a criteria of paramount importance in research on somatic hybridisation. It is also of great relevance to emphasize that the production of somatic hybrids between species which can be conventionally hybridized sexually would be of limited practical value.

Uptake of Macromolecules and Transfer of Genetic Information

DNA Uptake. In any plant breeding programme a major objective is the transfer of desirable genetic characters between the varieties and species. Apart from the

conventional hybridization which is an established approach, the techniques of tissue, cell, and protoplast culture have provided an additional method to accomplish
this goal. Protoplasts are especially a versatile material for the introduction of
foreign genetic material because of their wall-less nature. However, investigations
on the uptake of exogenous DNA by plant protoplasts are, relatively, in a state of
infancy (see review by Hess 1972). Hess et al. (1973) and Hoffmann and Hess
(1973) reported the uptake of exogenous DNA in protoplasts of *Petunia*. Ohyama
et al. (1972) reported that large amounts of bacterial DNA is taken up by protoplasts of soybean, carrot, and *Ammi visnaga*. Apart from DNA-mediated transformation through DNA uptake, another alternative method for genetic manipulation is by transplantation of cell organelles, especially nuclei. Potrykus and Hoffmann (1973) successfully transplanted isolated nuclei of *Petunia hybrida* into isolated protoplasts of *Petunia hybrida*, *Nicotiana glauca*, and *Zea mays*. The fate of
these transplanted nuclei is yet under investigation. These studies point towards a
new avenue in the field of cytogenetics, especially regarding nuclear-cytoplasmic
interactions. According to Smith (1974), "the broadening of genetic variability in
higher plants through the incorporation of foreign DNA is now a legitimate experimental goal which will undoubtedly take some time to adapt for practical
utilisation" (see also Bottino 1975).

Uptake of Virus Particles. Although plant tissue cultures have been used for studies
on the infection of tobacco mosaic virus (Chandra and Hildebrandt 1965; Kassanis
1967; Motoyoshi and Oshima 1968), the infection process is not synchronous and,
because of the rigid cell wall, the extent of infection is restricted to only damaged
cells. Protoplasts being cell-wall-less provide a good experimental system for studies on virus uptake.

Cocking (1966) and Cocking and Pojnar (1969) reported that tomato fruit protoplasts take up tobacco mosaic virus (TMV) particles by pinocytosis, and demonstrated the initial stages of infection by electron microscopic studies. Subsequently,
many workers have studied the virus uptake through leaf mesophyll protoplasts
(Aoki and Takebe 1969; Otsuki et al. 1972; Burgess et al. 1973; Coutts 1973). These
studies have shown that infection occurred when the purified TMV particles were
added to a protoplast suspension in the presence of poly-L-ornithine. Virus multiplication occurred in the cytoplasm of the protoplast, and the amount of intracel-
lular virus corresponded with that in the cells of infected plants. Virus multiplication was also followed by large-scale accumulation of ribosomal material near the
nuclear membrane. Thus, protoplasts could provide a powerful tool for elucidating
the processes underlying the infection, and the nucleo-protein replication (see also
Cocking 1970).

Bacterial Uptake and Nitrogen Fixation. An aspect of uptake which has been vigorously pursued by researchers in many laboratories relates to the attempts to induce
plant protoplasts to take up nitrogen-fixing bacteria, and to establish new symbiotic relationship between bacteria and plant cells. Davey and Cocking (1972) reported that *Rhizobium japonicum* can be introduced into pea leaf protoplasts during enzymatic digestion of the cell wall. This bacterial uptake occurred by invag-

inations of plasmalemma during plasmolysis. Davey et al. (1973) also successfully isolated legume root nodule protoplasts containing packets of bacteria. Attempts are being made towards fusing these protoplasts with non-legume protoplasts. Holsten et al. (1971) reported actual establishment of a symbiotic relationship between *Rhizobium* and soybean callus. The morphology of the in vitro symbiosis was analogous to the intact soybean root nodules. Uptake of free nitrogen-fixing bacteria like *Azotobacter* and the blue-green algae by non-leguminous protoplasts would be of great interest. Some useful work has recently been done in this direction. The establishment of mycorrhizal associations of the fungi with the roots of tree plants is well-known. For example, the fungus *Rhizopogon* is known to form mycorrhizal association with *Pinus radiata*. Giles and Whitehead (1975) demonstrated the transfer of nitrogen-fixing ability to the mycorrhizal fungus *Rhizopogon*. They incubated the protoplasts of the fungus *Rhizopogon* with nitrogen-fixing bacteria *Azotobacter vinelandii*, and the treated protoplasts were cultured on media containing penicillin which prevents the growth of *Azotobacter* but is lacking in nitrogen. The treated protoplasts grew on the nitrogen-free medium, and regenerated fungal colonies which showed acetylene reduction activity. Poly β-hydroxy-butyric acid, a storage product of *Azotobacter* cysts, was also identified in the fungal hyphae. In an extension of this work, Giles and Whitehead (1977 a, b) conducted further experiments to assess the potential for reassociation of modified strains of the mycorrhizal fungus *Rhizopogon* sp. capable of acetylene reduction activity in vitro with the roots of its host plant *Pinus radiata*. Reassociation was effected and acetylene reduction assays indicated that nitrogenase activity was present in the reassociated whole plants. Giles and Whitehead concluded that "the finding of acetylene reduction activity in plant roots reassociated with these modified strains and the increased nitrogen content of reassociated plants is a hopeful sign that this form of transfer of nitrogen fixation, using genetically modified mycorrhiza, may be successful." Cocking (1973) has rightly pointed out: "The agricultural and social implications of this sort of modification of non-legume food plants such as cereals would be profound – the need for plant biologists to work actively in this field is pressing."

Chloroplast Transplantation. Apart from DNA-mediated transformation through DNA uptake, another alternative method for genetic manipulation is by transplantation of cell organelles. Several investigators report that chloroplasts when transplanted survive, remain metabolically active, and retain their structural integrity. Nass (1969) observed that spinach chloroplasts divided normally when transplanted into the cytoplasm of animal cells. Giles and Sarafis (1971) also reported successful implantation of *Nitella* chloroplasts into hen's egg. These studies point to the autonomy of the chloroplasts. Potrykus (1973) also succeeded in introducing chloroplasts into the protoplasts of *Petunia*. Carlson (1973) took a step ahead and was successful in introducing functional chloroplasts into the protoplasts of an albino mutant *Nicotiana tabacum*, and in obtaining a whole-green plant containing chloroplasts of foreign origin. According to Carlson: "Perhaps chloroplast transfer will provide a method for plant breeders to incorporate highly efficient chloroplasts into important crop species."

Concluding Remarks

From the foregoing account it is apparent that rapid strides have been made in the field of protoplast research, and the areas of experimental investigation in which they are finding a use are becoming increasingly varied, providing a meeting ground for physiologists, geneticists, molecular biologists, and virologists.

Although somatic cell hybridization through protoplast culture is a new tool for plant cell genetics, still it would be a long time before this method is applied routinely in any plant breeding programme. Despite advances in fusion methods, the major problem is one of availability of a "selective technique" to permit preferential recovery of fused hybrid cells and, more important, to regenerate an entire hybrid plant. When these technical aspects required for parasexual hybridization are refined then, as Carlson (1973) puts it, "the potential offered by somatic hybridization may be expected to exceed the limitations imposed by the sexual processes, and extend the possibilities of combining widely divergent genotypes of plants."

The use of protoplasts in uptake studies has a tremendous potential. The findings that whole microorganisms, such as bacteria, are taken up by plant protoplasts suggest a novel way to establish new endosymbiotic associations, and to explore the possibilities of various kinds of transformation in higher plants including the gene transfer for nitrogen fixation. The induction of foreign chloroplasts into crop plants with inefficient photosynthetic ability would altogether produce new types of plants and, thus, revolutionize methods of plant improvement.

Haploid plants are of considerable interest and of value for the geneticist and the plant breeder, and the anther and pollen culture technique has provided a new source for obtaining the haploids. However, the limitations imposed by the anther culture technique are many, amongst which the critical factors are (1) the physiological age of the pollen at the time of culture, (2) the effect of anther wall on the development of pollen embryoids, and (3) the low frequency of haploids obtained. Besides these, plants of mixed ploidy are obtained because embryoids may originate not only from the pollen but also from the anther tissues. Microspore protoplasts being haploid offer an alternative experimental system in such situations. Being highly cytoplasmic with negligible vacuolation, microspore protoplasts are akin to meristematic somatic cell protoplasts (which fuse readily), and might also lend themselves to fusion studies. Future research would no doubt be focussed on this area (Bhojwani and Cocking 1972, Bajaj 1974a).

In short, the modification of the plant cell through genetic engineering or somatic cell hybridisation is the real challenge facing the biologists today. Although enormous amount of data have been gained through studies on isolated protoplasts, researchers in this field should not overlook the fact that "basically the real objective of this work is to broaden the base for genetic variability in plants" (see Smith 1971; Cocking 1973; Bajaj 1974b).

Isolation of Protoplasts. Usually, surface sterilization of leaves is done by a brief immersion in 70% ethanol followed by treatment with 2% sodium hypochlorite solution. However, exposing leaf material to treatment with extraneous chemicals is not suitable and might interfere in the yield of protoplasts that are isolated (Binding 1975). In some tobacco cultivars, for example, the leaves are extremely fragile and would be difficult to handle after sterilization. To overcome such problems Wilson et al. (1980) devised a new procedure for isolating contaminant-free protoplasts from unsterilized leaves. The leaves collected from plants growing in controlled environment chambers were rinsed twice in sterile water before digesting them in an enzyme mixture. Protoplasts were concentrated by centrifuging the filtrate twice in sterile 8% milktest Babcock bottles. According to the authors the advantages of using Babcock bottles are two-fold:

1. the protoplast layer can be more easily removed without including excess medium when a Babcock bottle is used in place of a centrifuge tube, and
2. Babcock bottles have a large volume and enable easy harvesting of densely packed protoplasts.

Regeneration in Protoplast Cultures. Ever since the totipotency of plant protoplasts was first demonstrated (Takebe et al. 1971), remarkable progress has been made and plants have been regenerated from protoplasts of several species especially belonging to Solanaceae. Isolation and culture of protoplasts of cereals and legumes – which constitutes one of the most important groups of plants – have proved difficult. The applicability of protoplast technology for crop improvement, viz., technique of somatic hybridisation and genetic manipulation, will remain restricted until the technology to regenerate plants reproducibly from protoplasts of crop species in large numbers is developed. Recently, there has been some work in this direction. V. Vasil and I. K. Vasil (1979, 1980) reported isolation of protoplasts from suspension cultures derived from immature embryos of pearl millet *(Pennisetum americanum)*. Two types of protoplasts were isolated: large, vacuolate protoplasts which never divided, and small, non-vacuolate protplasts derived from the embryogenic cells in suspension cultures, which divided and produced plantlets via somatic embryogenesis. Embryoids could be initiated from mesophyll protoplasts of Alfalfa, an important forage crop (*Medicago sativa*), and regenerated into plants (Kao and Michayluk 1980; Dos Santos et al. 1980), and isolated mesophyll protoplasts of cassava (*Mannihot esculenta* Crantz.) were induced to regenerate cell walls, divide and result in callus formation and occasional differentiation of shoots (Shahin and Shepard 1980). Cell suspension derived protoplasts of *Trifolium repens* (Gresshoff 1980) produced callus which had the potentiality to differentiate into complete plants. Other instances of regeneration of plants have been reported in protoplast cultures of *Salpiglossis sinuata* (Boyes and Sink 1981), *Hyoscyamus muticus* and *Nicotiana tabacum* (Wernicke and Thomas 1980), while organization of roots has been reported in *Brassica rapa* (Ulrich et al. 1980). Protoplast isolation and subsequent callus development has been observed in sugarcane (Evans et al. 1980a), *Coffea arabica* (Sondhal et al. 1980), *Rosa* "Paul's Scarlet" (Strauss and Potrykus 1980), soybean pods (Zieg and Outka 1980), and sweet potato (Bidney and Shepard 1980).

Somatic Cell Fusion, Hybrids and Selection System. Generally, fusion of two protoplasts is achieved by polyethylene glycol. However, a major disadvantage in all fusion techniques is the difficulty in obtaining rapid and synchronized fusion of a large number of cells. Zimmermann and Scheurich (1981) demonstrated that an electrical field pulse of very short duration but of high intensity can trigger fusion process between mesophyll cell protoplasts of *Avena sativa*. This occurs as a result of electrical breakdown of the membranes of cells that are in close contact with each other. Apparently, this is a clear-cut evidence for the role of the electrical field in the fusion process.

Comparison of the isoenzyme pattern in somatic hybrids with those of their parental plants has been one of the selection procedures. Lonnendonker and Schieder (1980) analyzed the amylase isoenzymes of four interspecific somatic hybrids, and observed that all the somatic hybrids had two bands representing the amylase activity of the species combined by protoplast fusion. The authors have indicated that amylase isoenzymes could be a reliable and easy method for the identification of somatic hybrids in *Datura*.

Fusion between protoplasts from cell cultures of *Sorghum bicolor* and leaf protoplasts of corn (*Zea mays*) has been achieved using PEG, and the fusion products divided and developed into 8 to 10-celled colonies (Brar et al., 1980a, 1980b). Somatic hybrids have been produced with *Datura innoxia* and the three *Datura* species, *D. sanguinea* and *D. candida* (Schieder 1980), however, the hybrids showed several abnormalities which was in contrast to earlier somatic hybrids between herbaceous *Datura* species (Schieder 1978). Gleba and Hoffmann (1980) reported, for the first time, the intergeneric-intertribal hybridization. They isolated somatic hybrid cell lines *Arabidopsis thaliana* + *Brassica campestris* ("Arabidobrassica") and obtained the regeneration of shoots and flowering plants. Cytological, biochemical and morphological analyses revealed that the genetic material of both the species was present in the resultant plants (see also Gleba and Hoffmann 1978; Gleba et al. 1978). Another example of somatic hybrid plants following fusion of protoplasts is *Nicotiana glauca* + *N. tabacum* (Evans et al. 1980b).

References

Akoi S, Takebe I (1969) Infection of tobacco mesophyll protoplasts by tobacco mosaic virus ribonucleic acid. Virology 39:439–448

Bajaj YPS (1972) Protoplast culture and regeneration of haploid tobacco plants. Am J Bot 59:647

Bajaj YPS (1974a) Isolation and culture studies on pollen tetrad and pollen mother cell protoplasts. Plant Sci Lett 3:93–99

Bajaj YPS (1974b) Potentials of protoplast culture work in agriculture. Euphytica 23:633–649

Bergmann L (1960) Growth and division of single cells of higher plants in vitro. J Gen Physiol 43:841–851

Bhojwani SS, Cocking EC (1972) Isolation of protoplasts from pollen tetrads. Nature: New Biol 239:29–30

Bidney DL, Shepard JF (1980) Colony development from sweet potato petiole protoplasts and mesophyll cells. Plant Sci Lett 18:335–342

Binding H (1974a) Cell cluster formation by leaf protoplasts from axenic cultures of haploid *Petunia hybrida* L. Plant Sci Lett 2:185–188

Binding H (1974b) Regeneration of haploid and diploid plants from protoplasts of *Petunia hybrida* L. Z Pflanzenphysiol 74:327–356

Binding H (1975) Reproducibly high plating efficiencies of isolated mesophyll protoplasts from shoot cultures of tobacco. Physiol Plant 35:225–227

Binding H, Behls R, Schieder O, Sopory SK, Wenzel G (1978) Regeneration of mesophyll protoplasts isolated from dihaploid clones of *Solanum tuberosum*. Physiol Plant 43:52–54

Bottino PJ (1975) The potential of genetic manipulation in plant cell cultures for plant breeding. Rad Bot 15:1–16

Boyes CJ, Sink KC (1981) Regeneration of plants from callus-derived protoplasts of *Salpiglossis*. J Amer Soc Hort Sci 106:42–46

Brar DS, Rambold S, Constabel F, Gamborg OL (1980a) Isolation, fusion and culture of *Sorghum* and corn protoplasts. Z Pflanzenphysiol 96:269–275

Brar DS, Gamborg OL, Constabel F (1980b) Somatic hybridisation of *Sorghum* and corn. In: Rao PS, Heble MR, Chadha MS (eds) plant tissue culture: Genetic manipulation and somatic hybridisation of plant cells. Proc Nat Symp, BARC, Bombay, pp 237–247

Bui Dang Ha D, Mackenzie IA (1973) The division of protoplasts from *Asparagus officinalis* and their growth and differentiation. Protoplasma 78:215–221

Bui Dang Ha D, Norreel B, Masset A (1975) Regeneration of *Asparagus officinalis* L. through callus cultures derived from protoplasts. J Exp Bot 26:262–270

Burgess J, Motoyoshi F, Fleming EN (1973) Effect of poly-L-orinithine on isolated tobacco mesophyll protoplasts: Evidence against stimulated pinocytosis. Planta 111:199–208

Carlson PS (1973) The use of protoplasts for genetic research. Proc Nat Acad Sci USA 70:598–602

Carlson PS, Smith HH, Dearing RD (1972) Parasexual interspecific plant hybridisation. Proc Nat Acad Sci USA 69:2292–2294

Chandra N, Hildebrandt AC (1965) Tobacco mosaic virus inclusion bodies in tobacco tissue cultures. Nature (London) 206:325–326

Chupeau Y, Morel G (1970) Obtention de protoplastes de plantes supérieures à partir de tissus cultivés in vitro. CR Acad Sci Paris 270:2659–2662

Chupeau Y, Bourgin JP, Claudine M, Morel G (1973) Les protoplastes de plantes supérieures. Etat actuel et perspectives. Bull Soc Bot Fr 120:175–188

Chupeau Y, Missonier C, Hommel MC, Goujand J (1978) Somatic hybrids of plants by fusion of protoplasts. Mol Gen Genet 165:239–245

Cocking EC (1960) A method for the isolation of plant protoplasts and vacuoles. Nature (London) 187:927–929

Cocking EC (1966) An electron microscopic study of the initial stages of infection of isolated tomato fruit protoplasts by tobacco mosaic virus. Planta 68:206–214

Cocking EC (1970) Virus uptake, cell wall regeneration and virus multiplication in isolated plant protoplasts. Int Rev Cytol 28:89–124

Cocking EC (1972) Plant cell protoplasts – Isolation and development. A Rev Plant Physiol 23:29–50

Cocking EC (1973) Plant cell modification: Problems and perspectives. In: Protoplastes et fusion de cellules somatiques végétales. Colloq Int CNRS, Paris, pp 327–341

Cocking EC, Evans PK (1973) The isolation of protoplasts. In: Street HE (ed) Plant tissue and cell culture. Blackwell Scientific Publications, Oxford, pp 100–120

Cocking EC, Pojnar E (1969) An electron microscopic study of the infection of isolated tomato fruit protoplasts by tobacco mosaic virus. J Gen Virol 4:305–312

Cocking EC, George D, Price-Jones MJ, Power JB (1977) Selection procedures for the production of inter-species somatic hybrids of *Petunia hybrida* and *Petunia parodii* II. Albino complementation selection. Plant Sci Lett 10:7–12

Constabel F, Kao KN (1974) Agglutination and fusion of plant protoplasts by polyethylene glycol. Can J Bot 52:1603–1606

Coutts RHS (1973) Viruses in isolated protoplasts, a potential model for studying nucleoprotein replication. In: Protoplastes et fusion de cellules somatiques végétales. Colloq Int CNRS, Paris, pp 353–365

Davey MR, Cocking EC (1972) Uptake of bacteria by isolated higher plant protoplasts. Nature (London) 239:455–456

Davey MR, Short KC (1973) The isolation, culture, and fusion of legume and corn flower leaf protoplasts. In: Protoplastes et fusion de cellules somatiques végétales. Colloq Int CNRS, Paris, pp 437–444

Davey MR, Cocking EC, Bush E (1973) Isolation of legume root nodule protoplasts. Nature (London) 244:460–461

Davey MR, Frearson EM, Withers LA, Power JB (1974) Observations on the morphology, ultrastructure, and regeneration of tobacco leaf epidermal protoplasts. Plant Sci Lett 2:23–27

Doon G, Hess D, Potrykus I (1973) Differentiation in calli originated from protoplasts of *Petunia hybrida*. Z Pflanzenphysiol 69:423–437

Dorion N, Chupeau Y, Bourgin JP (1975) Isolation, culture, and regeneration into plants of *Ranunculus sceleratus* L. leaf protoplasts. Plant Sci Lett 5:325–332

Dos Santos AVP, Outka DE, Cocking EC, Davey MR (1980) Organogenesis and somatic embryogenesis in tissues derived from leaf protoplasts and leaf explants of *Medicago sativa*. Z Pflanzenphysiol 99:261–270

Dudits D, Kao KN, Constabel F, Gamborg OL (1976) Embryogenesis and formation of tetraploid and hexaploid plants from carrot protoplasts. Can J Bot 54:1063–1067

Dudits D, Hadlaczky Gy, Levi E, Fejer O, Haydu Zs, Lazar G (1977) Somatic hybridisation of *Daucus carota* and *D. capillifolius* by protoplast fusion. Theor Appl Genet 51:127–132

Dudits D, Hadlaczky Gy, Bajszar Gy, Koncz CS, Lazar G, Horvath G (1979) Plant regeneration from intergeneric cell hybrids. Plant Sci Lett 15:101–112

Durand J, Potrykus I, Doon G (1973) Plantes issues de protoplastes de *Petunia*. Z Pflanzenphysiol 69:26–34

Eriksson T, Jonasson K (1969) Nuclear division in isolated protoplasts from cells of higher plants grown in vitro. Planta 89:85–89

Evans DA, Crocomo OJ, De Carvalho MTV (1980a) Protoplast isolation and subsequent callus regeneration in sugarcane. Z Pflanzenphysiol 98:355–358

Evans DA, Wetter LR, Gamborg OL (1980b) Somatic hybrid plants of *Nicotiana glauca* and *Nicotiana tabacum* obtained by protoplast fusion. Physiol Plant 48:225–230

Frearson EM, Power JB, Cocking EC (1973) The isolation, culture, and regeneration of *Petunia* leaf protoplasts. Dev Biol 33:130–137

Furner IJ, King J, Gamborg OL (1978) Plant regeneration from protoplasts isolated from predominantly haploid suspension culture of *Datura innoxia* Mill. Plant Sci Lett 11:169–176

Gamborg OL, Constabel F, Fowke L, Kao KN, Ohyama K, Kartha KK, Pelcher L (1974) Protoplast and cell culture methods in somatic hybridisation in higher plants. Can J Genet Cytol 16:737–750

Giles KL, Sarafis V (1971) On the survival and reproduction of chloroplasts outside the cell. Cytobios 4:61–74

Giles KL, Whitehead HCM (1975) The transfer of nitrogen fixing ability to a eukaryote cell. Cytobios 14:49–61

Giles KL, Whitehead HCM (1977a) The localisation of introduced *Azotobacter* cells within the mycelium of a modified mycorrhiza (*Rhizopogon*) capable of nitrogen fixation. Plant Sci Lett 10:367–372

Giles KL, Whitehead HCM (1977b) Reassociation of a modified mycorrhiza with the host plant roots (*Pinus radiata*) and the transfer of acetylene reduction activity. Plant Soil 48:143–152

Gleba YY, Hoffmann F (1978) Hybrid cell lines of *Arabidopsis thaliana* + *Brassica campestris*: No evidence for specific chromosome elimination. Molec Gen Genet 165:257–264

Gleba YY, Hoffmann F (1980) "*Arabidobrassica*": A novel plant obtained by protoplast fusion. Planta 149:112–117

Gleba YY, Shvydkaya LG, Butenko RG, Sytnik KM (1974) Cultivation of isolated protoplasts. Soviet Plant Physiol 21:486–492

Gleba YY, Butenko RG, Stynik KM (1975) Fusion of protoplasts and parasexual hybridisation in *Nicotiana tabacum*. Dokl Akad Nauk USSR 221:1196–1198

Gleba YY, Kohlenbach HW, Hoffmann F (1978) Root morphogenesis in somatic hybrid cell lines *Arabidopsis thaliana* + *Brassica campestris*. Naturwissenschaften 65:655

Glimelius K, Eriksson T, Grafe R, Müller AJ (1978) Somatic hybridisation of nitrate reductase-deficient mutants of *Nicotiana tabacum* by protoplast fusion. Physiol Plant 44:273–277

Gosch G, Bajaj YPS, Reinert J (1975 a) Isolation, culture and fusion studies on protoplasts from different species. Protoplasma 85:327–336

Gosch G, Bajaj YPS, Reinert J (1975 b) Isolation, culture and induction of embryogenesis in protoplasts from cell suspensions of *Atropa belladonna*. Protoplasma 86:405–410

Grambow HJ, Kao KN, Miller RA, Gamborg OL (1972) Cell division and plant development from protoplasts of carrot cell suspension cultures. Planta 103:348–355

Gregory DW, Cocking EC (1965) The large scale isolation of protoplasts from immature tomato fruit. J Cell Biol 24:143–146

Gresshoff PM (1980) In vitro culture of white clover: Callus, suspension, protoplast culture and plant regeneration. Bot Gaz 141:157–164

Grun P, Chu LJ (1978) Development of plants from protoplasts of *Solanum* (Solanaceae). Am J Bot 65:538–543

Handro W, Rao PS, Harada H (1972) Controle hormonal de la formation de cals, bourgeons, racines et embryons sur des explantats de feuilles et de tiges de *Petunia* cultivés in vitro. CR Acad Sci Paris 275:2861–2863

Handro W, Rao PS, Harada H (1973) A histological study of the development of buds, roots, and embryos in organ cultures of *Petunia inflata* R. Fries. Ann Bot 37:817–821

Harada H (1973) A new method for obtaining protoplasts from mesophyll cells. Z Pflanzenphysiol 69:77–80

Harris H (1970) Cell fusion (The Dunhan Lectures). Oxford England

Hayward C, Power JB (1975) Plant production from leaf protoplasts of *Petunia parodii*. Plant Sci Lett 4:407–410

Hess D (1972) Transformation an höheren Organismen. Naturwissenschaften 59:348–355

Hess D, Potrykus I, Doon G, Durand T, Hoffmann F (1973) Transformation experiments in higher plants: Prerequisites for the use of isolated protoplasts (isolation from mesophyll and callus cultures, uptake of proteins and DNA, and regeneration of whole plants). In: Protoplastes et fusion de cellules somatiques végétales. Colloq Int CNRS, Paris, pp 343–351

Hoffmann F, Hess D (1973) Die Aufnahme radioaktiv markierter DNA in Protoplasten von *Petunia hybrida*. Z Pflanzenphysiol 69:81–83

Holsten RD, Burns RC, Hardy RWF, Herbert RR (1971) Establishment of symbiosis between *Rhizobium* and plant cells in vitro. Nature (London) 232:173–176

Izhar S, Power JB (1979) Somatic hybridisation in *Petunia:* A male sterile cytoplasmic hybrid. Plant Sci Lett 14:49–55

Kameya T, Uchimiya H (1972) Embryoids derived from isolated protoplasts of carrot. Planta 103:356–360

Kao KN, Michayluk MR (1974) Methods for high-frequency intergeneric fusion of plant protoplasts. Planta 115:355–367

Kao KN, Michayluk MR (1980) Plant regeneration from mesophyll protoplasts of Alfalfa. Z Pflanzenphysiol 96:135–141

Kao KN, Keller WA, Miller RA (1970) Cell division in newly formed cells from protoplasts of soybean. Exp Cell Res 62:231–236

Kao KN, Gamborg OL, Miller RA, Keller WA (1971) Cell divisions in cells regenerated from protoplasts of soybean and *Haplopappus gracilis*. Nature (London) 232:124

Kao KN, Gamborg OL, Michayluk MR, Keller WA, Miller RA (1973) The effects of sugars and inorganic salts on cell regeneration and sustained division in plant protoplasts. In: Protoplastes et fusion de cellules somatiques végétales. Colloq Int CNRS, Paris, pp 207–213

Kao KN, Constabel F, Michayuluk MR, Gamborg OL (1974) Plant protoplast fusion and growth of intergeneric hybrid cells. Planta 120:215–227

Kartha KK, Michayluk MR, Kao KN, Gamborg OL, Constabel F (1974 a) Callus formation and plant regeneration from mesophyll protoplasts of rape plants (*Brassica napus* L. Cv. Zephyr.). Plant Sci Lett 3:265–271

Kartha KK, Gamborg OL, Constabel F, Kao KN (1974 b) Fusion of rapeseed and soybean protoplasts and subsequent division of heterokaryocytes. Can J Bot 52:2435–2436

Kassanis B (1967) Plant tissue culture. In: Maramorosch K, Koprowski (ed) Methods in virology, vol I. Academic Press, London New York, pp 537–566

Keller WA, Melchers G (1973) Effect of high pH and calcium on tobacco leaf protoplast fusion. Z Naturforsch 28:737–741

Koblitz H (1978) Recovery of plants from protoplasts of haploid long-term callus tissue of *Nicotiana tabacum* cultured in vitro. Biochem Physiol Pflanz 172:213–222

Kochba J, Spiegel-Roy P, Safran H (1972) Adventive plants from ovules and nucelli in *Citrus*. Planta 106:237–245

Konar RN, Nataraja K (1969) Morphogenesis of isolated flower buds of *Ranunculus sceleratus* L. in vitro. Acta Bot Neerl 18:680–699

Linsmaier KM, Skoog F (1965) Organic growth factor requirements of tobacco tissue cultures. Physiol Plant 18:100–127

Lonnendonker N, Schieder O (1980) Amylase isoenzymes of the genus *Datura* as a simple method for an early identification of somatic hybrids. Plant Sci Lett 17:135–139

Lorz H, Wernicke W, Potrykus I (1979) Culture and plant regeneration of *Hyoscyamus* protoplasts. Planta Medica 36:21–29

Maliga P, Lázár G, Joó F, Nagy AH, Menczel L (1977) Restoration of morphogenetic protential in *Nicotiana* by somatic hybridisation. Mol Gen Genet 157:291–296

Maliga P, Kiss ZR, Nagy AH, Lázár G (1978) Genetic instability in somatic hybrids of *Nicotiana tabacum* and *Nicotiana knightiana*. Mol Gen Genet 163:145–151

Melchers G, Labib G (1974) Somatic hybridisation of plants by fusion of protoplasts I. Selection of light-resistant hybrids of "haploid" light-sensitive varieties of tobacco. Mol Gen Genet 135:277–294

Melchers G, Sacristan MO, Holder AA (1978) Somatic hybrid plants of potato and tomato regenerated from fused protoplasts. Carlsberg Res Commun 43:203–218

Messerschmidt M (1974) Kallusbildung und Differenzierung aus isolierten Protoplasten von *Pharbitis* nil. Z Pflanzenphysiol 74:175–178

Motoyoshi F, Oshima N (1968) Multiplication of tobacco mosaic virus in suspension culture of tobacco cells. J Jap Microbiol 12:317–320

Murashige T, Skoog F (1962) A revised medium for rapid growth and bioassays with tobacco tissue cultures. Physiol Plant 15:473–497

Nagao T (1978) Somatic hybridisation by fusion of protoplasts. 1. The combination of *Nicotiana tabacum* and *Nicotiana rustica*. Jap Crop Sci 47:491–498

Nagata T, Takebe I (1970) Cell wall regeneration and cell division in isolated tobacco mesophyll protoplasts. Planta 92:301–308

Nagata T, Takebe I (1971) Plating of isolated tobacco mesophyll protoplasts on agar medium. Planta 99:12–20

Nagy JI, Maliga P (1976) Callus induction and plant regeneration from mesophyll protoplasts of *Nicotiana sylvestris*. Z Pflanzenphysiol 78:453–455

Nass MMK (1969) Uptake of isolated chloroplasts by mammalian cells. Science 165:1128–1131

Nitsch JP, Ohyama K (1971) Obtention de plantes à partir de protoplastes haploïdes cultivés in vitro. CR Acad Sci Paris 273:801–804

Ohyama K, Nitsch JP (1972) Flowering haploid plants obtained from protoplasts of tobacco leaves. Plant Cell Physiol 13:229–236

Ohyama K, Gamborg OL, Miller RA (1972) Uptake of exogenous DNA by plant protoplasts. Can J Bot 50:2077–2080

Otsuki Y, Takebe I, Honda Y, Matsui C (1972) Ultrastructure of infection of tobacco mesophyll protoplasts by tobacco mosaic virus. Virology 49:188–194

Pinto da Silva PG (1969) Isolation of protoplasts from higher plant cells. Naturwissenschaften 56:51

Poirier-Hamon S, Rao PS, Harada H (1973) Culture in vitro de protoplastes foliaires et de segments de tige d' *Antirrhinum majus* L.: Effets de divers phyto-hormones et milieux de culture sur la division et la differentiation cellulairies. Actes du 98 Congress National des Sociétés, Savantes, St. Etienne, vol II, pp 209–210

Poirier-Hamon S, Rao PS, Harada H (1974) Culture of mesophyll protoplasts and stem segments of *Antirrhinum majus* (Snapdragon): Growth and organization of embryoids. J Exp Bot 25:752–760

Potrykus I (1971) Intra- and interspecific fusion of protoplasts from petals of *Torenia baillonii* and *Torenia fournieri*. Nature: New Biol 231:51–58

Potrykus I (1973) Transplantation of chloroplasts into protoplasts of *Petunia*. Z Pflanzenphysiol 70:364–366

Potrykus I, Durand T (1972) Callus formation from single protoplasts of *Petunia*. Nature (London) 237:286–287

Potrykus I, Hoffmann F (1973) Transplantation of nuclei into protoplasts of higher plants. Z Pflanzenphysiol 69:287–289

Power JB, Cocking EC (1969) A simple method for the isolation of very large numbers of leaf protoplasts. Biochem J 111:33

Power JB, Frearson EM (1973) The inter- and intraspecific fusion of plant protoplasts: Subsequent developments in culture, with reference to crown gall callus and tobacco and *Petunia* leaf systems. In: Protoplastes et fusion de cellules somatiques végétales. Colloq Int CNRS, Paris, pp 409–421

Power JB, Cummins SE, Cocking EC (1970) Fusion of isolated plant protoplasts. Nature (London) 225:1016–1018

Power JB, Frearson EM, Hayward C, Cocking EC (1975) Some consequences of the fusion and selective culture of *Petunia* and *Parthenocissus* protoplasts. Plant Sci Lett 5:197–207

Power JB, Frearson EM, Hayward C, George D, Evans PK, Berry SF, Cocking EC (1976) Somatic hybridisation of *Petunia hybrida* and *P. parodii*. Nature (London) 263:500–502

Power JB, Berry SF, Frearson EM, Cocking EC (1977) Selection procedures for the production of interspecies somatic hybrids of *Petunia hybrida* and *Petunia parodii* I. Nutrient media and drug sensitivity complementation selection. Plant Sci Lett 10:1–6

Power JB, Sink KC, Berry SF, Burns SF, Cocking EC (1978) Somatic and sexual hybrids of *Petunia hybrida* and *Petunia parodii*. Hered 69:373–376

Raj B, Herr JM (1970) The isolation of protoplasts from the placental cells of *Solanum nigrum* L. Protoplasma 69:291–300

Rao PS, Harada H (1974) Hormonal regulation of morphogenesis in organ cultures of *Petunia inflata*, *Antirrhinum majus*, and *Pharbitis nil*. In: Plant growth substances. Hirokowa Publ Co, Tokyo, pp 1113–1120

Rao PS, Handro W, Harada H (1973a) Bud formation and embryo differentiation in vivo cultures of *Petunia*. Z Pflanzenphysiol 69:87–90

Rao PS, Handro W, Harada H (1973b) Hormonal control of differentiation of shoots, roots, and embryos in leaf and stem cultures of *Petunia inflata* and *Petunia hybrida*. Physiol Plant 28:458–463

Raveh D, Galun E (1975) Rapid regeneration of plants from tobacco protoplasts plated at low densities. Z Pflanzenphysiol 76:76–79

Reinert J, Hellmann S (1971) Mechanism of the formation of polynuclear protoplasts from cells of higher plants. Naturwissenschaften 38:419

Ruesink AW (1971) Protoplasts of plant cells. In: Methods in enzymology, vol 23. Academic Press, London New York, pp 197–209

Ruesink AW, Thimann KV (1966) Protoplasts from *Avena* coleoptile. Proc Natl Acad Sci USA 54:56–64

Schaeffer GW, Smith HH (1963) Auxin-kinetin interaction in tissue cultures of *Nicotiana* species and tumor-conditioned hybrids. Plant Physiol 38:291–297

Schenk RU, Hilderbrandt AC (1969) Production of protoplasts from plant cells in liquid culture using purified commercial cellulases. Crop Sci 9:629–631

Schieder O (1975a) Regeneration of haploid and diploid *Datura innoxia* Mill.: Mesophyll protoplasts to plants. Z Pflanzenphysiol 76:462–466

Schieder O (1975b) Selection of a somatic hybrid between auxotrophic mutants of *Sphaerocarpus donnelli* using the method of protoplast fusion. Z Pflanzenphysiol 74:357–365

Schieder O (1977) Attempts in regeneration of mesophyll protoplasts of haploid wild-type lines and those of chlorophyll-deficient strains from different Solanaceae.Z Pflanzenphysiol 84:275–281

Schieder O (1978) Somatic hybrids of *Datura innoxia* Mill + *D. discolor* Bernh. and of *Datura innoxia* Mill + *Datura stramonium* L. var Tatula L. Mol Gen Genet 162:113–119

Schieder O (1980) Somatic hybrids between a herbaceous and two tree *Datura* species. Z Pflanzenphysiol 98:119–127

Shahin EA, Shepard JF (1980) Cassava mesophyll protoplasts: Isolation, proliferation and shoot formation. Plant Sci Lett 17:459–465

Shepard JF, Totten RE (1977) Mesophyll cell protoplasts of potato: Isolation, proliferation, and plant regeneration. Plant Physiol 60:313–316

Sink KC, Power JB (1977) The isolation, culture, and regeneration of leaf protoplasts of *Petunia parviflora* Juss. Plant Sci Lett 10:335–340

Smith HH (1971) Broadening the base of genetic variability in plants. J Hered 62:265–276

Smith HH (1974) Model systems for somatic cell plant genetics. BioScience 24:269–276

Sondahl MR, Chapman MS, Sharp WR (1980) Protoplast liberation, cell wall reconstitution and callus proliferation in *Coffea arabica* L. callus tissues. Turrialba 30:161–165

Steward FC (1970) From cultured cells to whole plants: The induction and control of their growth and morphogenesis. Proc R Soc London 176:1–30

Strauss A, Potrykus I (1980) Callus formation from protoplasts of cell suspension cultures of *Rosa* Paul's Scarlet. Physiol Plant 48:15–20

Takebe I, Otsuki Y, Aoki S (1968) Isolation of tobacco mesophyll cells in intact and active state. Plant Cell Physiol 9:115–124

Takebe I, Labib G, Melchers G (1971) Regeneration of whole plants from isolated mesophyll protoplasts of tobacco. Naturwissenschaften 58:318–320

Ulrich TH, Chowdhury JB, Widholm JM (1980) Callus and root formation from mesophyll protoplasts of *Brassica rapa*. Plant Sci Lett 19:347–354

Vardi A, Spiegel-Roy P, Galun E (1975) *Citrus* cell culture: Isolation of protoplasts, plating densities, effect of mutagens and regeneration of embryos. Plant Sci Lett 4:231–236

Vasil IK, Vasil V (1972) Totipotency and embryogenesis in plant and tissue cultures. In vitro 8:117–127

Vasil V, Vasil IK (1974) Regeneration of tobacco and *Petunia* plants from protoplasts and culture of corn protoplasts. In Vitro 10:83–96

Vasil V, Vasil IK (1979) Isolation and culture of cereal protoplasts I. Callus formation from pearl millet (*Pennisetum americanum*) protoplasts. Z Pflanzenphysiol 92:379–383

Vasil V, Vasil IK (1980) Isolation and culture of cereal protoplasts II. Embryogenesis and plantlet formation from protoplasts of *Pennisetum americanum*. Theor Appl Genet 56:97–99

Wallin A, Eriksson T (1972) Plating of protoplasts from cell suspension cultures of *Daucus carota*. In: Protoplastes et fusion de cellules somatiques végétales. Colloq Int CNRS Paris, pp 301–307

Wernicke W, Thomas E (1980) Studies on morphogenesis from isolated plant protoplasts: Shoot formation from mesophyll protoplasts of *Hyoscyamus muticus* and *Nicotiana tabacum*. Plant Sci Lett 17:401–407

Wilson HM, Styer DJ, Conard PL, Durbin RD, Helgeson JP (1980) Isolation of sterile protoplasts from unsterilized leaves. Plant Sci Lett 18:151–154

Zieg RG, Outka DE (1980) The isolation, culture and callus formation of soybean pod protoplasts. Plant Sci Lett 18:105–114

Zimmermann U, Scheurich P (1981) Fusion of *Avena sativa* mesophyll cell protoplasts by electrical breakdown. Biochem Biophys 641:160–165

Plant Index

The plants listed in the Tables and References have not been indexed. As far as possible, common names are indexed with Latin names. + For somatic hybrid, × for sexual hybrid.

Subject Index

To locate pages for various topics discussed in this book, the reader is referred to the "Contents". Terms which occur throughout the book with high frequency, or are of unspecific or general nature, have not been fully indexed. The subject/s mentioned in Tables, and chemicals used in experimental studies have not been indexed.

Monographs on Theoretical and Applied Genetics

Editors: R. Frankel
(coordinating editor),
G. A. E. Gall, M. Grossman,
H. F. Linskens, D. de Zeeuw

Springer-Verlag
Berlin
Heidelberg
New York

Applied and Fundamental Aspects of Plant Cell, Tissue, and Organ Culture

Editors: J. Reinert, Y. P. S. Bajaj

1977. 181 figures. XVI, 803 pages
ISBN 3-540-07677-8

Contents: Regeneration of Plants, Vegetative Propagation and Cloning. – Haploids. – Cytology, Cytogenetics and Plant Breeding. – Protoplasts, Somatic Hybridization and Genetic Engineering. – Tissue Culture and Plant Pathology. – Cell Culture and Secondary Products. – Miscellaneous.

U. Lüttge, N. Higinbotham

Transport in Plants

1979. 1 portrait. 180 figures, 33 tables.
X, 468 pages
ISBN 3-540-90383-6

This book provides a comprehensive description of the transport of mineral and organic substances in plants. The basic concepts of plant biophysics, biochemistry, nutrient distribution, and physiology are presented and illustrated with well chosen examples. The text is organized in a logical progression from a discussion of the biophysical and thermodynamic aspects to discussion of:
- materials transported,
- construction and functions of membranes and cell walls,
- conceptual models of transport physiology, regulation and control,
- active transport,
- energy considerations,
- systems coupling,
- inter-cellular and inter-organ transport, and transport regulation in whole plants.

The material is well referenced, rigorous, and discusses topics at the forefront of modern research so that the book will be a valuable addition to the professional library. The development also makes this presentation ideal for new students in the field of plant physiology, biophysics and cell biology.

T. C. Moore

Biochemistry and Physiology of Plant Hormones

1979. 164 figures, 13 tables. XII, 274 pages
ISBN 3-540-90401-8

Biochemistry and Physiology of Plant Hormones is a comprehensive account of hormonal regulation of growth and seed plant development. The author summarizes current fundamental knowledge regarding the major kinds of hormones and the phytochrome pigment system, reflecting the steady output of important new discoveries in the field.

Chapter 1 introduces the reader to the growth and development of whole plants throughout ontogeny. This sets the stage for a consideration of hormonal regulation, specifically where it concerns auxins, gibberellins, cytokinins, abscisic acid and related compounds, ethylene, and phytochrome. Biochemical aspects of hormonal regulation are emphasized throughout the book.

Biochemistry and Physiology of Plant Hormones will be a valuable text and major reference for advanced students as well as researchers in biology, botany, and such fields of applied botany as agronomy, forestry, and horticulture.

Springer-Verlag
Berlin
Heidelberg
New York